ROUTLEDGE LIBRARY EDITIONS:
NUCLEAR SECURITY

Volume 22

THE PROLIFERATION PUZZLE

THE PROLIFERATION PUZZLE

Why Nuclear Weapons Spread (and What Results)

Edited by
ZACHARY S. DAVIS AND BENJAMIN FRANKEL

Routledge
Taylor & Francis Group
LONDON AND NEW YORK

First published in 1993 by Frank Cass

This edition first published in 2021
by Routledge
2 Park Square, Milton Park, Abingdon, Oxon OX14 4RN

and by Routledge
52 Vanderbilt Avenue, New York, NY 10017

Routledge is an imprint of the Taylor & Francis Group, an informa business

British Library Cataloguing in Publication Data
A catalogue record for this book is available from the British Library

ISBN: 978-0-367-50682-7 (Set)
ISBN: 978-1-00-309763-1 (Set) (ebk)
ISBN: 978-0-367-54353-2 (Volume 22) (hbk)
ISBN: 978-1-00-308908-7 (Volume 22) (ebk)

Publisher's Note
The publisher has gone to great lengths to ensure the quality of this reprint but points out that some imperfections in the original copies may be apparent.

Disclaimer
The publisher has made every effort to trace copyright holders and would welcome correspondence from those they have been unable to trace.

THE PROLIFERATION PUZZLE

Why Nuclear Weapons Spread (and What Results)

Zachary S. Davis *and* Benjamin Frankel
Editors

FRANK CASS

First Published 1993 in Great Britain by
FRANK CASS AND COMPANY LIMITED
Gainsborough House, 11 Gainsborough Road,
London E11 1RS, England

and in the United States of America by
FRANK CASS
c/o International Specialized Book Services, Inc.,
5804 N.E. Hassalo Street,
Portland, OR 97213-3644

British Library Cataloguing in Publication Data

Proliferation Puzzle: Why Nuclear Weapons
Spread (and What Results). – ("Security
Studies" Special Issue, ISSN 0963-6412)
I. Davis, Zachary S. II. Frankel,
Benjamin III. Series
355.825119

ISBN 0-7146-4108-1

Library of Congress Cataloging-in-Publication Data

The proliferation puzzle : why nuclear weapons spread / edited by
Zachary S. Davis and Benjamin Frankel.
p. cm.
Includes bibliographical references.
ISBN 0-7146-4108-1 (pbk.) : $37.50 (U.S.)
1. Nuclear nonproliferation. I. Davis, Zachary S., 1955–
II. Frankel, Benjamin, 1949–
JX1974.73.P78 1993
327'.74—dc20 93-28873
CIP

This group of studies first appeared in a Special Issue on The Proliferation Puzzle: Why Nuclear Weapons Spread (and What Results) of *Security Studies*2, nos. 3/4 (Spring/Summer 1993).

Typeset by Regent Typesetting, London
Printed in Great Britain by Antony Rowe Ltd., Chippenham

Contents

Nuclear Weapons Proliferation:
Theory and Policy

Benjamin Frankel and Zachary S. Davis

THE RELATIONSHIP between theory and policymaking is not a subject either theoreticians or policymakers are eager to establish or nourish. Both sides realize there is a connection between the two, but they remain skeptical that closer links would benefit either. Theory builders in the social sciences are reluctant to make the necessary compromises and concessions involved in translating their neat theories into policy recommendations; policymakers view most theory as too vague and unspecific.

Theory, however, is essential to policy. There is no way to comprehend reality in any meaningful fashion without the aid of a theory. There are different approaches to reality, but the difference is not between one which is theoretical and one which is not. Rather, the difference is between an approach which is theory-conscious and one which is theory-innocent. The former is aware of the assumptions and presuppositions it brings to the analysis of reality; the latter keeps its assumptions hidden, unacknowledged, and unexamined. The assumptions and presuppositions which undergird our understanding of the world, however, are not less powerful because they are hidden. As Kenneth Waltz writes,

> since empirical knowledge is potentially infinite in extent, without some guidance we can know neither what information to gather nor how to put it together so that it becomes comprehensible. If we could directly apprehend the world that interests us, we would have no need for theory. We cannot. One can find his way among infinite materials only with the guidance of theory.[1]

We thus cannot approach reality unencumbered by a system of meaning we impose on it in order to make sense of it.

The proliferation of nuclear weapons is a good subject for social science theorizing, and such theorizing is especially opportune now. First, the sample, that is, the number of countries whose decisions about nuclear weapons we may want to study, is large. Some argue that it is difficult to theorize about the spread of nuclear weapons because only a few countries have acquired them. These scholars, however, overlook the fact that many more countries could have built nuclear weapons but decided not to do so. Because the decisions to acquire or not to acquire nuclear weapons belong in the same domain, the understanding of why a certain state decided not to acquire nuclear weapons is helpful in understanding why another country chose to acquire them. Thus, rather than a sample of between five and nine states, the true weapons proliferation sample includes some forty states.

Second, the building of nuclear weapons is a complex task touching upon many of the subjects of study at the heart of social science and international relations. Nuclear weapons may be acquired as a hedge against external threat or for reasons of national prestige. They may also be built as a result of pressures by domestic coalitions among scientists, bureaucrats, and the military. They may be sought for defensive purposes or to support hegemonic aspirations. The destructiveness of these weapons and the short flight-time of their delivery vehicles raise questions about civilian command and control, especially in acute crisis situations. Further, because of their sheer destructiveness, nuclear weapons may blur the ready distinction between state-level characteristics and systemic effects.

Third, during the last two decades the building of nuclear weapons has been proscribed by the nonproliferation regime. The decisions countries have made about the acquisition of nuclear weapons can thus serve as a test of the effectiveness of this particular regime, and of regimes and regime theory in general.

Fourth, nuclear weapons were introduced at the same time bipolarity became the international order. It is argued that both nuclear weapons and bipolarity strengthen stability, and that the two have tended to reinforce each other. The world is moving away from bipolarity, however. What would be the effect of nuclear weapons in a multipolar order? How will the spread of nuclear weapons affect the distribution of capabilities among states? If nuclear weapons do spread to additional countries, will they enhance stability or exacerbate instability? Can the spread of these weapons be managed or controlled?

This volume is an effort to bring social science and international relations theory to bear on the question of nuclear weapons proliferation. There were two reasons why we thought it was important to put this volume together. We mentioned the first reason earlier: The disconnection between theory and practice is artificial and debilitating. The purpose of social science is to formulate hypotheses and devise theories that advance our understanding of society and aid in the fashioning of informed policy. Social scientists should articulate clearly the assumptions they bring to their analysis and enlighten policymakers about the tacit assumptions and implicit presuppositions they bring to their policymaking. The essays in this volume show how explicit hypotheses about the causes and consequences of nuclear weapons proliferation lead to a deeper understanding of the problem and to specific, theory-informed policy recommendations.

The second reason for putting this volume together is that we find ourselves now in a period of major transition. The old paradigm of international politics is disintegrating while the new one is yet to emerge. This is a time, in Thomas Kuhn's words, of revolutionary science. This profound transformation calls for a re-examination of the basic concepts and terms we use to understand the world and navigate in it. What is required, therefore, is not only a formulation of new policies but a re-evaluation of the explicit or implicit theories upon which these policies are based. This re-examination is the task of social science. In the essays collected in this volume different authors, relying on different social science approaches, offer different understandings of the problem of proliferation. This diversity of approaches typifies periods of paradigmatic shifts. It also reflects the fact that within international relations theory there are mounting challenges to realism and neorealism, which have been the dominant school within the discipline. This is the way it should be. Our understanding of the world is intimately and inextricably involved with the world. A transformation of the world calls for an examination of our understanding of it. One does not have to be a Marxist to agree with Marx that "One basis for life and another basis for science is a priori a lie."

This volume is part of a broader, on-going effort to apply social science and international relations theory to the question of nuclear weapons proliferation. The articles in this issue were presented as papers in a conference on Social Science Theory and Nuclear Weapons Proliferation held at the Center for International Security and Arms Control of Stanford University on 5–6 February 1993. Other papers presented at the conference will be published in future issues of the journal. We are especially grateful to Scott D. Sagan of Stanford University political science department and CISAC for sponsoring and hosting the conference. We gratefully acknowledge the financial support of CISAC, the Lawrence Livermore National Laboratory, and the Compton Foundation. We also thank Rosemary Hamerton-Kelly and Effie Toshav for helping make the conference a success.

NOTES

1. Kenneth N. Waltz, *Theory of International Politics* (Reading, Mass.: Addison-Wesley, 1979), 5.

I
Realist Approaches

Dividing Realism:
Structural Realism versus Security Materialism on Nuclear Security and Proliferation

Daniel Deudney

WITH THE END of the cold war and the resolution of the Soviet-American arms race, concern about the proliferation of nuclear weapons capability has taken a more prominent role in world politics and American foreign policy. The end of the cold war has also created novel proliferation challenges, as the newly formed states of Ukraine, Belarus, and Kazakhstan debate whether to retain nuclear weapons from the former Soviet Union as a deterrent against Russia or to dismantle them in accord with international pressures. The outcast countries of Iran, Iraq, and North Korea, continuing to strive for nuclear weapons capability, are on a collision course with the United States. With the expiration of the Non-Proliferation Treaty in 1995, and the growing debate over expanding the permanent members of the Security Council, the stage is set for a major change in the relationship between nuclear weapons and world public order.

Underlying these policy concerns is a very simple theoretical question: what impact does the diffusion of nuclear weapons have on international peace and security? The current policy of the United States and the nonproliferation regime are premised on the taken-for-granted assumption that the further spread of nuclear weapons is undesirable. How strong are the theoretical underpinnings for this well-established conventional wisdom? Nuclear proliferation has concerned leaders and publics since the beginning of the nuclear era, but international relations theorists have not given the topic extensive attention. The policy aspects of nuclear proliferation are addressed in a large and high quality literature, but no comparable theoretical literature exists. Furthermore, the theoretical work on nuclear proliferation that does exist raises as many questions as it answers.

A particularly powerful and provocative challenge to conventional wisdom has been posed by Kenneth Waltz. Contrary to the assumptions of the nonproliferation regime and the policies of leading states, Waltz argues that the spread of nuclear weapons might actually be beneficial because it would extend the stability and reduced incentives for war that has characterized relations between nuclear armed states over the last half century.[1] If Waltz is right, then the current policies might be unproductive, or perhaps even counterproductive, for achieving the goals that animate them. Since Waltz's structural realism is widely held to be one of the most authoritative contemporary statements of Realism, the challenge he raises to the conventional proliferation wisdom is particularly troubling. The long-standing policies of leading states and a leading version of Realism are in direct contradiction.

In order to solve the puzzle of nuclear proliferation, it is necessary first to understand the implications of nuclear weapons for the state and for the interstate system. Arguments about the causes and consequences of the spread of nuclear weapons derive at least implicitly from views of the relationship between nuclear explosives and the state-centric security order. Only those theories that provide an adequate understanding of the relationship between the state and state-system, and the nuclear forces of destruction can provide satisfactory answers to the nuclear proliferation puzzle.

This study analyzes nuclear proliferation from the standpoint of two distinct and competing Realist theories – structural realism and security materialism. Because of its concern with violence, states, and security, Realism contains the richest and most relevant theoretical tools for grappling with nuclear proliferation. Realism is not, however, one theory or even a tightly linked set of theories, but rather a family of related and competing theories. As developed by Waltz, structural realism attempts to explain the ways in which the ordering principles – or structure – of the international system shape political outcomes. Less well-known, security materialism posits that very basic features in the material conditions of destruction, constituted by the interaction of geography and technology, determine the number and character of viable security providing units. Both these theories are cast in very broad terms, and using them to generate expectations about nuclear weapons and nuclear proliferation is a complex and not entirely straightforward process. Producing expectations about nuclear proliferation from either structural realism or security materialism entails making additional assumptions about contested issues. Nevertheless, these two theories generate radically different interpretations of the relationship between nuclear explosives and the state, and hence of the impacts of further proliferation.

The main thrust of my argument is that structural realism has fundamental flaws in its understanding of the relationship between nuclear weapons, states, and security, and thus of nuclear proliferation, and that security materialism provides a superior understanding of these relationships. Specifically, Waltz's nuclear arguments contain two major flaws. First, his claims about nuclear weapons and nuclear proliferation are essentially ad hoc, do not derive from his structural realism, and are partially inconsistent with it. Second, Waltz lacks a theory of the state, both as institution and as a security mode, and his nuclear arguments entail the substitution of an idealized rational actor for the state.

In contrast, security materialism holds that nuclear weapons fundamentally compromise the basic security units that make up world politics, and that the statist institutions and orientations to power characteristic of the pre-nuclear world can lead to disaster if employed in the nuclear world. Security materialism holds that nuclear weapons have set in motion both a unit and a system change that will culminate in either a world state or a functionally differentiated security system in which the territorial states are circumscribed and then complemented by a strong nuclear regulatory system. This perspective helps explain the importance polities, both nuclear and non-nuclear, have attached to the nonproliferation regime. From this perspective, further proliferation intensifies the contradiction between statist forms of security governance and nuclear security imperatives, and is thus likely to shape the timing and character of nuclear adjustment.

In sum, I attack structural realism and its nuclear arguments from a primitive or paleo-realist direction. The problem with structural realism is not, as many critics have alleged, that it is too Realist because it places too much importance upon the role of violence capabilities and the state in world politics. Rather structural realism has drifted towards increasingly idealized and formalized conceptualizations of world politics and away from Realist fundamentals on the logics of violence capabilities and their role in shaping security institutions and systems. And instead of being too state-centric, it has lost the primitive Realist insights into the state as a mode of protection and the state as an institution with distinctive orientations to power and security. Whatever its merits for explaining the consequences of structure, structural realism does not attempt to explain system and unit change, and thus cannot illuminate nuclear realities.

The assessment of each of the competing Realist approaches to nuclear proliferation proceeds in three steps, from basic theoretical claims, to arguments about nuclear weapons generally, and then to arguments about nuclear proliferation. By beginning from their general claims and then progressing toward their more specific nuclear and then proliferation arguments, the underlying theoretical differences between security materialism and structural realism are revealed.

Structural Realism and Nuclear Weapons

Kenneth Waltz's "neorealist" or "structural realist" theory has been widely influential among academic international relations theorists.[2] Although Waltz does not deal extensively with nuclear weapons in *Theory of Inter-*

national Politics, his main exposition of structural realism, he does analyze nuclear weapons in several short pieces.[3] Waltz's brand of Realism has a wide following among international relations theorists, but his interpretation of nuclear weapons is a much more idiosyncratic blend of extremely unconventional and extremely conventional contemporary Realist views on the topic. To weigh Waltz's treatment of the impact of nuclear weapons on world politics we review the fundamentals of his structural realism, examine his analysis of nuclear weapons, assess how his views on nuclear weapons relate to his structural realist ideas, point to some weaknesses in his argument, and then examine modifications of structural realism proposed by Steve Weber.

Structural Realism

Waltz's theory of international relations is an effort to analyze the effects of structure on international outcomes. "Structure defines the arrangement, or ordering, of the parts of a system."[4] Waltz identifies three key dimensions of structure. First, there is the principle by which a system is ordered, which Waltz says must either be centralized and hierarchic, like domestic politics, or decentralized and anarchic, like international relations.[5] For Waltz, these two are the only ordering principles possible, and they are fundamentally dissimilar. In domestic politics, the units are hierarchically arranged. For Waltz, the "anarchy" of international relations is not complete disorder and chaos, but simply the absence of central government. Although Waltz acknowledges the existence of international organizations, he denies that they have significant effect, unless they are backed by states or are state-like in their capabilities.[6] Waltz's sharp distinction between domestic hierarchy and international anarchy systematizes a central insight of Hobbes' and Rousseau's international political theory.

The second element of Waltzian structure is the "specification of the functions of formally differentiated units."[7] When units are functionally differentiated, they exhibit a division of labor, with different parts performing different tasks. Units are functionally undifferentiated when they eschew specialization and each attempts to perform the full range of necessary tasks. Here too, Waltz draws a sharp distinction between domestic politics, where units are functionally differentiated, and international politics, where units are functionally undifferentiated. "States perform or try to perform tasks, most of which are common to other states; the ends they aspire to are similar. Each state duplicates the activities of other states at least to a considerable extent."[8]

The third dimension of structure for Waltz is the distribution of capabilities between the units, a variable long emphasized by Realist balance of power theorists. Waltz's structuralism aims to abstract as much as possible from the characteristics of the units that make up the system and to focus instead on how the parts are related to each other, but his third dimension of structure is an exception. Although capabilities are a unit attribute, he views the distribution of capabilities as a system feature. So central is the distribution of capabilities between actors for Waltz, that he seems to deny the significance of any other systemic feature.[9] It is changes in the distribution of capabilities that provide the basis for change in international politics: "The structure of a system changes with changes in the distribution of capabilities across the system's units."[10]

In part, Waltz's theory has had so much appeal because of the apparent scope of his theoretical argument. "Structures" he says, "never tell us all that we want to know. Instead they tell us a small number of big and important things."[11] Although Waltz's speaks of structure as the central variable in any genuine international theory, the scope of the phenomena explained by his theory is limited. He makes no attempt to explain either the origin of structures, or the causes of the changes in the units. And he also allows that much of state behavior and many system characteristics are caused by unit-level phenomena.

Nuclear Panglossism

Nuclear weapons are not central to Waltz's structural realism, but his recognition of their importance has been steadily increasing. In *Theory of International Politics*, Waltz gives only passing attention to nuclear weapons, and he generally belittles their importance as influences on international politics. "Nuclear weapons do not equalize the power of nations because they do not change the economic bases of a nation's power."[12] Furthermore, Waltz argues that nuclear weapons are relatively unimportant in international politics because they did not either cause or change the bipolarity of the system.

In subsequent writings, however, Waltz attributes much more importance to nuclear weapons and seems to hold that they have been more important than the "perennial forces of politics." Broadly speaking, Waltz makes two main arguments about nuclear weapons: their existence has greatly enhanced international security by greatly reducing the probability of war, and the further spread of nuclear would be desirable as a further cause of peace. Waltz's views of nuclear weapons are remarkably sanguine. Waltz is the Dr. Pangloss of the nuclear era: if we do not live in the best of all possible worlds, we live in something approaching it.

Generally holding that "nuclear weapons have been given a bad name,"[13] Waltz claims nuclear weapons have greatly strengthened the peaceful character of the international system. He thinks that "the probability of major war among states having nuclear weapons approaches zero."[14] Waltz argues that nuclear weapons have had the effect of greatly reducing the likelihood of war between states armed with nuclear weapons by reducing the possibility of a miscalculation. "With nuclear weapons, stability and peace rest on easy calculations." Waltz's overall characterization of the effects of nuclear weapons are unexceptional, except in the extremity of their optimism.

Waltz believes these basic realities of the nuclear era are simple and easily grasped, but that states have not fully grasped them. Waltz thinks "a basic strategic implication of nuclear weapons is yet to be appreciated."[15] Nuclear reality has not been fully reflected in the actual force structures and doctrines of the nuclear powers, because states still desire counterforce options and they fear that small differences in the number of nuclear weapons matter politically. Waltz labels the "window of vulnerability" fears propagated by Paul Nitze and others during the 1970s as "based on faulty assumptions, unfounded distinctions, and preposterous notions of how governments behave."[16a] Waltz also makes no mention of arms control, apparently reflecting a belief that its net impact upon the nuclear situation is negligible.[16b] In dismissing counterforce as obvious folly and arms control as trivial, Waltz is in effect saying the overwhelming preoccupations of American statecraft since the beginning of the nuclear age are wild goose chases. These preoccupations are for Waltz simply a "hangover from conventional days."[17] Waltz's qualified endorsement of nuclear proliferation also contradicts long-running American policy objectives. Contrary to those who claim that neo-realism serves as an ideological legitimation of American state policy,[18] Waltz's sweeping dismissal of the central concerns of American statecraft for the last half century demonstrates that his Realism maintains a healthy potential for critically judging state policy.

Non Sequiturs and Incongruities

Waltz's claims about the impact of nuclear weapons on the international system are authoritative in tone and sweeping in scope, but they have surprisingly little relation to his broader structural realist argument.

Waltz's structural realism emphasized the importance of the distribution of power as a powerful determinant of interstate outcomes. But in discussing nuclear weapons, Waltz insists that the balance or distribution of capabilities matters little: "Even with numbers immensely disproportionate, a small force strongly inhibits the use of a large one."[19] Once certain minimums are met, great inequalities in weapons deployment do not much matter, and the discrepancies in underlying economic potential are reduced greatly in significance.

Waltz's qualified endorsement of proliferation also seems to collide directly with his emphasis on distribution. One could expect a Realist theory assigning great importance to distribution to explain the nonproliferation regime as a means by which the powerful states protect their positions of relative strength. Waltz, however, argues the reverse: nuclear weapons are so unrelated to the distribution that their spread to other states would be desirable. Waltz's qualified endorsement of proliferation also seems to contradict his analysis about polarity and stability in structural realism. If, as Waltz holds, bipolarity is more stable than multipolarity, then it makes little sense to endorse nuclear proliferation, which should produce a more multipolar system.

Waltz's attempt to theorize about nuclear weapons in world security politics is also hampered by his treatment of technology. In *Theory of International Politics* he observed that "In shaping the behavior of nations, the perennial forces of politics are more important than the new military technology."[20] Here his view of the relative unimportance of nuclear weapons derives from his claims about the relative importance of political structure and technology. Yet nuclear weapons are a new military technology that Waltz thinks have had great influence. Because the independent variable in Waltz's structural realism is the ordering of the parts and distribution, his theory largely lacks the ability to analyze the effects of new technology, unless it alters the distribution of power.

Although admitting that nuclear weapons have had a profound impact on the overall character of the interstate system, Waltz does not view the advent of nuclear weapons as constituting a change in the structure of the international system. Rather, he says nuclear weapons constitute a "unit-level change in the extent and rapidity with which some states can hurt others," that has "system-wide effects."[21] Although not themselves structural nuclear weapons have "much diminished a structural effect."[22] In designating nuclear weapons as unit-level phenomenon, Waltz is essentially admitting that his theory of structural realism offers little insight into them. However penetrating Waltz's judgments about nuclear weapons may be, they are essentially unrelated to his theory of structural realism. Waltz's nuclear observations rest on his authority as a theorist rather than the power of his theory.

Waltz's relegation of nuclear weapons to the unit-level is unsatisfactory in two important ways. First, if nuclear weapons are a unit-level change of such influence that they overwhelm even structural causes of state behavior, then it would be logical to assume that they had some significant impact upon the identity of the units. Yet Waltz registers no such change in the units, and speaks of the units of both pre-nuclear and nuclear political systems as "states." The absence of a theory of the state in Waltz's neo-realism, noted by many critics,[23] thus leaves him ill-equipped to conceptualize possible changes in the unit driven by the unit-level change he believes nuclear weapons to be. However, it seems implausible, or at least odd, that a change at the unit-level capable of overwhelming structural effects would not also change the unit in some significant way. To draw out the implications of Waltz's observation that nuclear weapons are changes at the unit-level, a more general and robust theory of states as types of security providing units is necessary.

A second problem with Waltz's strategy of assigning nuclear weapons to the unit-level is that they are so unlike the other unit-level phenomena that he discusses.[24] The unit-level, as John Ruggie has observed, serves in Waltz's theory as a dumping ground for miscellaneous variables.[25] In describing the most important unit-level variables, Waltz gives a prominent place to ideology and domestic regime type. Ideology and regime type are institutional factors, subject to major human modification, while nuclear technology is a material factor based upon the realities of science and nature that institutions have little ability to escape.

In order to get out of this impasse, it is necessary to more broadly conceptualize the environment of states as being made of more than just other states. Instead of treating nuclear weapons as causes at the unit-level, it would be better to examine their effects at the unit-level. It would make more sense to treat them as parts of the material environment in which states exist, but over which individual states have limited control. Like geography, fundamental technological capabilities are constraints and opportunities that states must take largely as givens. Such underlying material realities are "structural" in the broad sense that they present actors with slowly changing powerful patterns of pressures and possibilities, but are not structural in Waltz's sense because they are not arrangements of parts.

Of course, Waltz's structural realism is not totally bereft of material dimensions. Like other Realists, Waltz emphasizes the importance of the *distribution* of capabilities. Waltz speaks about the importance of capabilities in shaping world political outcomes, but the only dimension of capability he analyzes is distributional, the relative capabilities of the actors in a particular system. Since the effects of nuclear weapons are clearly not distributional, Waltz is forced to speak of them as a sui generis phenomenon. The simple solution to this theoretical problem is to be found in analysis of the *composition* of power, the intrinsic ends to which a power asset can be put. In order to generalize and theorize about changes in the composition of power, we must look beyond Waltz's structural realism, but not beyond Realism.

Waltz's analysis of the effects of nuclear weapons also subverts his claims about the role of anarchy in shaping international relations. Nuclear-armed states remain in an anarchy in the formal sense that they are not subordinate to another juridical or administrative entity. But their patterns of interaction in the critical sphere of security are extremely orderly and highly certain, robbing interstate order of much of the disorder and unpredictability that justified conflating the technical sense of anarchy as absence of government with the more general sense of anarchy as chaos and disorder. In his classic analysis of the "third image" Waltz argued that war would occur because it was possible, but in the nuclear era this logic seems to have been reversed: nuclear war will not occur as long as it is possible. In Waltz's portrayal of nuclear deterrence, nuclear capability orders the state-system in much the same way that Hobbes' sovereign does civil society: peace is guaranteed for all actors, borders between the actors are preserved, and the actors face law-like certainty. Any force so powerful and pervasive deserves to be called the sovereign of the system. In Waltz's later formulations, nuclear weapons seem to function like an omnipotent but benevolent world sovereign. The sturdiness of the hedges that Waltz thinks have sprung forth from the bomb would surely be the envy of the architect of Leviathan.

The supreme ordering function of nuclear capability has further subversive implications for the state-system not explored by Waltz. He treats the overwhelming majority of the state arsenals as the residue of obsolete attitudes, as the product of illusion and institutional inertia. Waltz unflinchingly declares the main preoccupations of nuclear era statecraft to be much ado about little. But why stop here? Is it possible that the state itself, like the concern for counterforce, is also a "hangover from conventional days," an obsolete administrative conceit that nuclear weapons have robbed of substance and turned into a facade? If in fact the bomb provides so powerful an ordering principle, then the termination of juridical sovereignty might be little more than a formality, a bringing of form into conformity with substance. If the war-making "body" of the state is effectively paralyzed, then why accord such significance to its lingering "soul" of autonomy? If nuclear capability provides so many of the major features associated with an absolute sovereign, then it is also logical to conclude that the elaborate institutional apparatus premised upon autonomy and anarchy has been reduced to something of a facade, making it ripe for abolition.

In sum, Waltz's theoretical understanding of nuclear weapons is flawed both in its treatment of nuclear weapons as causes, or independent variables, and of the states and state-systems effected by them, or the dependent variables. In general, Waltz's analysis of both nuclear weapons as causes and of their effects is essentially ad hoc, with no attempt made to relate either cause or effect to more general classes of phenomena, either within or outside of structural realism. To explain the effects of nuclear weapons Waltz must step outside his broader theoretical framework without providing an alternative to it. In order to have a *theory* of the nuclear revolution rather than an illuminating case study, a more general logic relating levels and kinds of violence capability to different world security systems is needed. Ideally, the political effects of nuclear weapons should follow a pattern predicted by a theory formulated in sufficient generality to be applied fruitfully to outcomes in other eras as well.

In understanding nuclear weapons as causes, Waltz implausibly consigns them to the unit-level. A sounder approach would be to treat nuclear weapons as a part of the material environment external to, or "beneath," states. And to understand nuclear weapons as causes, it is also necessary to expand Waltz's emphasis upon capability distribution to include capability composition. In analyzing the effects of nuclear weapons, Waltz's lack of a theory of the state renders him unable to theorize about the ways in which units may have been changed, perhaps fundamentally, by nuclear explosives. At the system level, Waltz admits that nuclear weapons have overpowered the effects of structure, and admits that nuclear weapons have had powerful effects upon the overall international system, yet he provides no framework for classifying such systems, or criteria for measuring the completeness of the changes. However, the extremity of the effects of nuclear weapons in effectively paralyzing state-warmaking and robbing anarchy of its typical effects raises the possibility that an as-yet-unspecified system has replaced or is beginning to replace the system of states in anarchy.

Realism and the State as Ideal Rational Actor

If Waltz's two nuclear arguments have little relationship to his structural realism, then from whence do they derive? Waltz's nuclear arguments do not derive from his structural realism, but are rather a reductio ad absurdum of nuclear deterrence theory. Although Waltz does not have a theory of the state, or even a very full-bodied image of the state, he borrows crucial assumptions about the state from deterrence theory. These assumptions drive his entire analysis of nuclear politics. Stuctural realism did not assume that all states are rational, but rather held that those states that lacked the capacity to respond correctly to systemic imperatives gradually would be selected out of existence. But Waltz's "more may be better" argument assumes much more pervasive state rationality. Waltz's argument seems to assume that the progress of institutionalizing rationality and the progress of technology of destruction are closely linked.

On a wide front states fall short of what the revolution hypothesis would suggest, but this does not shake Waltz's confidence in their rationality. There is a jarring discrepancy between Waltz's dismissal of the concerns of states in counterforce, arms control, and nonproliferation, and his conviction that the lesson of the nuclear revolution will never be lost on states. This discrepancy is particularly pronounced in his arguments about nuclear proliferation. Waltz thinks "more may be better" because he holds that states will quickly grasp the simple realities of nuclear deterrence, and therefore will refrain from taking suicidal steps. But this view, as Joseph Nye has pointed out, assumes that prudence spreads along with technology:

> But the transferability of prudence assumes governments with stable command and control systems, the absence of serious civil wars, the absence of strong destabilizing motivations such as irredentist passions and discipline over the temptation for preemptive strikes during the early stages when new weapons capabilities are soft and vulnerable.[26]

One could multiply the list of disasters, follies, and disorders that have afflicted states in the past, and could befall them in the future, but the basic point is clear. In positing a happy outcome from proliferation, Waltz basically abstracts from all of the substantive attributes of states as concrete political entities and instead treats them as ideal rational actors.

Waltz's posture here is startlingly different than that found in most traditional Realist thinking. A central theme advanced by early postwar Realists such as Hans Morgenthau, Reinhold Neibuhr, and Martin Wight was that human beings were flawed, and that the organizations and projects of human beings could not ever fully escape these flaws. In Waltz's state, the follies and weaknesses of humans and their organizations that Realist theorists once emphasized have been left behind. Early Realist pessimism about human beings was aimed against the melioristic projects of Liberals, Internationalists, and Idealists. Although Waltz is silent on the topic of the perfectibility of human nature, he operates with a strikingly perfectionistic image of the state. In Waltz, we are confronted with a statist idealism. Waltz' views are "idealistic" both in the sense that he takes the existence of superstructures as given, but also in the sense that he has an idealized image of the state as an omni-rational entity. If one were to assume a more traditional Realist image of human fallibility, then one's confidence in the safety of the elaborate nuclear "fail deadly" deterrence order might not be so sanguine. Whatever its intrinsic strengths or weaknesses, it is doubtful that nuclear Panglossism is a Realist view.

The ultimate effect of Waltz's arguments about proliferation may be more damaging than helpful to the nuclear conventional wisdom that he obstensibly champions. Waltz's clever insight is to point out the discrepency between what works at the core among the greatest states, and what works at the periphery among the lesser states. The nuclear conventional wisdom codified in deterrence theory holds that peace among the superpowers has been greatly strengthened by the possession of nuclear weapons. But the nonproliferation regime is premised upon the view that nuclear possession among lesser states threatens peace and stability. Waltz's "more may be better" argument may be read, and seems to be intended to be read, as an attack upon the proliferation conventional wisdom. But his argument may also be read as an unconscious satire on deterrence theory, a reductio ad absurdum that calls into question our core theoretical beliefs by showing that they quickly lead to conclusions that contradict common sense and common practice. By showing us the gap between prevailing Realist theory and real security practice in the nuclear era, Waltz inadvertantly may have set us upon the path toward a more practical and realistic Realism.

Modified Structural Realism?

Steve Weber argues that the limitations of structural realism in understanding nuclear weapons can be overcome by modifying it.[27] Although Weber's analysis drives toward some of the conclusions of security materialism, he reaches his conclusions solely by modifying Waltz's structural realism. Weber also focuses solely on the relationship between structural realism and nuclear weapons, ignoring Waltz's own later remarks on nuclear weapons. Concluding that structural realism "leaves us empty-handed in our search for ways to think about nuclear weapons,"[28] Weber proposes a "revised realist framework" in which nuclear deterrence is understood as a "change in the organizing principle and joint custodianship is recognized as a new superpower function."[29]

Weber argues that nuclear weapons have produced a profound change in international relations that is structure-like in its character. He says that "while nuclear weapons per se did not alter the structure of the international system," nuclear deterrence has produced[11] a "system-wide condition that does constitute a particular kind of structural change."[30] Nuclear weapons, he says, have "engendered a structural change" because they have produced "a qualitative change not only in the destructiveness of war but also in the predictable consequences of war."[31] In this characterization of nuclear weapons Weber comes close to agreeing almost word for word with Waltz's own views. The key difference is that Weber wants to call this change "structural" while Waltz does not. Weber's argument is an improvement on Waltz's, but only a partial one. While Weber is willing to attach the label "structural" to nuclear weapons, he does not take to crucial next step and redefine or expand the meaning of the "structure" that nuclear weapons purportedly exemplify. Nuclear weapons clearly cannot be "structural" in Waltz's sense of the term, and so if the term "structural" is to be applied to them, some expanded or modified understanding of "structure," not provided by Weber, must be found.

The second part of Weber's proposed modification of structural realism is his claim that there is now a "significant degree of differentiation among the functions of states in the system," because the United States and the Soviet

Union now perform the new function of "joint custodianship" of the system.[32] Although it seems plausible that the adjustment of world politics to nuclear security imperatives might entail a change in the functional differentiation of the actors, Weber's concept of "joint custodianship" is unconvincing for two reasons. First, the idea that the leading states of the system can assume a special responsibility for the overall system is not significantly different from the traditional concepts of great power system management such as condominium, concert, and hegemonic coalition. Weber argues that the U.S.–Soviet custodianship is "fundamentally different" from the Concert of Europe, but the differences to which he points are incidental and exaggerated. In order to conceptualize the cooperative relationship between states caused by nuclear weapons in other than an ad hoc way, a more generalized typology of political forms existing between the stark Waltzian alternatives of hierarchy and anarchy is needed.

Weber's concept of nuclear functional differentiation also flounders on the problem of nuclear proliferation. If nuclear weapons are the cause of Weber's functional differentiation, then it is unclear why nuclear states other than the United States and the Soviet Union are not similarly distinguished. Furthermore, if this functional differentiation is based on the possession of nuclear weapons, then it could be expected to diminish with the further proliferation of nuclear weapons. Weber's argument thus produces the paradoxical conclusion that the functional differentiation produced by nuclear weapons would disappear as nuclear possession becomes universal, the point at which one would expect the impact of nuclear weapons to be greatest. The difference between nuclear and non-nuclear states marks the incompleteness of the diffusion of nuclear technology rather than functional differentiation. The basic problem with Weber's argument is that he conflates these two distinct phenomena. Despite these problems, the insight that attending to nuclear security imperatives will produce a functional differentiation of the units in world politics is important and deserves further development.

Security Materialism and Nuclear Weapons

The essentials of Waltzian structural realism are widely known, but security materialism is not so familiar to contemporary theorists. Therefore it will be necessary to describe the main features of this theory before it can be used to approach the question of proliferation. The aim here is not to vindicate or validate security materialism, but rather to describe its essential claims, to situate it within Realism, and then bring it to bear on the proliferation question.

Security Materialism

The branch of Realist theory that posits that the type, size, and number of the basic security providing institutions are determined by the material capabilities for destruction and protection can conveniently be labelled security materialism. Such arguments are often found in the geopolitical literature, along with many other types of claims. Because it is concerned with violence and security, security materialism is clearly a form of Realism. It is not, however, mainly a theory of international or interstate *relations*. Rather it is a set of propositions about the genesis, nature, and limits of particular forms or modes of security institutions, including states. Unlike many other forms of Realism, security materialism does not presume the state or the state-system, but rather attempts to explain why and when states and state-systems are the prevalent form of security institution. Stephen Krasner's recent claim that "[t]he ontological given for realism is that sovereign states are the constitutive components of the international system," does not hold true for this variety of Realism.[33] Thus the dependent variable in security materialist theory is not interstate relations, or even the state or state-system, but rather the world security order. Security materialism may thus be thought of as *generative realism* – it attempts to explain the genesis of security unit types.

The largest body of security materialist theory is found in the writings of the classical or global geopoliticans, notably Mahan, Mackinder, Haushofer, Spykman and a host of less well-known writers. Although the theoretical orientation of such writers was often muddled and heterogeneous, their main debate was clearly a variety of security materialism. They argued about how the new material forces produced by the industrial revolution (such as the railroad and the telegraph) interacting with the largest-scale geographic features of the earth would shape the character, number, and location of viable security units in the emergent global scale security system.[34] A somewhat less coherent body of even earlier writings sought to explain the genesis, character, and persistence of the European state-system as the product of the interaction a fragmented topography and a geography permitting extensive naval and maritime activity.[35]

Because of its focus upon very basic or deep structures' security materialist theory can be thought of as a *paleo-realism* in contrast to the dominant neo-realism of Waltz and Gilpin, and the fading classical realism of Morgenthau, Wight, and Bull. In Robert Gilpin's terms, security materialist theory is about system rather than systemic change.[36] And in contrast to Waltz's concept of structure as the ordering or arrangement of the units, security materialism is a theory about how the *deep structure* of material forces determines the type and number of the units. Security materialism attempts to understand large and infrequent changes in the constitutive units of security orders, and how they are driven by very basic alterations in the material forces of destruction that largely lie outside the control of humans. In contrast, most other Realist and particularly neo-realist theories seek to explain behavior among such units and within such systems. Because of this basic difference in what they are trying to explain, security materialist theories are complementary rather than competitive with Realist theories of interstate relations such as structural realism and hegemonic succession. These theories may, however, clash when they differ as to whether a particular change is, in Gilpin's terms, in the system or systemic.

In addition to attempting to explain the causes (rather than the consequences) of particular distributions of power, security materialist theory also has sought to understand how variations in the *composition* of power (that is, the intrinsic strengths and limits of different power assets) shape political outcomes. Most neo-realist thought operates with what might be called a "money pile" view of power assets, in which their essential commensurability is assumed. In contrast, security materialism, drawing from earlier geopolitical theory, employs what might be called a "tool-box" concept of power, in which different physical assets have intrinsic strengths and weaknesses. Security materialism attempts to show that significant variations in the composition of power shape security orders by determining the basic type of security unit that is feasible. Thus security materialism views the advent of nuclear weapons not primarily as a change in the distribution of power, or in the structural relationship of the units, but rather as a change in the composition of power. Nuclear weapons are understood as marking a qualitative change in violence capability that forces a qualitative change in security institutions.

Security materialist theories can be conveniently schematized using the conceptual apparatus of Marxian historical production materialism in which a base or infrastructure constituted by largely exogenous independent variables (in this case, the forces of destruction and protection, rather than of production), determines a superstructure of modes and political relations (in this case, protection modes and security orders, rather than class and society). In order to capture the features of security materialism relevant here, I briefly summarize its claims on technological determinism, historical periodization, and the state as a mode of protection.

Technological Determinism

A key feature of the security materialist model is that the interaction of geographical and technological factors is the driving or independent variable. Behind this move is the assumption that technology may in certain key respects be treated as a fact of nature to which the actors must adjust rather than something that they can call in and out of existence. In contrast, neo-realists such as Waltz and Gilpin maintain that technology is primarily derivative of human and political choice, and that technology is socially determined rather than socially determining.[37] Security materialism is thus a version of technological determination. Recent works on international relations and technology by Robert Art, Barry Posen, Matthew Evangelista and others have argued against "technological determinism." By this they mean that if a technical capability is possible or feasible, then it will be developed and/or used.[38] This image of technological determinism is, however, something of a strawman, for despite occasional rhetorical proclamations to this effect, it is hard to seriously suppose that every device which can be built will be built. In contrast, the security materialist model posits that the existence of a significant technology related to security, will either be built, or else some mechanism for its regulation will be established.[39]

The materialist argument depends upon the distinction between *scientific research*, *scientific knowledge*, *technology*, and *technics*. Broadly speaking, scientific research is a process of seeking and validating claims about nature. Scientific knowledge is the set of truths about nature discovered by scientific research. Technology, strictly speaking, is engineering "know-how," a practical knowledge of how natural principles can be put to work or harnessed to perform certain tasks. And technics are the actual tangible machines, apparatuses, and devices that are the product of labor and technology applied to natural resources. Scientific knowledge and technological "know-how" are intangible systems of symbols, while technics have materiality and can directly interact with nature. Another important difference is that scientific and technological knowledge diffuse, while technics have (or lack) mobility. Many of the effects commonly attributed to "technology" are really the effects of "technics."

Given these distinctions, there are three main ways in which scientific-technological phenomena can be thought of as standing, like non-human physical nature, as significantly outside human control. First and most importantly, the features of nature that science apprehends are not of human creation or subject to human alteration. The discovery of some new feature of nature is in reality a change in the human environment, much like the discovery of a new ocean or mountain range. Recent scholarship in the history of science has emphasized how social context determines the scientific enterprise, but there is an inescapable limit to the social determination of science. Humans may control, with increasingly conscious awareness, the rate and direction of scientific research, but they do not pick and cannot change the features of nature discovered by science. Scientific research finds natural realities that may or may not fulfill the social priorities of the research.

Second, the human environment, once changed by the arrival of scientific knowledge and technical know-how, cannot easily be returned to its previous state. Since scientific knowledge and technological know-how are both information, it is very difficult reverse or eradicate them. New realities of nature and know-how rarely disappear, unless supplanted by some newer version of scientific knowledge or technology.[40] Modern scientific knowledge and technology is widely disseminated and capable of being regenerated due to its systematic character. Barring a massive civilizational collapse or regression, there is little prospect that any substantial body of modern scientific or technical knowledge will disappear.

Third, technics are for human purposes like objects of nature. Their scope of potential consequence is set by their physical features interacting with their environment, rather than by the intent or purpose of their makers or users. Technics do not have to be built, and once they are built they can be destroyed. But while in existence, technics are part of the physical world, and their consequences are bound by natural physical laws as well as human choices as to how to employ them.[41]

To sum up on technology, security materialism posits that natural forces – particularly geography and technology – have an objective reality outside human intent or desire. These forces pose obstacles and opportunities for human institutions as they pursue the long-standing and basic goal of providing physical security. As these material realities change, human institutions are forced to adapt, sometimes successfully, and sometimes not. As humans struggle to cope with these realities, they come to be reflected in the main features of the world security order.

Historical Periodization

Unlike those varieties of Realism that emphasize recurrence and historical cycles, security materialism attempts to explain non-recurrent historical change resulting from the cumulative transformation of the human interaction with nature. Different eras are defined by the salience of different combinations of geographic and technological realities, and security providing institutions are conceived as being in essentially adaptive relationship to this material environment.

Security materialism identifies a fairly simple trajectory, in which the minimum size of the militarily viable terrain keeps expanding, and therefore the number of viable security communities keeps shrinking (given a finite earth). These scale shifts are driven by the simple and probably irreversible expansion in the growth of destructive power and the increase in the speed of weapons. Each successive scale shift has seen the superstructure of the state-system change – usually in a war – to take account of the new realities of security. These scale shifts have profoundly altered the world security order, and how the security-providing institutions have responded has profoundly affected the specifics of the scale shift.

This evolution in material constraints and opportunities has generated a succession in the type and number of security-providing institutions. In the pre-global era, many separate entities could achieve some semblance of security, relatively autonomous security systems were possible, and a world-wide state was impossible. During the global industrial era beginning in the 1850s and culminating during the Second World War, only a handful of entities of continental scope were capable of achieving security. As the world security system moved into equilibrium with the underlying material forces, the state mode of protection had proven its viability upon a new scale. Despite the anticipations of internationalists, the scope for interstate security organizations in the global industrial era dominated by continental superstates was limited to the marginal role codified in the UN Security Council mechanism.

Since the closing days of the Second World War, the *global* era has ended, and a new era *planetary* in scale and scope has begun with the advent of nuclear explosives, intercontinental ballistics, and planetary-scale information systems. The most important theoretical pay-off from security materialism is its insight into the implications of the planetary geopolitic, and particularly the impact of the master technic of nuclear explosives upon the state as a form of security institution.

The Real-State

Before it is possible to weigh the effects of nuclear explosives upon security providing institutions, forms of security-providing institutions must be specified. For such purposes, it is useful to think of the state as a "mode of protection," in the sense that there is a very distinctive, and widely-recurring statist approach to power and protection. In deference to rich tradition of European realpolitik, and in order to avoid the semantic confusions inherent in offering yet another definition or re-definition of the state in general, I will refer to the state mode of protection as the *real-state*. The real-state is a type of political institution with four inter-related features: [1] the *monopoly* of violence capability within a particular territorial space; [2] the *concentration* of control over that violence capability in the hands of a distinct organization, the state apparatus; [3] the relative *autonomy* of the state apparatus; and [4] the tendency to *employ* the capability at its disposal and thus to *couple* capability to outcomes. This concept of the state as a protection mode is held, with minor variations, by a wide range of Realist theorists.[42] According to this formulation, not every authoritative political system, state apparatus, territorially distinct polity, or recognized sovereign state is a real-state. Rather, real-states are governmental apparatuses that have other essential attributes and are capable of performing certain tasks – particularly the provision of security through the maintenance of a monopoly of violence capability in a particular territory.

Concentration and autonomy are closely linked to violence monopolies. For power to be monopolized in a meaningful fashion, it must be concentrated into the control of one organization. An organization with such a concentration of power will tend toward autonomy, both because its concentrated monopoly renders it intrinsically difficult to control, and because autonomy greatly facilitates effective and efficient employment of its concentrated power. To achieve autonomy, the state apparatus strives to avoid checks and limits imposed either from within its polity or from outside it. The tendency for a state apparatus to seek autonomy from external control is not confined to situations where violence capability lends itself to compartmentalization, but is most likely to be realized in such situations.

Real-states also have a constitutive tendency to *employ* the instrumentalities of violence at its disposal, and to *couple* political outcomes tightly to capability. The idea that real-state apparatuses have violence capabilities in order to employ them for political ends, particularly against other states, is so central to the political craft of the real-state that it is rarely explicitly articulated. Real-state apparats seek to squeeze as much political advantage out of weapons capability as possible.

The real-state also has a powerful ideology and a rich practitioners lore. The ideology of realpolitik and the closely related theory of classical or diplomatic realism provides the intellectual superstructure for the real-state. Realpolitik and classical realism provide real-state operatives with simplifying axioms to guide behavior, historical models to emulate, and normative claims to legitimate and justify their activities.[43] At its most basic level, realpolitik consists of ideological statism – the claim that the provision of security will and should always be provided by real-states. The concepts and practices of realpolitik and classical realism, like those of any ideology, are often cast in timeless and universally applicable terms, but are only accurate and useful in situations where the real-state exists.

To be a plausible guide to system change, security materialism must provide an explanation for the genesis and prevalence of the real-state. In very simple terms, the historical viability of the real-state rests upon the the degree to which violence capabilities lend themselves to territorial segmentation and compartmentalization. Economies of scale in the provision and delivery of security services determine the size at which units in the real-state mode emerge. In systems where more than one security unit is viable, balancing against outside threats is necessary. Also, for much of human history the volume and velocity of violence capability have been severely limited, and small relative differences have had a big impact upon security. Given these realities, polities have needed a distinct and autonomous state organization to mobilize and concentrate power and then employ it against outside threats.

The End of the Real-State

Security materialism applied to the nuclear era generates a view that I label *nuclear one worldism*. Nuclear one worldism contains two sets of claims: first, about the relationship between the real-state and nuclear explosives; and second, about the character of security governance congruent with nuclear security imperatives.

The heart of nuclear one worldism is an analysis of the impact of nuclear explosives upon the real-state. In general terms, nuclear weapons deny the possibility that a political order of less than universal scope can monopolize violence capability within its territory. This fundamentally challenges and compromises the real-state. Nuclear weapons are to the real-state what gunpowder was to the medieval fief – a technology that renders it militarily inviable. Prior to the advent of nuclear explosives, the scale requirements of military viability required that an entity be of approximately continental scale, as the World Wars demonstrated. With the advent of nuclear explosives, an entity must encompass the entire globe, or else monopolize all nuclear capability (in which case one state would be secure, and all the rest insecure). The scale imperatives of nuclear explosives mean that security can only be achieved through either the creation of a worldwide state or through the abandonment of the real-state approach to security and the creation of a security order that systematically paralyzes and bounds state and nuclear power.

Nuclear weapons render the real-state simultaneously superfluous and insufficient. Nuclear weapons ease some problems that the real-state mode struggled to solve, but they create others for which it is particularly ill-suited. Because they are so destructive, nuclear weapons largely solve the problem of balancing that animated the real-state. Nuclear weapons serve as *boosted balancers*. A relatively small number of them are enough to achieve balancing, and differences in their number beyond small numbers do not matter much. Because they provide so much destructive capability, mobilizing a sufficiency of violence capability greatly shrinks as a task for government. And because of related technological changes in the velocity of violence, the concentration of power in the state and its autonomy are both greatly enhanced. In short, nuclear weapons provide so much violence capability that they make outmoded institutions designed to cope with a world of violence scarcity.

While easing the problem of balancing, nuclear weapons create a new set of security needs that the real-state cannot meet. Security threats in a nuclear world stem from violence excesses – of overabundance and overrapidity. In general terms, this generates the need for a *separation* of power. The real-state is constitutively ill-prepared to achieve separation. The speed with which the superabundance of violence capability can be employed produces the need for institutions that artificially slow the process of employment. Whereas previously polities had to struggle to build institutions with sufficient autonomy and concentration, in the nuclear era autonomy and concentration are too extreme and so must be tempered. Instead of trying to connect their violence capabilities to the resolution of political disputes, polities must do the reverse – de-couple, insulate, and separate violence capability from political conflicts. Instead of war being the extension of politics, polities persist by the suspension of war.

The pursuit of security in the nuclear era also requires a reversal of approach to monopoly and consolidation. Whereas the real-state strives for security through monopolizing violence in its particular territory, security in the nuclear era requires polities to re-program their security institutions to permanently accept the impossibility of such monopolization. In the past the secure existence of separate units made unnecessary the operation of a general system to regulate violence, but in the nuclear era the existence of separate units is derivative of the general operation of a nuclear regulatory system.

In terms of geopolitical periodization, nuclear weapons mean that even the largest survivors of the previous scale consolidations have been rendered insecure. Whereas the shift from the pre-global to the global era altered the scale upon which states could achieve security, the shift from a global to a planetary geopolitic has thrown into question the territorial state as mode of protection. A planetary geopolitic presents a fundamentally new set of constraints and opportunities for the achievement of security to which the state mode of protection and statist realism are poorly matched. In this era, realpolitik is converted into an idealistic orientation, dangerously out of sync with real security imperatives. The adjustment of institutions to the new realities has been slow, in part because the previously quick mechanism of adjustment – war – has itself become detached from security objectives.

A simple analogy captures the impact of nuclear explosives upon the real-state. As Arnold Wolfers suggested, realpolitik theorists of states liken them to "billiard balls," knocking each other about, and only weakly constrained by the "cobwebs" of interstate norms. Nuclear weapons have transformed the billiard balls into "eggs." The "hard shelled" billiard balls of pre-nuclear times knocked each other about for political and security gain, but collision has become suicidal for the inescapably vulnerable eggs of the nuclear world.

Where the billiard ball states crashed through the weak webs of international norms and institutions, the fragile nuclear egg states sit paralyzed.

Nuclear Contradiction

Nuclear weapons have rendered the real-state approach to security inviable, but they have not yet generated a full substitute. Nuclear weapons do not immediately eliminate state apparats with realpolitik orientations, only their viability as security providers.

The concept of *contradiction* is the key to understanding this schizophrenic situation. In social theory, a contradiction is a "maladjustment" or an "incompatibility" between different aspects or parts of social and political systems. Contradiction plays an important role in Realist theories of balance of power and hegemony,[45] as well as in Marxism. Much of the explanatory power and practical appeal of these theories stems from their ability to analyze disjunctures between social systems and underlying material circumstances. The concept of contradiction thus gives Realist and Marxist theories a powerful *critical* dimension. By identifying contradictions Realist and Marxist theorists are able to criticize policy choices and institutions, and identify alternatives more congruent with underlying power realities.

The concept of contradiction can also be employed to illuminate situations in which changes in the forces of destruction have obsolesced a protection mode. A key criteria for judging whether such a contradiction exists is the degree to which dominant security approaches generate counterproductive consequences when they are brought to bear on security issues, and the degree to which security derives from the ad hoc departures from the dominant approaches rather than from their application.

With the effective abolition of the real-state, the remaining state apparats with realpolitik orientations are thrown into a state of contradiction with the existing forces of destruction. They have become drained of their previous functionality, and become facades, unrelated, except negatively, to the actual achievement of the security goals that motivated their creation. In this condition of contradiction, realpolitik attitudes and institutions threaten rather than enhance security, and disaster is avoided by suspending and departing from statist approaches. As the ability of states to monopolize violence-capability in a particular territory becomes unobtainable in a nuclear world, there emerges a fundamental misfit between statist approaches to power and security and the constraints and opportunities for achieving security. Like natural monopolists suddenly thrown into one market by a technological advance, states in a nuclear world must achieve some form of consolidation in order to avoid ruinous competition.

The Logic of Nuclear Adjustment

Security materialism and nuclear one worldism generate two distinct images of adjustment, of institutional configurations congruent with the security opportunities and constraints of the nuclear era. Both varieties of nuclear one worldism hold that a more consolidated security providing institution is necessary, but differ on its character, and the path to it.

Operating with an implicit security materialist framework, early nuclear one worlders sketched statist solutions. Figures as diverse as James Burnham, Hans Morgenthau[46] and Robert Hutchins thought only a world *state* with a monopoly of the means of destruction could solve permanently the problem of nuclear explosives. In the world state solution nuclear capability is *concentrated* and then *employed* to maintain peace between the disarmed or dismantled states. With the *multi-state* system no longer able to provide security, they looked for salvation in an *omni-state*.

Projecting the state form to a world-scale was bolder in practice than in theory. This agenda resulted from the extrapolation of pre-nuclear images of peace into the nuclear era, rather than from an analysis of the novel and distinctive security conditions created by nuclear explosives. In retrospect, the expectation that nuclear weapons would stimulate the emergence of a world state was either incorrect or premature. But the key argument of the nuclear one worlders on the increasing scale of military viability is not so much wrong as incomplete. Their diagnosis is correct to a first approximation. And the outcome of a world state emergent from a nuclear war remains the simplest, or default version of nuclear one worldism.

The prediction of territorial consolidation failed to materialize because of the simultaneous presence of another, seemingly opposite, trend in the relationship between violence capability and territorial security from conquest. The contemporary composition of the powers of destruction and protection are bifracated: *annihilation is simple*; *conquest is hard*. It is nearly impossible to protect territory from annihilation; but it is easier than ever to prevent conquest. This means that territorial control and the upper reaches of destructive capability have, after millennia of close connection, largely parted company. The bifracation of contemporary violence capability reinforces the viability of territorial security units in some ways, while compromising them in other ways. These underlying material forces thus pull world order in opposite directions at once – away from centralized territorial control but towards the consolidated management of territory-negating nuclear capability.

Considered together, these two features create a novel configuration of security threats and possibilities for governance structures. The old problem of aggression is now much less acute. Peace can be kept without elaborate or powerful international institutions to prevent *interstate aggression*. As the problem of interstate territorial aggression has declined, the solution of "collective security" has declined in relevance and appeal. But the high costs of conquest do not eliminate the threat of annihilation posed by nuclear explosives. This threat, however, can be addressed without creating a government structure that aggregates of territories (and thus populations) world-wide. By side-stepping the aggregation of territories, nuclear security institutions do not need to address the many other intractable problems afflicting and differences dividing peoples. This points to the possibility for an institution that contains and paralyzes nuclear capability, but does not aggregate all territory into one state. The disposition of territory-negating capability (which are distinctive and specialized) rather than checks upon territory aggregating capabilities becomes the central and circumscribed problem for interstate security institutions.

This modified image of the nuclear one world diagnosis implies a modified image of the character of a world order in which the reality of nuclear capability has been reflected adequately in institutions. This less developed and more complicated image of a world order in adjustment to nuclear realities is one in which the state-system is augmented and complemented with a nuclear containment system rather than abolished and replaced. In this image of adjustment nuclear capability is *separated* from state control and *paralyzed*. This image of adjustment is implicit in the architecture of the Acheson–Lillienthal Plan and has been kept alive (but never given adequate theoretical critical mass), by Leonard Beaton,[47] John Barton,[48] and Jonathan Schell.[49] Because it envisages a major institutional center of gravity between the states, but is not itself a state, conceptualizing such an alternative in the largely statist vocabulary of "international" relations theory is difficult. This image of adjustment as separation and paralysis rather than concentration and employment is an extrapolation of the observed elements of state adjustment in the nuclear era, and takes seriously that the state mode, rather than the size of states, has been rendered obsolete.

In more general terms, the logic of sustainable nuclear governance is essentially *republican* in character. By this I mean institutions in which devices such as balance, separation and mixture serve to limit, check, and constrain power. Whereas statist institutions seek to mobilize and wield power, republican ones seek to constrain, bind, and paralyze both the instruments of violence and the actors that might employ them.[50] In the nuclear era, security and survival depend upon the systematic de-coupling and dis-employment of power capabilities from interstate conflict resolution. In a nuclear world, security derives from the paralysis of states rather than from the exercise of state power, and from the acceptance of the impossibility of territorial violence monopolization rather than from its pursuit. Thus we can measure adjustment by the degree to which apparats shed their realpolitik orientations and become effectively bounded by series of institutionalized constraints. This runs fundamentally against the grain of statist approaches to power.

Such republican tendencies characterize both the first steps of adjustment visible in world politics and more robust systems of nuclear control needed to close the gap between security objectives and nuclear realities. As they respond to nuclear security imperatives, states as security-providing institutions have been increasingly republicanized. They are first paralyzed, and then increasingly displaced, complemented, and enveloped by larger, less-centered, and less-sovereign institutions whose logic is republican. The norm of nuclear no use, interstate arms control agreements, and the co-ordinate design of nuclear forces constitute important, but incomplete, institutional adjustments to nuclear security imperatives. These are crucial first steps down a path whose essential logic is republican rather than statist. These positive steps remain incomplete, and they coexist uneasily with nuclear doctrines and force structures that are continuing sources of insecurity and evidence for continuing institutional maladjustment.

A more complete and sustainable institutional adjustment to nuclear security imperatives would result from intensifying the republicanization of nuclear governance. Full adjustment need not be a world state, but rather could be a republican union dedicated to the comprehensive control of nuclear capability and the internal re-configuration of states to constrain quick executive action. In such a situation of adjustment, the state-system is not abolished, but rather limited and complemented by security institutions of a non-statist logic. Such a situation does not mark the overthrow of the balance of power, but rather its complication by the other republican institutional forms of separation and mixture. The resulting mix of territorial states and a nuclear republican union is more complex than a world order composed of a series of real-states, but is more in alignment with real security imperatives. In contrast to Steve Weber's view that some states serve functions in the system that others do not, a world political system adjusted to nuclear security imperatives would be functionally differentiated in a quite different way: territorial states will serve some security functions, while other non-state institutions serve others.

Our analogy of billiard balls and eggs can be extended to capture this republican image of nuclear adjustment. After the invention of nuclear explosives turns billiard balls into eggs, the continuing quest for security drives fragile states to create cushioning and protective structures not unlike an "egg carton." Where the billiard states crashed through the weak webs of international norms and institutions, the fragile nuclear egg states are first paralyzed, and then begin creating an egg carton, a system of mutually protective norms and institutions. Materialism illuminates the transformation of billiard balls into eggs, republicanism the logic of egg cartons. With the emergence of a protective republican union, the separate units of the world political system do not disappear, nor are they absorbed into one superstate. And the emerging protective egg carton – the nuclear republican security union – is not a state and it does not transform states back into billiard balls.

Security Materialism and Nuclear Proliferation

Security materialism applied to the nuclear era produces nuclear one worldism, in both a simple world-statist and a complex republican form. What does this model imply about the proliferation of nuclear weapons? Here, as before, the main thrust of the argument is deductive, an effort to generate from the general model a set of more specific expectations and insights about nuclear proliferation. Nuclear one worldism makes three main points about nuclear proliferation, [1] on the motives for and pacing of nuclear proliferation, [2] on the appeal of the nonproliferation regime and its consequences for world order, and [3] on the consequences of nuclear proliferation.

First, security materialism and nuclear one worldism suggests that, ceteris paribus, polities might not be powerfully attracted to nuclear possession. If, as nuclear one worldism postulates, the acquisition of nuclear explosives fundamentally calls into question the real-state mode of protection, then polities can be expected to eschew acquisition. In the nuclear one worldist view, to possess nuclear weapons is to be possessed by them. Nuclear weapons add to the violence capabilities of a state, but not in ways (in most circumstances) that add to its security and expand its choices. Nuclear one worldism suggests that the more polities know about nuclear realities, the less they will want to get possessed by them.

This expectation resonates with an important feature of the nuclear era – the fact that the spread of nuclear weapons has been so slow. In the early days of the nuclear era, it was assumed widely that states would acquire nuclear weapons capability as soon as they were able. These expectations followed from the simple assumptions that states seek to maximize power assets, and that nuclear weapons are power assets. In retrospect, these expectations have proven to be incorrect. World politics has not been marked by the extensive acquisition of nuclear weapons, even by states with the ability to acquire them. Since the 1950s, what Albert Wohlstetter dubbed the "nuclear overhang" – the gap between the number of states with the ability to fabricate nuclear weapons, and the number of states actually possessing nuclear weapons – has grown progressively larger with each passing year.[51] Because it hypothesizes that polities will eschew nuclear possession, nuclear one worldism provides a simple and straightforward explanation for this "nuclear overhang" phenomenon.

A second and related conclusion of nuclear one worldism concerns the nonproliferation regime. Modified or complex nuclear one worldism generally postulates that security governance in the nuclear era will tend to take the form of a republican union in which states and nuclear capabilities are both limited, bound and paralyzed. Such a security order would entail restraints on the number of actors with access to nuclear capability, norms of non-possession, and an elaborate system for the monitoring of nuclear capability. In short, security materialism and modified nuclear one worldism postulates that something like the non-proliferation regime will emerge. The current nonproliferation regime constitutes only a weak version of one facet of a full-fledged nuclear republican security union. Nevertheless, this regime does exist, and a substantial political effort has gone into its creation, expansion, and maintenance.

It is important to note that only security materialism and nuclear one worldism, among the varieties of Realist theory, provides a straightforward explanation for the nonproliferation regime. For Waltz, the existence of this regime is a complete enigma and anomaly. An alternative simple Realist explanation for the regime is that it represents the interests of the strong nuclear possessors against the weak nuclear aspirants. But this explanation implies that the strong possessors coerce or bribe the weak aspirants, and is confounded by the intense support that the regime has among many non-possessor states.

The third contribution of security materialism to understanding the phenomenon of proliferation concerns its consequences. Although the modified version of nuclear one worldism postulates that a republican union is the optimum the form of sustainable nuclear governance, it does not guarantee that the path to this outcome will be smooth and disaster-free, or that this outcome will prevail over the simple or default outcome of a world nuclear war or state. When all is said and done, security imperatives may not prevail over statist and bellicist tendencies. Nuclear weapons possibly pose a challenge too great for human governance. History may end in the ash heap.

Nuclear proliferation intensifies the contradiction between the institutions of security and the forces of destruction. Given the nuclear one worldist proposition that the size of a securable space has expanded to world-wide scale, or become "unitary," the number of viable nuclear security units is also expected to follow suit, and thus become "unipolar." The extent of contradiction can be gauged by the discrepancy between the number of securable terrains, and the number of security institutions. Thus, a world with three nuclear armed states is more in contradiction than a world with two, and so on.

As the contradiction between the institutional superstructure and the material forces intensifies, the pressure for a correction or major institutional change increases. (Should this contradiction intensify or remain high for an extended period without a correction or major institutional adjustment, then the basic proposition of nuclear one worldism is called into doubt.) In simple terms, the more state actors have nuclear weapons, the more likely it is that nuclear weapons will be used, causing the elimination of some actors, and/or triggering an institutional consolidation. Also, as the number of states with nuclear explosives increases, the number of potential nuclear adversaries increases as well. This, in turn, can be expected to increase the incentives of all states to construct a consolidated nuclear governance system. In short, nuclear proliferation can be expected to intensify the contradiction between the organization of the security system as a multi-state-system and the unitary security potential, and thus accelerate its resolution.

Notes

An earlier version of this article was presented at the Conference on Social Science Theory and Nuclear Weapons Proliferation, held at the Center for International Security and Arms Control, Stanford University, 5–6 February 1993. The author would like to thank Michael Desch, Benjamin Frankel, Avery Goldstein, and Scott Sagan for helpful comments on the earlier draft. Parts of this article are derived from *Pax Atomica: Planetary Geopolitics and Republicanism*, forthcoming.

1. Kenneth Waltz, *The Spread of Nuclear Weapons: More May Be Better*, Adelphi Paper no. 171 (London: International Institute of Strategic Studies, 1981).
2. Kenneth Waltz, *Theory of International Politics* (New York: Random House, 1979).
3. See in particular: Waltz, *More May Be Better*; Kenneth Waltz, "Reflections on *Theory of International Politics*: A Response to My Critics," in Robert O. Keohane ed., *Neorealism and Its Critics* (New York: Columbia University Press, 1986); and Kenneth Waltz, "Nuclear Myths and Political Realities," *American Political Science Review* 84, no.3, (September 1990): 731–45.
4. Waltz, *Theory of International Politics*, 81.
5. Ibid., 88.
6. "Supranational agents able to act effectively, however, either themselves acquire some of the attributes and capabilities of states, as did the medieval papacy in the era of Innocent III, or they soon reveal their inability to act in important ways except with the support, or at least the acquiescence, of the principal states concerned with the matters at hand." Ibid., 88.
7. Ibid., 82.
8. Ibid., 96.
9. "Students of international politics make distinctions between international-political systems only according to the number of their great powers." Ibid., 17.
10. Ibid., 97.
11. Waltz, "A Response to My Critics," 329.

12. Waltz, *Theory of International Politics*, 181.
13. Waltz, "Nuclear Myths," 731.
14. Ibid., 738.
15. Ibid., 738.
16a. Ibid., 735.
16b. Waltz's view here resembles Bernard Brodie's late thought on arms control: arms control may help avoid wasteful spending, but does not alter security fundamentals. Bernard Brodie, "The Objectives of Arms Control," *International Security* 1, no. 1 (Summer 1976).
17. Ibid., 737. For a more extended examination of the persistence of pre-nuclear strategic approaches in the nuclear era, see: Robert Jervis, *The Meaning of the Nuclear Revolution: Statecraft and the Prospect of Armageddon* (Ithaca: Cornell University Press, 1989).
18. See for example Richard Ashley, "The Poverty of Neorealism," *International Organization* 38, no. 2 (Spring 1984): 225–86.
19. Waltz, "Nuclear Myths," 734.
20. Waltz, *Theory of International Politics*, 173.
21. Waltz, "A Response to My Critics," 327.
22. Ibid., 327.
23. As R. B. J. Walker observes, "And not the least significant characteristic of Waltz's analysis is the absence of any serious theory of the state." "Realism, Change, and International Political Theory," *International Studies Quarterly* 31, no. 1 (March 1987): 78.
24. This incongruence is noted by Joseph S. Nye, Jr., "Neorealism and Neoliberalism," *World Politics* 30, no. 2 (January 1988): 235–51.
25. John Gerard Ruggie, "Continuity and Transformation in World Politics: Toward a Neorealist Synthesis," *World Politics* 35, no. 2 (January 1983): 261–85.
26. Joseph S. Nye, Jr., "Maintaining a Nonproliferation Regime," in George H. Quester, ed., *Nuclear Proliferation: Breaking the Chain* (Madison: University of Wisconsin Press, 1981), 32.
27. Steve Weber, "Realism, Detente, and Nuclear Weapons," *International Organization* 44, no. 1 (Winter 1990): 55–82.
28. Ibid., 62.
29. Ibid., 65.
30. Ibid., 55.
31. Ibid., 63.
32. Ibid., 64.
33. Stephen Krasner, "Realism, Imperialism, and Democracy: A Response to Gilbert," *Political Theory* 20, no. 20 (February 1992): 39.
34. For an extended reconstruction, see: Daniel Deudney, "Global Geopolitics: Materialist World Order Theories of the Late Nineteenth and Early Twentieth Centuries" (unpublished Ph.D. dissertation, Department of Politics, Princeton University).
35. For a reconstruction, see: Daniel Deudney, "The Natural Republic of Europe, ca 1500–1900: The Geopolitics of a Balancer State-System" (Paper presented at the annual meeting of the International Studies Association, Vancouver, British Columbia, March 1991).
36. Robert Gilpin, *War and Change in World Politics* (New York: Cambridge University Press, 1982), 39–43.
37. The view most prevalent among contemporary international relations scholars is stated by Robert Gilpin: "... whereas environmental factors such as climate and geography lie outside of state control, the technological environment is man-made, and a society will develop technological capabilities in order to gain an advantage over other states." *War and Change in World Politics*, 55. For

further elaboration on this view of technology see Robert Art and Kenneth Waltz, "Technology, Strategy, and the Uses of Force," in Art and Waltz, eds., *The Use of Force* (Boston: Little Brown, 1971).

39. Should a technology significant for security emerge and not generate either deployment or an institutional adjustment to constrain employment, then the security materialist model would be falsified.
40. History provide examples of forgetting, such as the loss of knowledge about siege machines in Western Europe after the fall of the Roman Empire, and the loss of the formula for Byzantine "Greek Fire." Loss of knowledge was much more likely to occur in pre-modern times when it often was a craft secret held by a few individuals and was unconnected to systematic science.
41. An example of this is provided by Arnold Toynbee, who noted that the Roman system of roads built to expand military mobility were as useful to the invading Germanic tribes as they were to their builders. Arnold Toynbee, *Universal States*, vol. 7A of *A Study of History* (Oxford: Oxford University Press, 1947), 91 and 703.
42. For a classic statement see John Herz, *International Politics in the Atomic Age* (New York: Columbia University Press, 1959).
43. As Richard Ashley has pointed out, much of classical realist literature is not a set of social science propositions, but is rather "the ethnomethodology of the modern tradition of statecraft." Richard Ashley, "The Poverty of Neorealism," *International Organization* 38, no. 2 (Spring 1984): 265.
44. In the "billiard-ball model" states "represent a closed, impermeable and sovereign unit, completely separated from all other states." Arnold Wolfers, *Discord and Collaboration: Essays on International Politics* (Baltimore: Johns Hopkins University Press, 1962), 3–24. This image is similar to John Herz's "hard-shelled" vs "permeable" units. *International Politics in the Atomic Age* (New York: Columbia University Press, 1959).
45. For example, in Realist theories of hegemony, a contradiction occurs when an interstate system is still organized around a hegemon whose power is no longer sufficient to maintain its position of privilege and pre-eminence in the system. As Gilpin puts it, there can be "a disjuncture between the basic components of the existing international system and the capacity of the dominant state or states to maintain the system." *War and Change*, 157.
46. For discussion of Morgenthau's areas of agreement with the nuclear one worlders, see James Speer, "Hans Morgenthau and the World State," *World Politics* 20, no. 2 (January 1968); and Richard Rosecrance, "The One World of Hans Morgenthau," *Social Research 48*, no. 4 (Winter 1981).
47. Leonard Beaton, *The Reform of Power: A Proposal for an International Security System* (New York: Viking, 1972).
48. John Barton, "The Proscription of Nuclear Weapons: A Third Nuclear Regime," in David Gombert et al., *Nuclear Weapons and World Politics*, (New York: Council on Foreign Relations, 1977).
49. Jonathan Schell, *The Abolition* (New York; Knopf, 1984).
50. In contrast to the contemporary view that republics are either democracies, democracies with participating and virtuous citizens, or simply kingless political orders, I argue that republican political orders distinctively aim to constrain violence capability and authority without centralization. Republicanism can thus be thought of as the part of the broader Liberal tradition concerned with security and violence, or *security liberalism*, and the side of the realist tradition concerned with complex stable quasi-state systems, or *realist institutionalism*.
51. Albert Wohlstetter et al., *Swords from Plowshares: The Military Potential of Civilian Nuclear Energy* (Chicago: University of Chicago Press, 1979), 15–16.

The Brooding Shadow:
Systemic Incentives and Nuclear Weapons Proliferation

Benjamin Frankel

THE END of the cold war and the demise of the bipolar international order have been greeted with enthusiasm and hope. Many have expressed the belief that a more pacific, tranquil world would now replace the cold war tensions and crises. In this article I argue that a structural analysis of international relations does not lend support to these earnest hopes. Specifically, an examination of the differences between bipolar and multipolar orders shows that nuclear arms proliferation will likely intensify in the 1990s and beyond, and that the owners of these newly acquired weapons will likely brandish them more openly to advance their political objectives.[1]

This intensified proliferation of nuclear weapons will be facilitated by both continuity and change. Familiar technological trends will continue: The spread of technology will enable more countries to manufacture nuclear weapons while countries already in possession of such capabilities will more readily augment them. The more important incentive for intensified proliferation, however, is the change in the international political system. The change that will most directly influence the acceleration of nuclear proliferation is the move away from bipolarity. The end of bipolarity means that superpower guarantees – the most effective instrument to moderate the effects of systemic characteristics – will be reduced and weakened. As a result, the international system will revert to a more unvarnished form of anarchy in which systemic attributes such as the security dilemma and self-help will be accentuated.[2] The accelerated proliferation of weapons of mass destruction will be an early and noticeable consequence of this change.

In this article I present, in an explicit and accessible fashion, a theory of nuclear weapons proliferation. At the core of this theory is a hypothesis postulating a causal relationship between the structure of the international system and the spread of nuclear weapons. I claim that bipolarity inhibits the spread of nuclear weapons while multipolarity induces their proliferation. Departing from Kenneth Waltz's proposition that in the international system the constituent units do not have a specification of role, I argue that in a bipolar system the superpowers do have a specific role: their preponderance allows them to attenuate systemic characteristics. I use the examples of nuclear weapons proliferation and the creation of NATO to show that a concerted effort by a superpower in a bipolar system reduces the influence of systemic characteristics such as the security dilemma and reliance on self-help. If we represent the security situation of states as points on a 'security curve' extending from utter insecurity to comfortable security, then the relief from systemic effects that a superpower provides its clients amounts to more than a move on the security curve: such relief is a move of the curve itself.

The theory I advance here, like any other theory, has its limitations. Critics will claim that my explanation of the proliferation of nuclear weapons leaves out factors such as international regimes, domestic structures, bureaucratic politics, organizational processes, national ethos and political culture, technological system-building, the psychology of individual leaders, the cognitive processes and perceptions of these leaders, the consequences of learning and adaptation, and more. In addition, some may argue that changes in polarity will not be the only or even the most important source of instability in the post-cold war period and that, therefore, to the extent that increased instability will result in accelerated proliferation, that proliferation should be correlated with other variables rather than with changes in polarity. To list what the theory omits, however, is not a valid criticism of the theory.[3] I know that the factors I left out influence a state's decision on nuclear weapons. I argue, however, that these factors are secondary to structural incentives and that in any event it is impossible to build a theory with too many independent variables.[4] To formulate a theory is to pick and choose from the "infinite materials" reality offers. Good theories elucidate by simplifying: they are parsimonious, incorporating only a few independent variables.[5] There are inherent difficulties in testing theories in the social sciences,[6] and the nature of the social world makes predictions especially difficult.[7] These difficulties notwithstanding, the future offers a valuable testing ground for the validity of theories.[8] Therefore, I offer explicit predictions about nuclear weapons proliferation in the future as a test for the explanatory and predictive power of structural analysis.

My discussion aims to contribute to international relations theory by applying an important theoretical explanation to a significant scholarly and policy question. It also aims to contribute to informed policymaking on proliferation. It is impossible to formulate a policy without an appeal to theory. The only issue is whether such an appeal is made explicitly or implicitly.[9] I argue that policies informed by structural analysis of international relations would be more effective in curbing runaway proliferation.

I do not take a position on whether or not the spread of nuclear weapons increases or decreases international stability. John Mearsheimer and Stephen Van Evera, among others, argue that the managed or controlled proliferation of nuclear weapons to key states – Ukraine, for example – would have a stabilizing effect. Even supporters of managed proliferation, however, agree with most students of the subject that a runaway, uncontrolled proliferation to many more states would probably not be beneficial. This is the position I take here.

The article proceeds as follows. In the first section I argue that as the technological difficulties in acquiring nuclear weapons diminish, the proliferation of such weapons becomes a matter of political decisions. To answer the question why more states in the post-bipolar world will make the political decision to acquire nuclear weapons, we must rely on a theory of international relations. Such a theory consists of our assumptions about how the world works: its nature, ordering mechanism, and the forces operating in it. In the second section, therefore, I briefly outline a structural analysis of international politics, highlighting those elements of the theory that bear more directly on the question of proliferation. In the third section I discuss the structure of post-Second World War international politics to show the causal relationship between bipolarity and the slow pace of nuclear proliferation. In section four I briefly discuss several cases of nuclear weapons proliferation and nonproliferation to illustrate that the availability of a superpower's security guarantee and the assessment of the strength and credibility of that guarantee were decisive in states' decisions about nuclear weapons. I show that in cases in which a superpower's guarantee was believed to be strong and credible, states decided not to acquire nuclear weapons even when facing a disadvantageous security situation. To provide another example of a superpower's ability to ameliorate systemic characteristics, I examine, in section five, the creation of NATO. I show that without U.S. intervention to attenuate the effect of systemic attributes, the creation of NATO would have encountered nearly insurmountable obstacles. In section six I summarize the lessons of nonproliferation and the creation of NATO. In the final section I show why multipolarity is conducive to the accelerated proliferation of nuclear weapons. I also argue that strict interpretation of the nuclear Non-Proliferation Treaty (NPT) and deep cuts in American and Russian arsenals will cause nuclear weapons proliferation to intensify further.

Proliferation: Technology and Politics

Technology is not the obstacle to acquiring nuclear weapons it once was. As one recent study concludes, technology is no longer a barrier to weapons proliferation, but merely a hurdle.[10] Technological progress and the dissemination of knowledge make building nuclear weapons a more manageable task now, and they also facilitate the development of related technologies such as ballistic missiles which augment a country's nuclear capability.[11] A recent Aspen Strategy Group Report concluded that the nuclear, chemical, and ballistic missile proliferations reinforce each other, creating a "dangerous synergism ... When taken together, the interaction of all three forms of proliferation is greater than the sum of the individual changes in each area."[12]

Nuclear weapons proliferation, however, is not driven by technology, although it requires technological capabilities to be sustained. The spread of nuclear weapons is determined by international politics, that is, the relationship among states, the constituent units of the international political system.[13] This relationship is governed by the structure of the international system. It is, therefore, the structure of the international system and the manner in which it is ordered that determine the rate and scope of nuclear weapons proliferation. Certain international structures inhibit proliferation while other international structures make it more likely. The fact is, "units differently juxtaposed and combined behave differently and in interacting produce different outcomes."[14]

What are the likely ramifications of the end of the cold war for nuclear weapons proliferation? It is not easy to answer this question because the introduction of nuclear weapons coincided with bipolarity becoming the international order. The only experience we have with nuclear weapons is as an integral part of a bipolar system. The postwar period also provides the only modern experience with bipolarity, which means that the only direct knowledge we have of a bipolar system has its two dominant powers equipped with nuclear arms. This coincidence makes it difficult to distinguish systemic features such as bipolarity from unit-level characteristics such as the quality of armaments states possess.[15] The distinction is made more difficult by the fact that the effects of nuclear weapons and bipolarity reinforce each other.[16] Without making this distinction, however, we will be unable to separate those features of the postwar period attributable to the effects of bipolarity from features attributable to the effects of military technology. The analytical separation of system and unit-level effects is important for predictions about nuclear weapons proliferation in the 1990s and beyond.

For our purpose it is not necessary to discuss the reasons for the changes in the international system or whether these changes would result in a familiar form of multipolarity. What is important is to show that a multipolar international structure is more conducive than a bipolar one to nuclear weapons proliferation. It can then be argued that the pace and scope of nuclear proliferation will intensify as the international system moves away from bipolarity toward some form of multipolarity. In the absence of historical experience, the conclusion that multipolarity is conducive to intensified proliferation of nuclear weapons has to be logically deduced. This logical deduction holds whether the international system is in fact moving towards multipolarity or not.

International Politics: A Structural Analysis[18]

It is not possible to analyze nuclear weapons proliferation without making theoretical judgements about international relations. Accordingly, I make explicit the theory on which I base my analysis of nuclear weapons proliferation. Below I offer a brief outline of structural analysis as developed by Waltz and others, stressing those aspects of the theory that bear more directly on my discussion of proliferation.

Structural analysis posits that the organization of units within a system, the way in which they are arranged and ordered, affects the way they function. The differences in the behavior of states are correlated more closely with their position in the system than with the differences among them in terms of ideology, type of regime, and economic system. The system does not determine what a state does, but rather constrains and disposes. It encourages behavior by providing incentives and disincentives. Behavior conforming with system requirements is rewarded while behavior that does not is punished. Different systems offer different clusters of incentives and disincentives. International systems change periodically. Structural changes in the international system occur in one of two ways: changes in the system's arrangement or changes in the distribution of capabilities among the system's units.[19]

Changes in the System's Arrangement[20] The system's arrangement is the principle according to which the units of the system are ordered. The ordering principle of domestic political systems is hierarchical, with relationship of subordination and superordination among the system's units. The ordering principle of the international system is anarchical, with relationship of coordination among the units. The relationships among states express relative strength, not entitlement. If one-worlders had their way, the world would become hierarchical: states would concede their sovereignty to a central authority and a world government would be formed, a government with powers such as monopoly over the use of force. Parts of the system would then be entitled to command while others would be required to obey. Hierarchy would replace anarchy. This would be a structural change because the principle of arrangement of the system changed.

The principle of action in the international system is self-help. States have to take care of themselves because other states would not do it for them. The international system exemplifies what Emile Durkheim called mechanical society. In a mechanical society,[21] the constituent units resemble each other because each unit performs functions the other units perform. There is a duplication of effort among the units, not a specialization of function or a division of labor. Because of the constituent units' similarity, self-sufficiency, and independence, mechanical societies are loosely linked. The opposite of mechanical society is a society in which there is an organic solidarity among the units. There is a differentiation of function and specification of role among the different units, and this division of labor strengthens the bonds among them. Modern societies are organic societies, with members performing specialized tasks. Because of the constituent units' dissimilarity and dependence on each other, organic societies are closely integrated.

Some students of international relations argue that growing economic interdependence among states, interpenetration of markets, and the growing importance of transnational actors such as multinational corporations, have moved the international system, or at least its more developed segments, closer to being an organic society. There is no agreement over the merit of these arguments or their meaning,[22] but if the international system were to change from mechanical to organic society, it would be a structural change because one principle of order would be replaced by another.

Change in the Distribution of Capabilities A system may also change through change in the distribution of capabilities among its constituent units. This change alters the number of major actors in the system, and a reduction or an increase in the number of major actors in the system is a structural change. The Second World War and the process of decolonization brought about the demise of Germany, Japan, Britain, and France as major global actors. This, and the introduction of nuclear weapons, changed the distribution of capabilities across the parts of the system, reducing the number of major global actors to two. Bipolarity replaced multipolarity, a structural change.[23]

The change brought about by the end of the cold war is also a structural change because it is caused by a change in the distribution of capabilities across the system's units. This change increases the number of major actors in the system and moves it from bipolarity toward multipolarity. The principle of arrangement of the system – anarchy – has not changed, however. The system that is now emerging is characterized by anarchy and a form of multipolarity.

Anarchy and the Use of Force The international system is an anarchical system in which a state's most important goal is to secure its survival.[24] All other goals a state wishes to pursue are conditioned on its survival. States secure their survival by accumulating military force they can use, singly or in combination with other states, against states wishing them ill. In an anarchical system force is the final arbiter. The centrality of force does not mean that states are constantly at war with each other. War is costly, so states have reason to settle their disputes by less expensive means. This, however, does not change the fact that a state with access to greater amounts of force has more control over its security than a state with access to lesser amounts of force.[25] The importance of force means that in anarchical systems units are arranged in accordance with the power they possess, not the authority or legitimacy they claim to have. International relations are relations of strength, and it is a state's strength that determines its position in the international system. The key question in an anarchical system is how a state maintains its own security and risks the security of others. As Robert W. Tucker says, authority in the international system is but an expression of capability.[26]

Internal and External Balancing The purpose of a state's effort to increase the amount of military force available to it is to balance the power of states it perceives as hostile. Balancing is achieved by internal or external means. A state may decide to mobilize its internal resources to increase its military strength or it may choose to counter a threat by forming a coalition with other states which feel similarly threatened. In certain situations a different strategy is adopted: if a state perceives itself to be under threat and it is unable to mobilize sufficient internal resources or find coalition partners, it may opt for bandwagoning over balancing. It may cozy up to the potential aggressor in an effort to remove any reason he may have to attack. Balancing, however, is more common than bandwagoning.[27]

Bipolarity and Stability The inherent complexity of multipolar architecture dooms multipolar systems to instability, making them susceptible to crisis and war.[28] As Alvin Saperstein concisely states, "stability decreases as the system complexity increases."[29] In multipolar systems the number of possible conflictual relationships among the powers in the system is great. The flexibility of alliances creates a situation in which there is no firm or enduring distinction between friend and foe. States cannot be certain who would side with whom over what issues and for how long. The relatively large number of powers in the system makes it difficult to have confidence in the commitment of allies while it is still too small to allow for ready compensation if an ally defects or shifts alliance.[30]

Checking an ambitious expansionist state is complicated and such states tend to get an advantage in trying to realize their aspirations. Because no single power is preponderant, there is a diffusion of responsibilities among states, causing responses to disruptions in the balance of power to be late and not always effective. In a multipolar world countries rely on external means of coalition building to keep the balance, but the process of alliance formation is susceptible to miscalculation and misjudgment that result in failure to balance the aggressive state properly. Even after a balancing coalition is formed, its members may continue to bicker among themselves over the nature of the threat, the best way to counter it, and the balance of costs and risks each member should bear. Estimating the collective power of a coalition is a complicated matter, and uncertainty about the relative strength of opposing coalitions causes crisis and war.[31]

A bipolar system is not hobbled by these crippling flaws. With only two great powers in the system, balancing is achieved by internal means of mobilizing domestic resources, a more reliable method than coalition building. There is no doubt who is responsible for maintaining the balance and the lines between friend and foe are clearly marked. Because of their global reach, the superpowers in a bipolar system are interested in developments far from home if such developments bear on their power relationship, limiting the potential for destabilizing developments on the periphery. As a result, a bipolar world is more stable than a multipolar world.[32]

Scholars are divided on the issues of the stability of a bipolar system relative to a multipolar one and on whether the world is moving toward multipolarity. Some argue further that lessons drawn from history about the advantages of one order or another are irrelevant because the behavior of states has profoundly changed. An argument is offered that some characteristics of the post-cold war system, especially increasing economic interdependence, decrease in nationalistic fervor, the spread of democratic regimes, and greater awareness of the costs of war, will enhance stability and peace. Proponents of this view make a strong case with regard to Western Europe.[33] Most of them agree, however, that outside Western Europe the situation is more complex and that familiar patterns of international politics are likely to prevail there.[34] The distinction between Western Europe and the rest of the world means that politics among states in which the vast majority of mankind lives will continue to be conducted more or less as it has been conducted in past, and that explanations of state behavior based on past experience will be applicable to them. The countries of Eastern Europe and the former Soviet Union lie in close proximity to the countries of Western Europe. Regardless of the changes that have taken place in the relationships among the Western European countries, they will be exposed to the consequences of the traditional interstate politics in the East. And without taking sides in the on-going debate about the importance of the Third World to the United States,[35] it is reasonable to assert that the spread of nuclear weapons and ballistic missiles to countries in the Third World will make more Third World countries relevant to the United States because of the increased ability of these states to threaten U.S. interests and make more costly a U.S. effort to protect such interests. The countries of Western Europe and North America may thus be viewed as an island of new politics in a sea of old politics. The lessons we draw from history are thus relevant to the future.

Structure and Stability in the Postwar Period

The long peace[36] that followed the Second World War was not the fruit of extra-mundane intervention but the result of a specific combination of structural arrangements and unit-level features. The world benefited from a lucky combination of bipolarity and caution-inducing nuclear weapons in the hands of the superpowers.[37] Different combinations, different structural arrangements, would have yielded different results.[38] The end of the cold war means the end of bipolarity and its replacement by a form of a multipolarity.[39] Historical evidence suggests that the end of the postwar bipolar arrangement is not likely to make the world more pacific. Indeed, the opposite may well be the case.[40]

Bipolarity and Nuclear Weapons Proliferation

Since 1945 only eight or nine states have equipped themselves with nuclear weapons. Many more countries with the technological capability to do so – one estimate puts the number at forty[41] – have opted instead for nuclear abstinence. This remarkably slow pace of nuclear proliferation is the direct result of the postwar bipolar arrangement and its underlying logic.

A state's decision to build nuclear weapons is a result of its perception of the security equation it faces.[42] A state measures the immediate and potential threats to its security against the means it has or can develop to thwart these threats. Each set of means has associated costs and these costs, too, are measured against the severity of the threat. Mobilizing domestic resources in order to augment defense means the diversion of resources from other uses, such as social programs. Forming an alliance with other states means giving up a measure of political freedom. The more serious the threat, the higher the cost a state will be willing to bear to meet it. For example, during the 1950s and 1960s Sweden, Switzerland, and Yugoslavia considered developing nuclear weapons. The three concluded that their security concerns did not justify the development of nuclear weapons and the costs associated with these weapons and that their security problems could be met below the nuclear threshold.[43]

In the postwar bipolar order, the superpowers' disposition was an important factor in the decision of states whether or not to develop nuclear arms.[44]

An alliance with one of the nuclear superpowers meant that that superpower was willing to use its nuclear forces on behalf of the minor coalition partner. This willingness on the part of a superpower obviated the minor partner's need to develop independent nuclear forces. Such superpower-client arrangements are encouraged by the logic of bipolarity because the superpowers consider a war or a threat of war anywhere of concern to them if it might lead to significant losses or gains for either of them.[45]

We can now refine the generic formulation we used earlier to say that in a bipolar world, a state's relationship with one of the superpowers is an important element in its security equation.[46] State N's decision to build nuclear weapons is the result of N's assessment of the strength and credibility of a superpower's commitment to N's security relative to the security threat N is facing.[47] If a state believes it is facing a severe security predicament, a superpower's commitment would have to be strong and credible to persuade that state not to build nuclear weapons. If a state believes it is facing a lesser security challenge, a lower level of superpower's commitment would suffice to persuade it not to build nuclear weapons. With the exception of states acquiring nuclear weapons for the prestige they confer, the key to nuclear proliferation in a bipolar world is thus the manner in which the superpowers exercise their role as security providers.[48] If they exercise it with sensitivity and authority, the spread of nuclear weapons would be slow. If there are doubts in the minds of potential security consumers regarding the credibility and effectiveness of the superpowers as providers of security, these consumer states would prefer to rely on their own devices.[49] Germany and Japan abstained from building nuclear forces not because they lacked the capability but, as Betts says, "As long as their security is enhanced more by association with the United States than by autonomous nuclear forces, their interests make their capability irrelevant."[50] This applies with equal force to all nuclear-capable states and their relationships with one of the two superpowers in a bipolar world. If a state's belief that its great-power ally is not sure to retaliate on its behalf is an important reason for it to build nuclear weapons, it follows that the strongest means by which the superpower can persuade a country to forego nuclear weapons is to guarantee its security.[51] The early predictions about the rapid spread of nuclear weapons did not materialize because they failed to take into account the logic of bipolarity.[52] As Nye writes, "The credibility of the nuclear umbrella extended by Washington and Moscow over their allies is a major reason why proliferation has been much slower than Kennedy feared."[53]

A cursory look at a few cases from the history of nuclear proliferation shows that states' confidence in the strength of a superpower security commitment was a key element in the decision whether or not to build nuclear forces.

Superpower Guarantees and Nuclear Weapons Proliferation

In this section I discuss the decisions on nuclear weapons made by states that decided to acquire them and states that decided to abstain from acquiring them. Because of space limitations the discussion is brief and sketchy. In the notes the reader will find references to sources providing detailed discussions of each of the cases.

Proliferation Checked

Japan Japan had historical reasons not to build nuclear weapons.[54] More important than painful memories, however, was the fact that it had no compelling security incentives to develop nuclear weapons because the security arrangements between the United States and Japan, including the United States-Japan Mutual Security Pact of 1960, have provided for Japan's security.[55] Japan's close relationship with the United States kept in check any inclination the Japanese might have had to develop an indigenous nuclear force. With the protection offered by the American nuclear guarantee, Tokyo saw no need to develop its own nuclear armaments.[56] Japan security concerns centered on a possible nuclear blackmail or attack by the Soviet Union or China, conventional threats against the islands, a unified and hostile Korea, and interference with Japanese shipping on which the nation depends. Japan's policy for meeting those threats was to enter into an alliance with the United States as a means of protection against external aggression. Japan, in Reiss's words, "resolved its postwar security dilemma by reliance on the military strength of the United States."[57]

Developments in Asia and the Pacific in the 1970s heightened Japan's anxiety about its security. The Soviet Union built up its forces in the Pacific to a point where one-quarter of the Soviet ground forces and combat aircraft were deployed in the Far East, one-third of the Soviet Navy was in the Pacific Ocean, and Soviet naval and land units in the Pacific accounted for one-third of the Soviet nuclear missile force. Soviet power projection capabilities were enhanced by the use of the former U.S. facilities at Cam Ranh Bay and Da Nang.[58] In 1978 the Soviets positioned armored forces and artillery on the disputed islands of Kunashiri, Etorofu, and Shikotan.

The Soviet buildup was made more worrisome by what was perceived as a U.S. disengagement from Asia. The United States withdrew its forces from Vietnam, reduced its forces in South Korea, and reduced the size of the Seventh Fleet. American reluctance to commit itself abroad was evident elsewhere. Congress blocked attempts by the administration to provide aid for the pro-Western forces in the Angolan civil war and the Carter administration was divided over the issue of how to solve the conflict in the

Horn of Africa on terms favorable to the United States. In 1977 President Carter announced his plan to complete the withdrawal of American troops from Korea, and in 1979 the administration was viewed as encouraging the departure from power of the Shah of Iran and Somoza in Nicaragua.

The Japanese reaction to these developments is instructive. The increase in Soviet power, the erosion of American commitment to the Far East, and the growingly uncertain international environment, strengthened Japan's resolve that the best means to safeguard its security was to reaffirm its ties to the United States.[59] Consultations between the United States and Japan yielded a plan for a comprehensive military cooperation between the two countries. The understanding covered such matters as coordination of advanced logistics and intelligence gathering and the elaboration of joint military response in the event of an armed attack on Japan. Following the talks, the naval cooperation between the two nations, which had already included joint antisubmarine and minesweeping operations, was increased to include the participation of the Maritime Self-Defense Forces in the U.S. Navy's exercise in conjunction with forces from Canada, Australia, and New Zealand. Joint air and ground exercises also increased in number and complexity. In the early 1980s the Japanese government agreed to sell or license military technology to the Pentagon and the U.S. private sector, while the United States began to weigh the advantages of the two countries joining in the development of advanced technology weapons.

The cost of nuclear weapons was not an issue in Japan because the Japanese economy made it possible for the country to build nuclear weapons without materially interfering with national priorities.[60] Domestic opposition stemming from the country's painful experience certainly played a role in the government's decision, but the security relationship with the United States was the most important factor disposing Japan against building nuclear weapons.[61] Faced with an increasingly uncomfortable security situation, Japan recognized it did not need to build its own nuclear weapons because, in Reiss's words,

> ... a viable alternative already existed. The American nuclear umbrella deterred Soviet and Chinese nuclear forces, while U.S. conventional strength addressed lesser threats to Japanese interests. The Japanese-American relationship ... provided Japan with an adequate – and when compared with the nuclear alternative – vastly superior, guarantee of its security.[62]

South Korea South Korea's history and its security situation made it sensitive to U.S. attitudes and policies.[63] Indeed, the central foreign policy objective of South Korea since the Korean War had been to tie America's military strength to the defense of the ROK.[64] The importance South Korea placed on its security relationship with the United States, and fear of a rupture in the relationship, provided the major disincentive to constructing a South Korean nuclear force.[65] The United States provided South Korea with large quantities of advanced weaponry and these provisions were important in maintaining the military balance between South and North Korea. These provisions also symbolized the U.S. alliance connection to South Korea.[66] The U.S. withdrawal from Vietnam, the elaborate detente relationship with the Soviet Union, and the normalization of relationship with China, led South Korea increasingly to doubt the firmness of American commitment to South Korean security. The result was predictable: a growing South Korean interest in developing its own nuclear weapons. South Korea did not ratify its membership in the NPT until April 1975 and public statements by South Korean officials at the time of ratification appeared to condition that ratification on U.S. guarantee against nuclear threats and aggression.[67] Two months after ratifying its membership in the NPT, however, South Korea was revealed to have negotiated with France for the purchase of a reprocessing plant. The United States pressured South Korea into canceling its purchase by telling South Korea that the deal would jeopardize the American-Korean security relationship. South Korea relented and canceled the deal because of the value it placed on the presence of American troops south of the 38th Parallel.[68]

The message South Korea's conduct conveyed was that South Korea was capable of constructing nuclear weapons but that it would abstain from doing so as long as America's nuclear commitment to the country was not removed.[69] President Park Chong-hui told American reporters that if U.S. support weakened, South Korea would build nuclear weapons.[70] There was thus no doubt, writes Dunn, that if the U.S.-South Korean alliance connection were to be weakened and its most immediate manifestation – the presence of American troops – withdrawn, South Korea would be likely once again to take steps towards a nuclear weapons capability.[71] If there were any doubts they were removed by South Korean reaction to President Carter's March 1977 plan to remove American troops from the peninsula. Two months after Carter's announcement, South Korean officials stated that if the United States would go through with the plan, South Korea would build its own nuclear weapons.[72] The withdrawal plan was suspended in July 1979 but the turmoil it had caused demonstrated the link between American security guarantees to South Korea and the latter's abstention from developing nuclear weapons. Aware of Washington's sensitivity to nuclear weapons proliferation, Seoul used the threat to develop nuclear weapons as a reverse leverage on the United States to make sure American security commitments to South Korea were not diluted.[73]

The March 1993 threat by North Korea to withdraw from the NPT underlined the role played by U.S. security commitments to South Korea and Japan in preventing a nuclear arms race in Northeast Asia. The prevention of such an arms race, David Sanger reported from South Korea, "has always depended on Seoul and Tokyo remaining unshakably confident that the United States' support, and its nuclear umbrella, extended over them."[74] South Korean civilian and military officials, reacting to the North Korean threat, let it be known that "if Seoul lost confidence that the United States was capable of protecting it against the North's threats, South Korea would restart a nuclear effort, and Taiwan could be close behind. Some believe even Japan, if surrounded by nuclear powers, would be forced to rethink its deeply ingrained commitment not to develop nuclear weapons."[75] Han Sung Joo, the South Korean foreign minister, summarized the mood in Seoul following the North Korean threat, saying "We have stripped ourselves, while North Korea is getting the bomb, and Japan has all the nuclear materials it needs and then some. There is something of a clamor to reconsider [South Korea's commitment not to develop nuclear weapons]."[76]

Taiwan Taiwan provides another example of a state accelerating or slowing down its nuclear program in response to its perception of the strength of a superpower commitment. In 1987 Taiwan secretly began to construct a small scale plutonium extraction unit.[77] The unit would have enabled Taiwan secretly to extract plutonium from a reactor's spent fuel in order to use it in nuclear weapons. Taiwan's 1987 attempt was similar to its effort in 1976 secretly to build a plutonium extraction unit. The timing of these clandestine operations was not arbitrary. Both the 1976 and the 1987 incidents followed periods of a deepening Taiwanese apprehension about the U.S. commitment to the island's independence and security. The 1976 effort was the culmination of Taiwanese attempts throughout the early 1970s to acquire reprocessing technology. Those attempts paralleled the process of normalization of U.S.-China relations, a process accentuated by the visit Richard Nixon and Henry Kissinger paid to Beijing in May 1972.[78]

Following the 15 December 1978 announcement of the establishment of formal diplomatic relations between the United States and China, Taiwan's doubts about the firmness of the U.S. commitment deepened even more. Dunn writes that had the U.S. Senate not taken the lead in providing new security assurances after full normalization, it is quite likely that the Taiwanese government would have faced an intense pressure for resumption of Taiwan's earlier nuclear weapon activities.[79] Further erosion in U.S. security commitments to the island nation could easily heighten incentives to acquire nuclear weapons, both for defense against invasion and as a symbolic replacement for U.S. security and political connections.[80] Spector agrees, writing that Taiwan's 1987 decision to build an extraction facility was made amidst growing concerns about the country's future security, fueled by Washington's increasing ties, even under the conservative Reagan administration, to the People's Republic of China and the concomitant weakening of American commitment to Taiwan's independence.[81] The connection between the quality of the American commitment to Taiwan's security and the latter's decision to build nuclear weapons was noticed in a 1974 CIA report, which concluded that

> Taiwan conducts its small nuclear program with a weapon option in mind ... Taipei's role in the world is changing radically, and concern over the possibility of complete isolation is mounting. Its decisions will be much influenced by U.S. policies in two key areas – support for the island's security and attitudes about the possibility of a nuclear-armed Taiwan.[82]

Other Cases Other cases may be mentioned.[83] Canada cooperated closely with the United States in the Manhattan project. In 1946, however, it decided not to build the bomb because Canadian security was credibly assured by the United States.[84] Sweden, too, was motivated in its decision by U.S. policies. Although Sweden was not a member of NATO, Reiss argues that its strategic value and geographical proximity to the NATO countries ensured that the American nuclear guarantee extended to Sweden in much the same manner as it did to Washington's European allies.[85] The Federal Republic of Germany, in possession since the early 1950s of the technical-industrial capacity to produce nuclear weapons, decided in 1954 not to do so. There were historical reasons for Germany to renounce nuclear weapons, but more importantly, underlying Germany's continued nuclear abstinence has been the fact that Germany was provided military security by the Atlantic Alliance.[86] Syria's decision not to go beyond its nascent nuclear infrastructure was influenced by the nuclear guarantees extended to it by the Soviet Union. Syria's defense minister Mustafa Tlas and Soviet officials asserted that the Soviet Union would assist Syria militarily, including using tactical nuclear weapons, if the latter were to be attacked by Israel.[87] The Soviet willingness to threat the use of its nuclear weapons on behalf of its Arab clients became apparent during the 1973 Yom Kippur War.[88] Providing a nuclear umbrella to a client state also served the nonproliferation objectives of the provider because, in countering the Israeli nuclear threat, the Soviets reduced Syria's incentives to seek an independent nuclear capability.[89]

Proliferation Unchecked

Some states perceive their security situation to be so dire that no outside assurance would persuade them to abjure the bomb.[90] The security situation of these states is exacerbated by the psychological effects of questions about their very legitimacy. The sense of illegitimacy leads to profound insecurity and distrust which tend to be confirmed by the political isolation in which these states find themselves. This anxiety is further reinforced by traumatic history and cataclysmic visions that have become part of the national ethos and as such may be beyond assuagement. The putative nuclear Masada and nuclear Laager[91] confirm this formulation: Relative to these proliferators' security problems, superpower commitments would likely have been judged insufficient even if offered (in the cases of both Israel and South Africa no such superpower guarantees were offered[92]). However, even in instances in which a superpower's security guarantee may not be sufficient to persuade a state to abstain from acquiring nuclear weapons, the superpower may still be able to influence the state's decision whether to deploy the weapons openly. Israel is a case in point.

Israel Israeli leaders, in deciding the develop an indigenous nuclear force, were motivated primarily the traumatic experiences of the 1940s. Tom Segev observes that "The most fateful decisions in Israeli history, other than the founding of the state itself – the mass immigration of the 1950s, the Six-Day War, and Israel's nuclear project – were all conceived in the shadow of the Holocaust."[93] Israeli leaders were also persuaded in their decision by a realistic appreciation of Israel's security equation. The decisive steps in Israel's nuclear history were taken in mid- and late-1960s, with the 1967 Six Day War serving as the catalyst.[94] That war is remembered for the conceptual and operational brilliance of Israel's destruction of the Arab states' air forces and the magnitude of the defeat Israel inflicted on Egypt, Syria, and Jordan. Israelis, however, remember the sense of abandonment and solitude that characterized their situation during the three weeks immediately preceding the outbreak of the war. Between 15 May and 6 June 1967 Egypt marched its army into Sinai and blockaded the Straits of Tiran; Syria concentrated its forces on the Golan Heights, overlooking Israel's northern sector; and Jordan allowed Iraqi and Egyptian forces to enter its territory and move close to the Israeli border. The Arab leaders accompanied the encirclement of Israel with dire and explicit predictions concerning Israel's fate. There were expressions of concern for Israel in Western capitals but little concrete action was offered to assist it.

The perilous three weeks of *hamtana* (wait) left their mark on the nation's psyche. The inconclusive War of Attrition in 1968–70, the surprise of the 1973 October War, the improved performance of the Arab armies in it, and the growing economic and political influence of Arab states in the wake of the oil embargo, indicated that the political, economic, and military trends favored the Arab states. More importantly, during this period Israel had been unable to attain a formal security guarantee from a superpower or a membership in a military alliance.[95] Israel's decision to build nuclear weapons was thus an expression by Israel of profound skepticism over the international community's willingness to come to its aid in a crisis, and signalled Israel's determination to rely on its own capabilities instead.[96]

Israel's history and its security situation drove it to build an independent nuclear force but its relationship with the United States dictated the manner in which this process was pursued.[97] Israeli leaders' belief in the U.S. commitment to deter direct Soviet military employment against Israel on behalf of the Arabs, a commitment demonstrated by the American nuclear alert during the 1973 war, may have induced Israel to continue to equivocate on the nuclear issue.[98] Israel's equivocation may also reflect Israel's awareness of the choices an open Israeli nuclear deployment would impose on the U.S. administration and Congress. Successive Congressional legislative efforts relating to nuclear nonproliferation had to be rewritten so as not to force a confrontation with Israel.[99] The United States may have been unwilling to provide Israel with formal security guarantees, and Israel may have been unwilling to abjure the bomb even if such guarantees were offered, but U.S.-Israeli relations provide an example of a superpower tacitly using its influence with a client to keep the client's nuclear posture opaque, mitigating the more onerous effects an overt nuclear posture might have.

Pakistan, India There are other examples of states developing nuclear weapons because of their lack of confidence in a superpower's commitment to their security, but keeping such development opaque because of the importance they attach to their relationship with the superpower.[100] Pakistan is cited as a case in which a more forthright and unequivocal American commitment to assuage that country's security concerns may have checked nuclear proliferation.[101] Even as Pakistan pursued its nuclear weapons program, however, the economic and military aid it was receiving from the United States was influential in taming Pakistani public statements about Pakistan's nuclear potential and intentions. This may be changing. The end of the war in Afghanistan and the end of the cold war have reduced Pakistan's strategic importance to the United States. Pakistani clandestine efforts to obtain weapons-related materials in the West also came to light in recent years. Congress has increasingly voiced its displeasure with Pakistan's nuclear behavior, and U.S. economic and military aid to Pakistan may be the first casualty of this displeasure.[102]

India's story is also instructive because its failure to secure a nuclear guarantee from either of the superpowers following China's 1964 nuclear detonation provided it with an incentive to embark on its nuclear program.[103] India's humiliating defeat in its border war with China in October–November 1962 and China's explosion of a nuclear bomb in October 1964 heightened India's security concerns. While leading Indian politicians recognized the value of a superpower's guarantee to their country's security, they were also afraid that such a guarantee would compromise their country's non-aligned international status.[104] In any event, Indian efforts to obtain a superpower's guarantee against China's bomb failed.[105] Subsequent Indian diplomatic efforts to promote an international nonproliferation agreement that would reduce the disparities between countries with and without nuclear weapons also failed.

Events in 1971 provided India with a strong impetus to acquire the bomb. In July, after mediation efforts by Pakistan, Henry Kissinger traveled to Beijing. A month later, India, fearing a tilt in the regional balance of power against her, signed a twenty-year friendship treaty with the Soviet Union. In December of 1971 India sent its troops to help the Bengalis in their bitter war against Pakistani units in east Pakistan. India defeated the Pakistani army, helping create the independent Bangladesh. However, the American-Chinese-Pakistani alliance and the Indian-Soviet treaty polarized the subcontinent, raising doubts about India's ability to maintain its cherished nonalignment under the new circumstances. The development of an Indian bomb was an attractive solution. After the explosion of May 1974, India's nuclear activities became more muted. The benefits of demonstrating India's technical ability were garnered and India's willingness to defend its non-aligned status was demonstrated. Also important in India's decision to slow its nuclear weapons activities despite early reservations, was the fact that the treaty with the Soviet Union provided India with some insurance against Chinese aggression.[106]

The Case of NATO

The superpowers' security guarantees played a decisive role in inhibiting the proliferation of nuclear arms. In restraining this proliferation, the superpowers demonstrated their ability to reduce systemic effects. There are other examples of how superpowers conduct in a bipolar system ameliorates and softens the mechanical nature of international society. The role the United States played in fostering Western European defense cooperation illustrates how a superpower may wield its enormous power in a way that effects structural outcomes.

The Western alliance has been from its inception a dual-purpose alliance.[107] One purpose was to protect U.S. allies from the Soviet Union. The other purpose was, in Ronald Steel's words, "to reassure them that [the United States] would keep watch over the Germans ... [so that Europe's] ... endemic rivalries would be muted and its ability to harm itself and others limited."[108] The Western Alliance had to offer the states of Western Europe the firm containment of Germany before it could get them to attend to the containment of the Soviet Union.

On 26 October 1950 the French government introduced the Pleven Plan for a Western European defense confederation. It envisioned a Western European military force under a Western European secretary of defense and a Western European assembly. West Germany would participate in the confederation, but without a national army or a general staff. The plan, called the European Defense Community (EDC), stipulated that the strength of the West German contingent would not exceed one-fifth of the entire Western European force. The EDC draft agreement was ready by May 1952, but the French, who signed it on the 27th of that month, continued to insist on firm American and British guarantees to contain potential German threats against European nations. On 30 August 1954 the French National Assembly rejected the EDC plan because the French were not convinced of British and American readiness to counter West German resurgence. It is important to recall that the French rejected this Western defense arrangement in the face of what appeared to be a growing Soviet threat and a perception of an intensifying global communist challenge to the West. The apprehension about Germany was stronger than the fear of the Soviet Union.

Within weeks of their rejection of the EDC, the French accepted a communal defense arrangement with Germany by agreeing to the admission of a sovereign and rearmed West Germany as a full member of the North Atlantic Treaty Organization (NATO) and the Western European Union (WEU). The major difference between the EDC, on the one hand, and the WEU and NATO, on the other hand, was the participation of Britain and the United States. Britain and the United States were not members of the EDC, but Britain was a member of the WEU and both Britain and the United States were members of NATO. With the EDC, France and the smaller Western European countries had to face West Germany on their own. With NATO, they were protected by the United States. Josef Joffe, in his book *The Limited Partnership*, makes a strong and persuasive case that it was the specific role the United States played that made it possible to form the alliance. It is worth quoting him at some length. He writes,

> It was only the virtually permanent intrusion of the United States that changed the terms of state interaction to the point where the Europeans no longer had to conduct their business in the brooding shadow of violence ... By committing its power to the protection of Western Europe as well as itself against the Soviets, the United States swept aside the rules of the self-help game that had governed and regularly brought grief to Europe in centuries past.[109]

To say that the assumption by a superpower of security responsibilities in Western Europe "changed the terms of state interaction" and "swept aside the rules of the self-help game" is to say that by investing its enormous power in Western European security matters, the United States attenuated structural effects. Joffe further writes that

> States go to war for many reasons ... but an irreducible reason is built into the very nature of the sovereign state system whose members must ultimately rely on themselves to protect their security and independence. In an anarchic self-help system, the very search for security turns into the root cause of insecurity, since each nation's quest poses a threat to the other. ... the perverse dynamic of self-reliance explains why war breaks out even in the absence of pernicious motives such as greed, ambition, and aggrandizement.
>
> ... Yet, by underwriting the security of the West European nation states, by sparing them the necessity of self-reliant choice in matters of defense, the United States banished the systemic cause of conflict at the root of so many of Europe's past wars ...

The United States' preponderance – a unit-level feature – enabled it to provide Western European nations with guarantees of systemic significance. America's role in Europe was the indispensable condition for cooperation among Western European nations because it softened, if not removed, the security dilemma from the Western European state system. As Joffe writes, " ... [B]y extending its guarantee, the United States embedded another girder into the postwar order. The lasting integration of the United States into Europe's state system dispatched the prime structural cause of conflict among states – the search for an autonomous defense policy."[111] U.S. commitment attenuated the mechanical nature of Western European politics. As a result of the American guarantee, the international relations of Western Europe "came to resemble domestic politics in one crucial respect: insecurity, the underlying condition of statecraft, and force, its ultima ratio, were banished from the system."[112]

Preponderance as a Form of Specialization

The Lessons of Nonproliferation

There are fewer cases of the proliferation-inducing combination of a lack of confidence in a superpower's commitment and the development of indigenous nuclear capabilities than cases of the proliferation-retarding combination of confidence in a superpower's guarantee and nuclear self-abnegation. This is not surprising: In a bipolar world the superpowers play the role of security providers nimbly and adroitly because the structural logic of bipolarity rewards such behavior. They extend their security guarantees far and wide when they perceive a threat to the power relationship between them, bringing worried states under their protective umbrella.

The reasons for the slow pace of nuclear proliferation during the last four decades are clear: In a bipolar world, the two superpowers provided security, including nuclear guarantees, to their client states. The security needs of most would-be proliferators were satisfied by these superpower commitments. The only instances of nuclear weapons proliferation were cases in which superpower guarantees were not relevant or in which more than security concerns were involved. Pariah states such as Israel and South Africa believed external guarantees, even if offered, would be inherently insufficient. Pakistan, too, harbored a deep sense of insecurity, accentuated by its dismemberment in 1971. In deciding to build their independent nuclear forces, France and India, in varying degrees, were motivated by considerations of prestige. Superpower guarantees are irrelevant to states seeking the prestige of owning nuclear weapons because such guarantees symbolize the inferior position these states wish to escape. Where security concerns were the main motivation for wanting the bomb, however, and where superpowers' guarantees were available, proliferation was kept in check.

Walt argues that the global commitments the United States assumed after the Second World War – commitments I see as decisive in slowing down the pace of nuclear proliferation – were less the result of the underlying logic of bipolarity than a consequences of U.S. misperceptions and miscalculations. The United States undertook such commitments because its leaders exaggerated both the Soviet threat and the "Finlandization" tendencies of allies.[113] Walt's argument weakens the case for bipolarity as an inhibitor of nuclear proliferation, but it lends support to the claim that superpowers' security guarantees, regardless of why they are given, are decisive in retarding proliferation. We may thus accept that in bipolarity the superpowers' preponderance and reliance on internal mobilization allow them to be indifferent to events on the periphery, and that balancing tendencies would cause states to balance against an aspiring hegemon even without a superpower assuming more than its proportional share of the balancing task. We may then conclude that the United States overcommitted itself to allies and invested more than was necessary for the effective containment of the Soviet Union. Still, U.S. global engagement, even if motivated be an exaggerated view of the stakes and the adversary, allowed other states a measure of security that led them to abstain from acquiring their own nuclear forces. The end of the cold war will cause the United States to adopt a more sober view of its national interest and a more circumscribed understanding of its global role, leading to reduced U.S. commitments and security guarantees to allies. This reduction, regardless of its motives, will increase pressures for nuclear proliferation.

The Lessons of NATO

A discussion of NATO will not be complete without mentioning the fact that the cooperation among European states fostered by U.S. involvement was not limited to security matters. Indeed, the expectations of peaceful relations were a necessary condition for the establishment of the European Economic Community. If the European believed that there was a chance that they would have to fight each other, they would not have allowed their economies to grow so interdependent.[114] Those critics of U.S. foreign policy who offered the promotion of international economic cooperation as an alternative to containment[115] overlooked the causal relationship between bipolarity and cooperation among states. Increased economic cooperation among states cannot be portrayed as an alternative to bipolarity and containment: The relatively high degree of such cooperation during the last three decades was a byproduct of bipolarity and its logic.[116] It is useful to quote Walt on this point:

> [The critics'] view understates the role that containment plays in facilitating cooperation among the advanced industrial countries. If the Soviet threat declined dramatically or the U.S. abandoned containment, the other industrial powers would worry more about their positions relative to one another. From this perspective, economic cooperation among the industrial powers has been facilitated by the fact that the Soviet threat and the strategy of containment created a stable security environment and strong incentives for political collaboration in the West. Thus, declining cooperation among the industrial powers may be due as much to decreased concern about the Soviet Union as to the erosion of U.S. economic supremacy.[117]

Recall that a system is defined by three elements: the principle according to which it is organized; the differentiation of its units and the specification of their roles; and the distribution of capabilities across the system's units. The second element does not apply to the international system because it is composed of like units different from each other by capability, not function.[118] Waltz points out that because of the undifferentiated character of its units, the international political system resembles what Durkehim called mechanical society.

Bipolarity, created by the distribution of capabilities across the system's units, modifies the mechanical nature of the international political system. The preponderance of the two superpowers is a form of specialization because it allows them to do things other states cannot and would not do. In Waltz's words, "Great tasks can be accomplished only by agents of great capability."[119] Specifically, the superpowers can take actions that reduce structural effects. Walt writes that "America's global presence helps safeguard its allies from one another; they can concentrate on balancing the Soviet Union because they do not need to worry about other threats";[120] Joffe writes that the security guarantees the United States gave Western Europe changed "the terms of state interaction," swept "aside the rules of the self-help game," banished "the systemic cause of conflict," and dispatched "the prime structural cause of conflict among states – the search for an autonomous defense policy." Both, in other words, write about the United States reducing structural effects.

Other students of the effects of structure on alliance formation also draw attention to the beneficial systemic influence of bipolarity. Discussing how domestic politics or faulty ideas may cause overreaction or underreactions to disrupt the balance of power, Thomas Christensen and Jack Snyder write that " ... bipolarity is an ameliorator ... bipolarity mitigates the structural causes of such problems."[121] There is a qualitative difference between even very powerful states and states capable of assuming tasks of systemic significance. In a bipolar world, the superpowers may assume such tasks.

What we see in the post-Second World War world, then, is a division of labor, with the superpowers specializing in offering other states services that enabled those state to escape systemic effects. Removing the security dilemma from an international subsystem is such a service. By offering Western European states security guarantees, the United States solved the security dilemma there. It may be argued that the superpowers' strength means only that they possess more of what all other units possess, but there is a point beyond which quantity becomes quality. Purposefully wielding an overwhelming power is akin to a specialized skill.

The specialization represented by the superpowers' overwhelming strength only modifies, not changes, the fact that the international system comprises like units. It also does not change the fact that states are sovereign to decide for themselves how to cope with challenges and problems by choosing, for example, between mobilization of internal resources and coalition formation when faced with the need to balance an adversary. The exertion of a preponderant strength by the superpowers, however, may provide relief to client states from systemic effects, as evidenced by states' tacit recognition and acceptance of the division of labor in the security area. This recognition led many states to behave in a manner not expected of units in a mechanical society. Rather than insist on nuclear weapons self-sufficiency, many states entrusted the performance of nuclear weapons-related tasks to a superpower. The conduct of the superpowers in bipolarity thus allowed states a measure of escape from the harsh logic of the security dilemma.

Multipolarity and the NPT

In a multipolar world powers will not be able or willing to shoulder the responsibilities borne by the two superpowers in a bipolar system. Without a superpower umbrella to shield them, more states will rely on the mobilization of internal resources, including building their own nuclear forces, to secure their survival, with all the associated destabilizing consequences.[122]

The flexibility of alliance formation is cited as a virtue of a multipolar system, but this flexibility militates against the provision of the kind of security guarantees available in a bipolar world. I suggested earlier that state N's decision to build nuclear weapons or abstain from building them is the result of N's assessment of the strength and credibility of a superpower's commitment to N's security relative to the security threat N is facing. The rigidity of the bipolar system and the fact that there are no peripheries because of the superpowers' perception of their competition as a zero-sum game allow clients a high degree of confidence in a superpower guarantee, if one is given. But in a multipolar world it is not always clear who is siding with whom and for how long. This uncertain quality of alliances would make states reluctant to entrust their fate to another state and its guarantees. In addition, it is not clear that middle powers in a multipolar world would offer nuclear umbrellas to their clients the way the superpowers did. France's behavior is indicative.[123] Talks about devolution in Europe have usually been accompanied by references to the French and British nuclear forces replacing American forces in providing the extended deterrence for fellow European nations. There is no more persistent theme in French strategic thinking, however, than the rejection of sharing France's nuclear powers. French officials may obliquely refer to *sanctuarisation élargie*, as General Guy Méry did in 1976, but the thrust of most declarations on the subject reinforces the message of French unwillingness to extend nuclear deterrence.[124]

The major reason for the reluctance of middle powers to extend nuclear deterrence guarantees is rooted in the size and structure of their nuclear forces.[125] What allows one country to threaten the use of its nuclear forces on behalf of another is the possession of a large and invulnerable nuclear force and developed counterforce capabilities.[126] In theory, such capabilities allow the nuclear guarantor to use nuclear weapons discriminately against the aggressor to signal its resolve to defend its allies but also its desire to avoid an all-out nuclear conflagration. Counterforce capabilities afford this tit-for-tat option because they can be tailored to hit military targets with little collateral damage.[127] If the collateral damage is kept low, intrawar deterrence would hold and neither side would have an incentive to escalate the war to the level of destroying the adversary's population centers. In theory, then, an extended nuclear guarantee is credible if a nuclear guarantor can use his nuclear forces in a manner that will not provoke the adversary to launch everything he has and bring about the guarantor's own demise. Countries do not commit suicide on behalf of other countries.

Middle nuclear powers do not have such forces and are not likely to possess them in the future. French and British nuclear forces provide an example of the arsenals middle powers are going to have. The purpose of such arsenals is to protect the homeland against nuclear attack by threatening heavy but less than total damage to an attacker's population (its authors call this posture "pure" or "proportional" deterrence, while others describe it, more colloquially, as a "tear off an arm" strategy). They are meant to be used when all else has failed and there is nothing more to lose. These are not forces that can credibly provide extended deterrence to other states.[128]

The decline in superpower security guarantees and increase in the number of states relying on their own nuclear devices will revive the anxieties associated with the security dilemma, contributing to a weakening of stability. Some point to the Non-Proliferation Treaty (NPT) and the non-proliferation regime it stands for as a possible barrier against rapid proliferation in the post-cold war world. The consequences of the end of bipolarity for the nonproliferation regime, however, are not reassuring, as the structural changes in international politics expose the flawed conceptual foundations of the NPT. The post-Second World War bipolarity helped mask that flaw. With the passing of bipolarity, the NPT's feeble foundations may cause the whole edifice to collapse.

Article VI of the NPT requires the nuclear weapons states to take steps toward nuclear disarmament. In what Joseph Nye calls a "basic intuitive compact" between nuclear abstainers and nuclear weapons states, the former agreed not to acquire nuclear weapons while the latter agreed to negotiate the dismantling of their nuclear arsenals. The failure of the nuclear weapons states to live up to their commitments under Article VI has been predictably criticized as an expression of hypocrisy and discrimination by the rich and powerful against the weaker states. Some say this failure may doom the renewal of the NPT, scheduled for 1995.[129]

It may be a cruel irony, but the causal relationship between vertical and horizontal nuclear proliferation is the exact opposite of that assumed by Article VI of the NPT. It is precisely the large size and sophistication of the superpowers' nuclear forces, and the avowed willingness of the superpowers to threaten to use them on behalf of their clients, that have prevented a more pervasive nuclear weapons proliferation.[130] The intuition of the authors of the basic intuitive compact – the NPT – was wrong, their logic flawed. As Nye writes, "... [T]he key role played by [the superpowers'] alliance guarantees in stemming proliferation indicates that [the superpowers'] Article VI obligations cannot be interpreted as simple disarmament."[131] Deep reductions in the nuclear arsenals of the two superpowers will lead to the intensification of nuclear proliferation, not to its arrest. Edward Luttwak is correct to refer to deep reductions in U.S. strategic systems as counter-nonproliferation policy.[132] Betts agrees, writing that to approve "the retrenchment of U.S. commitments and the decline of American military hegemony outside Europe, is necessarily to accept that there are severe limits to what we are willing to do to contain nuclear proliferation."[133]

Vertical nuclear proliferation inhibits horizontal nuclear proliferation. Vertical nuclear disarmament induces horizontal nuclear proliferation. Nye writes that the superpowers' "high levels of weaponry have instilled prudence in their relationship and permitted their allies the eschew the nuclear option."[134] Some states criticize the superpowers for failure to reduce their nuclear arsenals but there is no reason to believe this failure will lead these states to build nuclear weapons of their own. States acquire nuclear weapons because of their perception of their security situation, not out of spite. Deep reductions in the superpowers' arsenals, and the concomitant diminution in their ability to offer security guarantees to other states, will change these states' security situation and provide them with an incentive to build their own nuclear weapons.[135]

Deep cuts in the superpowers' strategic arsenals may provide incentives for proliferation in yet another way because there is an inverse relationship between the size of the arsenals of the superpowers and the benefits states may derive from possessing a small nuclear force of their own. The small nuclear forces second-generation proliferators are likely to build would increase in importance the smaller the superpowers' forces become. The value of owning a hundred nuclear weapons in a world in which the superpowers possess tens of thousands of weapons each is different from owning a hundred nuclear weapons in a world in which the superpowers possess only a few hundreds weapons each. The benefits of a small nuclear force and the resulting incentives to acquire one thus increase in the absence of the superpowers' overwhelming numerical advantage in weapons.

States' accession to the NPT should be understood in this light. Subscribing to the NPT is a contingent, not transcendent act. It does not connote an existential renunciation of nuclear weapons. Rather, it indicates an understanding by a member state that under present conditions it can do without nuclear weapons. The key to this understanding is the availability of superpower security guarantees. The division between nuclear 'have's' and 'have nots' has thus not been imposed from above by the nuclear weapons states, but also has been supported from below by the nonnuclear weapons states.[136] Nye puts it nicely when he writes that "most states are likely to accept some ordered inequality in weaponry because anarchic equality appears more dangerous."[137] Betts correctly observes that to assert that the prevention of horizontal nuclear proliferation depends on reductions in the superpowers' large nuclear-arms capabilities is analytically wrong, prescriptively quixotic, and a dangerous misdirection of effort.[138]

The NPT, like any other international treaty, reflects the interests of the states which signed it. States comply with a treaty's stipulations when doing so serves their interests. When their interests are served by doing what a treaty forbids, they do it regardless of what the treaty mandates.[139] A state's compliance with a treaty means that it would likely have followed the course of action mandated by the treaty even if there were no treaty. An exception would be an international regime, within which a state may behave in manner that cannot be accounted for by the state's narrow, short-run interests. It is doubtful, however, that a regime exists in the security area.[140] The absence of a regime does not mean that states do not cooperate or exhibit self-restraint. It does mean, however, that states' cooperation and self-restraint are a result of their self-interest. As Jervis writes, "If patterns of international relations can be explained by the distribution of military and economic power among the states, then the concept of regime will not be useful."[141] Nuclear abstinence on the part of nuclear-capable states can be explained by reference to their pursuit of their self-interest within a bipolar world, bipolarity being the result of the distribution of military and economic power among the states.

All other things being equal, it is better to have the NPT than not have it. Treaties have an undeniable symbolic value and patterns of cooperation among nations, even if motivated by self-interest, should be encouraged and nurtured. We should not have more confidence in the NPT ability to prevent nuclear proliferation, however, than we do in the UN ability to keep the peace. There are several technical components to the nonproliferation system, such as agreements among nuclear materials and missile supplier states, which make it marginally more difficult for states to build nuclear weapons for themselves. Pakistan, South Africa, and Iraq show that a nation determined to acquire the weapons can circumvent these barriers. Israel built most of its nuclear infrastructure before the NPT but continued to engage in activities proscribed by the treaty to enhance its nuclear weapons capability. India built its bomb with plutonium obtained by irradiating an indigenous source of natural uranium in a Canadian-built reactor, covered by bilateral safeguards, a reactor operated with U.S.-supplied heavy water. India, Israel, Pakistan, and, until recently, South Africa, are not members to the NPT. Iraq is, however, a fact that did not prevent it from trying to build nuclear weapons.[142] After its initial attempt was thwarted by an Israeli air raid in 1981, Iraq redoubled its efforts. In any event, the NPT regime was not airtight within a bipolar order, and it is difficult to see how it will grow stronger under multipolarity.

Nervous Times, Hard Choices

The Aspen Strategy Group report refers to the dangerous synergism created by the proliferation of nuclear and chemical weapons and ballistic missiles in the 1990s, and the interactive, reinforcing nature of this proliferation. The changes entailed by the passing of the bipolar structure, however, are more important in accelerating nuclear arms proliferation. The end of the cold war and the resulting new international structure are creating conditions conducive to nuclear weapons proliferation of increasing scope and pace. The combination of technological capabilities and political changes creates a more dangerous synergism than the one referred to in the Aspen report. As Betts wrote more than fifteen years ago, "No less than a function of diffusion of technology, proliferation will now be a function of the postcontainment era's diffusion of power."[143] Unchecked nuclear proliferation and multipolarity will create an unnerving combination. Jervis writes about former Soviet clients who now may be at once more desperate because they lack a powerful superpower patron and more autonomous because they lack a superpower to restrain them.[144] This observation also applies to erstwhile American allies.

The consequences of using nuclear weapons may be unprecedented, but the incentives for acquiring them are wholly traditional. To believe that the changes in the international system in the 1990s will not be accompanied by an intensified nuclear arms proliferation, we must postulate a fundamental change in the way states behave and a sharp break with historical patterns. These departures are possible, but not likely, certainly not on a universal scale.

There is hope, however, in the fact that most proliferators are motivated by traditional security concerns. A detailed discussion of the policy implications of my argument is outside the scope of this article, but a brief sketch of the principle that should guide that policy is in place. The thrust of my discussion is to urge a recognition of the intimate relationship between the ability of the superpowers to offer security to their allies and the pace of nuclear weapons proliferation. The United States built its arsenal and extended its security commitments primarily to contain the Soviet Union. The undeniable, if initially unintended, side-effect of the U.S. effort has been the remarkably slow pace and scope of nuclear weapons proliferation. The main recommendation my discussion suggests is that before American policymakers dismantle the cold war-inspired U.S. military structure, they should keep in mind the beneficial effects it had on nuclear proliferation. If they believe that a runaway proliferation of nuclear weapons is inimical to U.S. interests, they should be careful not to destroy the most effective means to keep such proliferation in check. The reader will find examples of specific policies derived from this recognition of the dual purpose of U.S. military might in Art and Quester and Utgoff.[145] I agree with most of what they suggest.

Electoral pressures militate against long-term, steady policymaking on foreign policy and defense issues. Still, if policymakers responsibly resist the urge to become complacent as a result of the presumed benefits of the end of the cold war, the United States may be able to ameliorate somewhat the more disagreeable attributes of the post-cold war world politics. If, against the odds, the United States is successful in this endeavor, we may yet enjoy a welcome sequel to the long peace.

NOTES

I have benefited from the comments of Sean M. Lynn-Jones, Eric Mlyn, and Neil Joeck on earlier drafts of this article. I am grateful for the spirited criticism offered by the participants in the conference on Social Science Theory and Nuclear Weapons Proliferation held at the Center for International Security and Arms Control, Stanford University, 5–6 February 1993. Special thanks to Scott Sagan, Jerry Mullins, Peter Lavoy, Glenn Chafetz, and Zachary Davis.

1. Earlier versions of some sections of this article have appeared in Benjamin Frankel, "An Anxious Decade: Nuclear Proliferation in the 1990s," in Benjamin Frankel, ed., *Opaque Nuclear Proliferation: Methodological and Policy Implications* (London: Frank Cass, 1991), 1–13; and Benjamin Frankel, "International Political Changes and Nuclear Proliferation in the 1990s,"in Eric H. Arnett, ed., *Science and International Security* (Washington, D.C.: American Association for the Advancement of Science, 1990), 85–113.
2. For good discussions of the security dilemma see Robert Jervis, "Cooperation under the Security Dilemma," *World Politics* 30, no. 2 (January 1978): 167–214; and Glenn H. Snyder, "The Security Dilemma in Alliance Politics," *World Politics* 36, no. 4 (July 1984): 461–95. John Herz coined the term in his *Political Realism and Political Idealism* (Chicago: University of Chicago Press, 1951).
3. "Theory, after all, is mostly omissions." Kenneth N. Waltz, "Realist Thought and Neorealist Theory," *Journal of International Affairs* 44, no. 1 (Spring 1990): 31. See Stephen Walt's criticism of Robert Kaufman in Stephen M. Walt, "Alliance, Threat, and U.S. Grand Strategy: A Reply to Kaufman and Labs," *Security Studies* 1, no. 3 (Spring 1992): 465.
4. As Gaddis writes, "One might – at least as a thought experiment – construct a model capable of simulating all of international relations in all of their complexity. But the model would have to be of such complexity as to render it indistinguishable from that which was being modeled, which would rather

defeat its purpose." John Lewis Gaddis, "The Cold War's End Dramatizes the Failure of Political Theory," *Chronicle of Higher Education*, 22 July 1992, p. A44.

5. Kenneth N. Waltz, *Theory of International Politics* (Reading, Mass.: Addison-Wesley, 1979), 5. See also Stephen Van Evera, "Hypothesis, Laws, and Theories: A User's Guide," ver. 2.4, rev. April 1991, in Benjamin Frankel, "*Security Studies*: A Guide to Authors," ver. 1.4 (Washington, D.C.: November 1992).
6. On the need for hard tests of theories in the social sciences see Waltz, *Theory of International Politics*, 15–16; Waltz, "Reflections on *Theory of International Politics*: A Response to My Critics," in Robert O. Keohane, ed. *Neorealism and its Critics* (New York: Columbia University Press, 1986), 335–36; Robert O. Keohane, *International Institutions and State Power: Essays in International Relations Theory* (Boulder, Colo.: Westview, 1989), 8 and 167; Joseph M. Grieco, *Cooperation among Nations: Europe, America, and Non-Tariff Barriers to Trade* (Ithaca: Cornell University Press, 1990), 2; Oran R. Young, "The Effectiveness of International Institutions," in Ernst-Otto Czempiel and James N. Rosenau, eds., *Governance Without Government: Order and Change in World Politics* (Cambridge: Cambridge University Press, 1992), 165–66.
7. On the difficulty of making predictions in the social sciences see Robert Jervis, "The Future of World Politics: Will It Resemble the Past?" *International Security* 16, no. 3 (Winter 1991/92): 39–46. See also Nazli Choucri, "Key Issues in International Relations Forecasting," in Nazli Choucri and Thomas W. Robinson, eds., *Forecasting in International Relations. Theory, Methods, Problems, Prospects* (San Francisco: W. H. Freeman, 1978), 3–22; and Thomas W. Milburn, "Successful and Unsuccessful Forecasting in International Relations," in Choucri and Robinson, *Forecasting in International Relations*, 79–91. For good examples of theory-based predictions see John J. Mearsheimer, "Back to the Future: Instability in Europe after the Cold War," *International Security* 15, no. 1 (Summer 1990): 5–56; and Stephen Van Evera, "Primed for Peace: Europe After the Cold War," *International Security* 15, no. 3 (Winter 1990/91): 7–57.
8. See Gunther Hellmann and Reinhard Wolf, "Neorealism, Neoliberal Institutionalism, and the Future of NATO," *Security Studies* 3, no. 1 (Autumn 1993), forthcoming; Keohane, *International Institutions and State Power*, 16; and Grieco, *Cooperation among Nations*, 227–32.
9. As Stephen Walt writes, "It is impossible to formulate any strategy without making theoretical judgements; the only issue is whether these judgements will be made explicit or left vague and untested." Walt, "Alliance, Threat, and U.S. Grand Strategy," 466. See also Walt, "The Search for a Science of Strategy: A Review of *Makers of Modern Strategy*," *International Security* 12, no. 1 (Summer 1987): 141–42.
10. Peter D. Zimmerman, "Technical Barriers to Nuclear Proliferation," *Security Studies* 2, nos. 3/4 (Spring/Summer 1993): 345–56.
11. Reflecting on the cumulative effect of these trends, Leonard Spector writes that the determined drive of countries towards and across the nuclear threshold "has been accompanied by a second trend: horizontal proliferation is going vertical, as the regional states that have already crossed that barrier strive to increase their stocks of weapons material, develop advanced delivery systems, and acquire thermonuclear options." Leonard S. Spector, "Nuclear Proliferation in the 1990s: The Storm after the Lull," Appendix 1, *New Threats: Responding to the Proliferation of Nuclear, Chemical, and Delivery Capabilities in the Third World*, an Aspen Strategy Group Report (Lanham, Md.: University Press of America, 1990);48–49. Spector has described the different technological and political paths different countries have chosen to acquire the capabilities to build nuclear weapons. See accounts of countries' nuclear capabilities and detailed histories of their nuclear programs in his *Nuclear Ambitions* (Boulder, Colo.: Westview,

1990); "Nuclear Proliferation in the 1990s: The Storm after the Lull," *The Undeclared Bomb* (Cambridge, Mass.: Ballinger, 1988); *Going Nuclear* (Cambridge, Mass.: Ballinger, 1987); *The New Nuclear Nations* (New York: Vintage, 1985); and *Nuclear Proliferation Today* (New York: Vintage, 1984).

12. *New Threats: Responding to the Proliferation of Nuclear, Chemical, and Delivery Capabilities in the Third World*, 2.
13. In Keeley's words, "Proliferation is a political problem ... the problems of the nonproliferation regime draw on and reflect fundamental dynamics of any political order: they are based on deep rooted problems of politics." James F. Keeley, "Legitimacy, Capability, Effectiveness and the Future of the NPT," in David B. DeWitt, ed., *Nuclear Non-Proliferation and Global Security* (New York: St. Martin's, 1987), 30. Betts writes: "Proliferation does not have a life of its own; it is a political problem as much as a technical one." Richard K. Betts, "Paranoids, Pygmies, Pariahs & Proliferation," *Foreign Policy*, no. 26 (Spring 1977): 163. A revised version of Betts's article is published in *Security Studies* 2, nos. 3/4 (Spring/Summer1993): 100–123; see also George H. Quester, "Some Conceptual Problems in Nuclear Proliferation," *American Political Science Review* 66, no. 1 (March 1972): 490–97. See Waltz's discussion in *Theory of International Politics*, 88–99, 116–23; and in Waltz, "War in Neorealist Theory," in Robert I. Rotberg and Theodore K. Raab, eds., *The Origins and Prevention of Major Wars* (New York: Cambridge University Press, 1990), 42–43. For a discussion of countries that decided not to acquire the bomb for other than technological reasons, see Lewis A. Dunn, "Four Decades of Nuclear Non-Proliferation: Some Lessons from Wins, Losses, and Draws," *New Threats: Responding to the Proliferation of Nuclear, Chemical, and Delivery Capabilities in the Third World*, Appendix 8, pp. 234–37; and Dunn, *Controlling the Bomb: Nuclear Proliferation in the 1980s* (New Haven: Yale University Press, 1982), 11–14.
14. Waltz, *Theory of International Politics*, 81. For good discussions of the influence of international structures on nations' domestic policies and foreign policy behavior, see Peter Gourevitch, "The Second Image Reversed: The International Sources of Domestic Politics," *International Organization* 32, no. 4 (Autumn 1978): 881–912; Gourevitch, "International Trade, Domestic Coalitions, and Liberty: Comparative Responses to Crisis, 1873–1896," *Journal of Interdisciplinary History* 8 (1977): 281–313; Gourevitch, *Politics in Hard Times: Comparative Responses to International Economic Crises* (Ithaca: Cornell University Press, 1986); Ronald Rogowski, "Political Cleavages and Changing Exposure to Trade," *American Political Science Review* 81, no. 4 (December 1987): 1121–37; and Jack Snyder, "International Leverage on Soviet Domestic Change," *World Politics* 42, no. 1 (October 1989): 1–30.
15. Jack S. Levy, "The Causes of War: A Review of Theories and Evidence," in Philip E. Tetlock, Jo L. Husbands, Robert Jervis, Paul S. Stern and Charles Tilly, eds., *Behavior, Society, and Nuclear War*, vol. 1 (New York: Oxford University Press, 1989), 209–333; and Charles W. Kegley, Jr. and Gregory A. Raymond, "Must We Fear a Post-Cold War Multipolar System?" *Journal of Conflict Resolution* 36, no. 3 (September 1992): 573–85, 579.
16. Kenneth N. Waltz, "What Will the Spread of Nuclear Weapons Do to the World?" in John Kerry King, ed., *International Political Effects of the Spread of Nuclear Weapons* (Washington, D.C.: Government Printing Office, 1979), 176.
17. For criticism of the argument that the world is moving toward multipolarity because the United States is in decline, see Joseph S. Nye, Jr., *Bound to Lead: The Changing Nature of American Power* (New York: Basic Books, 1990); Nye, "Against 'Declinism'," *New Republic*, 15 October 1990, pp. 12–13; and Samuel

P. Huntington, "The U.S. – Decline or Renewal?" *Foreign Affairs* 67, no. 2 (Winter 1988/89): 76–96.

18. Structural analysis, or neorealism, abstracts from the characteristics of the system's constituent units, their behavior, and their interactions, concentrating instead on how the units are positioned or arranged in the system. See Waltz, *Theory of International Politics*, esp. 79–80, 99; Waltz, "War in Neorealist Theory," 42–43; and Waltz, "Reflections on *Theory of International Politics*: A Response to My Critics," 325–30, 341–44. Keohane's volume contains useful discussions of neorealism.
19. "International structures vary only through a change of organizing principle or, failing that, through changes in the capabilities of the units." Waltz, *Theory of International Politics*, 93; see also 100–1.
20. "A structure is defined by the arrangement of its parts. Only changes of arrangement are structural changes." Waltz, *Theory of International Politics*, 80.
21. See discussion in Waltz, "Reflections on *Theory of International Politics*: A Response to My Critics," 323–25.
22. Waltz, "Reflections on *Theory of International Politics*: A Response to My Critics," 325–26; Waltz, *Theory of International Politics*, 93–97. See discussion of interdependence in Waltz, "The Myth of National Interdependence," in Charles P. Kindelberger, ed., *The International Corporation* (Cambridge: MIT Press,1970), 205–23.
23. "Reducing a multi- to a tri- or bipolar world would change the system's structure. Wars that eliminate enough rival great powers are system-transforming wars"; and " ... a change of system ... [would occur] ... by building a world hegemony or by promoting an area of great power status by helping it find political unity ... "; and " ... international political structure [is defined] by counting states." (*Theory of International Politics*, 199, 201, 99); and "Students of international politics make distinctions between international-political systems only according to the number of their great powers. The structure of a system changes with changes in the distribution of capabilities across the system's units." (*Theory of International Politics*, 97).
24. For criticism of claims about the anarchic nature of world politics see Helen Milner, "The Assumption of Anarchy in International Relations Theory: A Critique," *Review of International Studies* 17, no. 1 (January 1991): 67–85.
25. "To interdict the use of force by the threat of force, to oppose force with force, to influence the policies of states by the threat or use of force: These have been and continue to be the most important means of control in security matters." Waltz, *Theory of International Politics*, 209.
26. See discussion in Robert W. Tucker, *The Inequality of Nations* (New York: Basic Books, 1977), 3–15.
27. See discussions in Stephen M. Walt, *The Origins of Alliances* (Ithaca: Cornell University Press, 1987), ch. 2; Waltz, *Theory of International Politics*, ch. 6; Snyder, "The Security Dilemma in Alliance Politics"; and Barry R. Posen, *The Sources of Military Doctrine: France, Britain, and Germany Between the World Wars* (Ithaca: Cornell University Press, 1984), 60–65.
28. There is no agreement on this issue. The argument that the bipolar system is more stable than the multipolar one is made in Kenneth N. Waltz, "The Stability of a Bipolar World," *Daedalus* 93, no. 3 (Summer 1964): 881–909; Waltz, "War in Neorealist Theory," 44–48; Waltz, *Theory of International Politics*, 163–76; Waltz, *The Spread of Nuclear Weapons: More May Be Better*, Adelphi Paper no. 171 (London: International Institute for Strategic Studies, 1981), 2–3; Waltz, "What Will the Spread of Nuclear Weapons Do to the World?" 169–71; Manus I. Midlarsky, "Hierarchical Equilibria and the Long-Run Instability of Multipolar Systems," in Manus I. Midlarsky, ed., *Handbook of*

War Studies (Boston: Unwin Hyman, 1989), 55–81; and Mearsheimer, "The Future of Europe," 6–9, 13–14. For an argument for the benefits of multipolarity see Karl W. Deutsch and J. David Singer, "Multipolar Power Systems and International Stability," *World Politics* 16, no. 3 (April 1964): 390–406. On the increased risks of war present in highly polarized systems see Michael D. Wallace,"Alliance Polarization, Cross-Cutting, and International War, 1815–1964," *Journal of Conflict Resolution* 17 (1973): 575–604; and Woosang Kim, "Power, Alliance, and Major Wars, 1816–1975," *Journal of Conflict Resolution* 33 (1989): 255–73. See also the discussions in Richard N. Rosecrance, "Bipolarity, Multipolarity, and the Future," *Journal of Conflict Resolution* 10, no. 3 (September 1966): 314–27; and Jack S. Levy, "The Causes of War: Contending Theories," in Charles W. Kegley, Jr. and Eugene R. Wittkopf, eds., *The Global Agenda*, 3rd ed. (New York: McGraw–Hill, 1992), 59–69.

29. Alvin M. Saperstein, "The 'Long Peace' – Result of a Bipolar Competitive World?" *Journal of Conflict Resolution* 35, no. 1 (March 1991): 68–79.
30. For a good summary of the arguments about bipolarity and multipolarity see Van Evera, "Primed for Peace," 33–40.
31. Geoffrey Blainey, *The Causes of War*, 3rd ed. (New York: Free Press, 1988), 113–14, 117–18.
32. Waltz, *Theory of International Politics*, ch. 8, pts. 2, 3; and Waltz, "What Will the Spread of Nuclear Weapons Do to the World?" 169–70.
33. See Van Evera, "Primed for Peace"; Jervis, "The Future of World Politics: Will It Resemble the Past?"; Kegley and Raymond, "Must We Fear a Post-Cold War Multipolar System?"; and Glenn Chafetz, "The End of the Cold War and the Future of Nuclear Proliferation: Alternative to the Neorealist Perspective," *Security Studies* 2, nos. 3/4 (Spring/Summer 1992): 127–58. For an early argument about the development of pluralistic security communities, see Karl Deutsch et al., *Political Community and the North Atlantic Area: International Organization in the Light of Historical Experience* (Princeton: Princeton University Press, 1957).

    Gaddis is skeptical, writing that "These are pleasant things to believe, but there is remarkably little historical evidence to validate them." Gaddis, "The Long Peace: Elements of Stability in the Postwar International System," in John Lewis Gaddis, *The Long Peace: Inquiries into the History of the Cold War* (New York: Oxford University Press, 1987), 224. See discussion in Mearsheimer, "Back to the Future," 40–51. See also Josef Joffe, *The Limited Partnership: Europe, the United States, and the Burdens of Alliance* (Cambridge, Mass.: Ballinger, 1987), esp. ch. 5; and Joffe, "NATO and the Limits of Devolution" (Paper presented at "NATO at 40: Confronting a Changing World," a Cato Institute Conference, Washington, D.C., 3–4 April 1989). There are many historical studies from which one may draw support for the argument that the emerging international system will be conducive to instability. See, among others, Blainey, *The Causes of War*; A. W. DePorte, *Europe Between the Superpowers: The Enduring Balance*, 2nd ed. (New Haven: Yale University Press, 1986); Stephen R. Rock, *Why Peace Breaks Out: Great Power Rapprochement in Historical Perspective* (Chapel Hill: University of North Carolina Press, 1989); and Richard Ned Lebow, *Between Peace and War: The Nature of International Crisis* (Baltimore: Johns Hopkins University Press, 1981).
34. Thus Jervis writes that "While Eastern Europe and the Third World are not likely to simply recapitulate the West's history from which so many of our theories of international politics are derived, neither should we expect a basic change from the familiar ways in which nations relate to each other." Jervis, "The Future of World Politics: Will It Resemble the Past?" 73.
35. See Stephen Van Evera, "Why Europe Matters, Why the Third World Doesn't:

America's Grand Strategy After the Cold War," *Journal of Strateguc Studies* 13, no. 2 (June 1990): 1–51; Van Evera, "American Intervention in the Third World: Less Would Be Better," *Security Studies* 1, no. 1 (Authumn 1991): 1–24; Michael C. Desch, "The Keys that Lock Up the World: Identifying American Interests in the Periphery," *International Security* 14, no. 1 (Summer 1989): 86–121; and Steven R. David, "Why the Third World Matters," *International Security* 14, no. 1 (Summer 1989): 50–85.

36. See Gaddis, "The Long Peace," 216. For support of Gaddis's argument, see Saperstein, "The 'Long Peace'– Result of a Bipolar Competitive World?" 68. For criticism of the definition of the cold war period as especially peaceful, see Robert Jervis, "Will the New World Be Better?" in Robert Jervis and Seweryn Bialer, eds., *Soviet-American Relations after the Cold War* (Durham: Duke University Press, 1991), 7–19; and Michael Brecher and Jonathan Wilkenfled, "International Crises and Global Instability: The Myth of the 'Long Peace'," in Charles W. Kegley, Jr., ed., *The Long Postwar Peace* (New York: HarperCollins, 1991), 85–104.

37. "Never since the Treaty of Westphalia in 1648," Waltz writes, " ... have great powers enjoyed a longer period of peace than we have known since the Second World War." He emphasizes the role nuclear weapons played in securing the long peace. Kenneth N. Waltz, "Nuclear Myths and Political Realities," *American Political Science Review* 84, no. 3 (September 1990): 744.

38. There are scholars who challenge the claim that there is a correlation between polarity and propensity for war. See Ted Hopf, "Polarity, the Offense-Defense Balance, and War," *American Political Science Review* 85, no. 2 (June 1991): 475–93; Jervis, "The Future of World Politics: Will It Resemble the Past?" 40–42; Bruce Bueno de Mesquita, "Risk, Power Distributions, and the Likelihood of War," *International Studies Quarterly* 25 (1981): 541–68, 564; Charles W. Ostrom, Jr. and John H. Aldrich, "The Relationship between Size and Stability in the Major Power International System," *American Journal of Political Science* 22 (1978): 743–71, 765; Jack S. Levy, "The Polarity of the System and International Stability: An Empirical Analysis," in Alan Ned Sabrosky, ed., *Polarity and War* (Boulder, Colo.: Westview. 1985), 41–66; and Frank Wayman, "Bipolarity, Multipolarity, and the Threat of War," in Sabrosky, *Polarity and War*, 115–44. On the distinction between system stability and resource stability see Manus I. Midlarsky, *The Onset of World War* (Boston: Allen & Unwin, 1988), 48–52; and Emerson M. S. Niou and Peter C. Ordeshook, "Stability in Anarchic International Systems," *American Political Science Review* 84, no. 4 (December 1990): 1207–34.

39. For a compelling argument see Mearsheimer, "Back to the Future." An abridged version, titled "Why We Will Soon Miss the Cold War," appeared in *Atlantic Monthly* (August 1990): 35–50.

40. Mearsheimer, "Back to the Future." See also Mearsheimer, "Back to the Future, Part II: International Relations Theory and Post-Cold War Europe," *International Security* 15, no. 2 (Fall 1990): 194–99; and Mearsheimer, "Back to the Future, Part III: Realism and the Realities of European Security," *International Security* 15, no. 3 (Winter 1990/91): 219–22. Cf. Samuel P. Huntington, "No Exit: The Errors of Endism," *The National Interest*, no. 17 (Fall 1989): 3–11; John Lewis Gaddis, "One Germany – in Both Alliances," *New York Times,* 21 March 1990, p. A27; John Lewis Gaddis, "Coping with Victory," *Atlantic Monthly* (May 1990): 49–60; Alice M. Rivlin, "New World, New Dangers," *Washington Post*, 10 April 1990, p. A23; and Benjamin Frankel, "A New Continent Emerges, Full of Old Uncertainties," *Los Angeles Times*, 29 October 1989, p. M2. For more optimistic assessments of the post-cold war world see Van Evera, "Primed for Peace"; and Kegley and Raymond, "Must We Fear a Post-Cold War Multipolar System?" 581–83.

41. Joseph S. Nye, Jr., "Maintaining a Nonproliferation Regime" (unpublished paper, June 1980), as quoted in Waltz, *The Spread of Nuclear Weapons: More May Be Better*, 29. Nye refers to "40 odd states that possess nuclear technology" in Joseph S. Nye, Jr., "NPT: The Logic of Inequality," *Foreign Policy*, no. 59 (Summer 1985): 126. According to IAEA figures, by the end of the 1970s nuclear research or power programs existed in over forty-five non-nuclear weapons states. Cited in Mitchell Reiss, *Without the Bomb: The Politics of Nuclear Nonproliferation* (New York: Columbia University Press, 1988), 23. Cf. Dunn, *Controlling the Bomb*, 11–18.
42. Or, as Betts writes, "A nation's decision to manufacture nuclear weapons depends on its incentives, capabilities, and disincentives." Richard K. Betts, "A Diplomatic Bomb for South Africa?" *International Security* 4, no. 2 (Fall 1979): 91.
43. Dunn, *Controlling the Bomb*, 13–14. Cf. Leonard Beaton and John Maddox, *The Spread of Nuclear Weapons* (London: Chatto and Windus, 1962), 151–67.
44. In many cases superpower involvement also played a role in dampening conflicts in the Third World. See Jervis, "The Future of World Politics: Will It Resemble the Past?" 59–60.
45. Waltz, *Theory of International Politics*, 170–71; Waltz, *The Spread of Nuclear Weapons: More May Be Better*, 3; Waltz, "The Stability of a Bipolar World," 882.
46. For a vigorous argument for this position see George H. Quester and Victor A. Utgoff, "U.S. Arms Reductions and Nuclear Proliferation: The Counterproductive Possibilities," *Washington Quarterly* 16, no. 1 (Winter 1993): 129–40.
47. For a discussion of the role of American security guarantees in retarding nuclear proliferation see Robert J. Art, "A Defensible Defense: America's Grand Strategy after the Cold War," *International Security* 15, no. 4 (Spring 1991): 28–30. For examples from the pre-nuclear era of big powers offering alliance arrangements to small powers in order to restrain and moderate the defense policies of these smaller powers, see Paul Schroeder, "Alliances, 1815–1945: Weapons of Power and Tools of Management," in Klaus Knorr, ed., *Historical Dimensions of National Security* (Lawrence, Kan.: University of Kansas Press, 1976), 227–62.
48. Dunn, "Four Decades of Nuclear Non-Proliferation: Some Lessons from Wins, Losses, and Draws," 235. See also Michael J. Brenner, *Nuclear Power and Non-Proliferation: The Remaking of U.S. Policy* (New York: Cambridge University Press, 1982). Brenner's discussion of the perception by other countries of U.S. nuclear conduct is good, especially his discussion of the Carter administration's policies.
49. Waltz does not see this as necessarily a bad thing. "Politically, allies who are unsure of one another's support have reason to work harder for the sake of their own security." Waltz, "Nuclear Myths and Political Realities," 739.
50. Betts, "Paranoids, Pygmies, Pariahs & Proliferation," 174.
51. Waltz, "What Will the Spread of Nuclear Weapons Do to the World?" 167; Waltz, *The Spread of Nuclear Weapons: More May Be Better*, 28. Cf. Dunn, *Controlling the Bomb*, 3, 16; and Gerard C. Smith and Helena Cobban, "A Blind Eye to Nuclear Proliferation," *Foreign Affairs* 68, no. 3 (Summer 1989): 69.
52. In 1963 President John F. Kennedy expressed the fear that by the 1970s an American president may face "a world in which fifteen or twenty or twenty-five nations may have [nuclear] weapons." *Public Papers of the Presidents of the United States: John F. Kennedy, 1963* (Washington, D.C.: Government Printing Office, 1964), 280; Cf. predictions cited in Reiss, *Without the Bomb*, 15.
53. Nye, "NPT: The Logic of Inequality," 126. Cf. Lincoln P. Bloomfield, "Forward," in Reiss, *Without the Bomb*, vii. Robert Art agrees, writing that "American actions, including the stationing of troops overseas, have played a

critical role in retarding nuclear weapons spread among key states." Art, "A Defensible Defense," 29.

54. For background discussion of Japan's security situation, see Yukio Satoh, *The Evolution of Japanese Security Policy*, Adelphi Paper no. 178 (London: International Institute for Strategic Studies, 1982); Saburo Kato, "Japan: Quest for Strategic Compatibility," in Robert M. Lawrence and Joel Larus, eds., *Nuclear Proliferation, Phase II* (Lawrence, Kan.: University Press of Kansas, 1974), 183–208; Kiichi Saeki, "Japan's Security in a Multipolar World," in *East Asia and the World System*, pt. 2, Adelphi Paper no. 92 (London: IISS, 1972), 21–29; Wayne Wilcox, "Japanese and Indian National Security Policies in the Asia of the 1970s: The Prospects for Nuclear Proliferation," in *East Asia and the World System*, pt. 2, Adelphi Paper no. 92 (London: IISS, 1972), 30–39; Masataka Kosaka, "Japan's Nuclear Options," in Geoffrey Kemp et al., eds., *The Superpowers in a Multinuclear World* (Lexington, Mass.: D. C. Heath, 1973), 91–105; John E. Endicott, *Japan's Nuclear Option* (New York: Praeger, 1975); Joseph A. Yager, "Japan," in Joseph A. Yager, ed., *Nonproliferation and U.S. Foreign Policy* (Washington, D.C.: Brookings, 1980), 9–46.
55. Dunn, *Controlling the Bomb*, 12, 64–66. See also George H. Quester, *The Politics of Nuclear Proliferation* (Baltimore: Johns Hopkins University Press, 1973), 117–21.
56. Reiss, *Without the Bomb*, 254.
57. Ibid., 128.
58. For a discussion of these developments, see Alvin H. Bernstein, "The Soviets at Cam Ranh Bay," in Benjamin Frankel, ed., *In the National Interest: A National Interest Reader* (Lanham, Md.: University of America Press, 1990), 167–83.
59. Reiss, *Without the Bomb*, 130.
60. Kunio Muraoka, *Japanese Security and the United States*, Adelphi Paper no. 95 (London: IISS, 1973), 26.
61. Reiss, *Without the Bomb*, 136.
62. Ibid., 135.
63. For background information on South Korea's security situation, see Reiss, *Without the Bomb*, 78–86; Stuart E. Johnson, *The Military Equation in Northeast Asia* (Washington, D.C.: Brookings, 1979); Young-sun Ha, *Nuclear Proliferation, World Order, and Korea* (Seoul: Seoul National University, 1983); Joseph A.Yager, "The Republic of Korea," in Yager, ed., *Nonproliferation and U.S. Foreign Policy*; Franklin B. Weinstein, "Korea and Arms Control II," in John Barton and Ryukichi Imai, eds., *Arms Control II, A New Approach to International Security* (Cambridge, Mass.: Oelgeschlager, Gunn and Hain, 1981); Franklin B. Weinstein, "The United States, Japan, and the Security of Korea," *International Security* 2, no. 2 (Fall 1977); and Frank Gibney, "The Ripple Effect in Korea," *Foreign Affairs* 56, no. 1 (October 1977): 160–74.
64. Reiss, Without the Bomb, 98.
65. Ibid., 253.
66. Lewis A. Dunn, "Some Reflections on the 'Dove's Dilemma'," in George H. Quester, ed., *Nuclear Proliferation: Breaking the Chain* (Madison: University of Wisconsin Press, 1981), 184.
67. Reiss, *Without the Bomb*, 92.
68. Betts, "Paranoids, Pygmies, Pariahs & Proliferation," 179. Cf. Reiss, *Without the Bomb*, 101.
69. Reiss, *Without the Bomb*, 93.
70. Rowland Evans and Robert Novak, "Korea: Park's Inflexibility," *Washington Post*, 12 June 1976, p. A19; Richard Halloran, "Korean Chief Linked to Illegal Lobbying," *New York Times*, 1 November 1976, p. 22; Betts, "Paranoids, Pygmies, Pariahs & Proliferation," 180; Reiss, *Without the Bomb*, 93–94.

71. Dunn, "Some Reflections on the 'Dove's Dilemma'," 184. Cf. Reiss, *Without the Bomb*, 85, 98; and Young-sun Ha,"Nuclearization of Small States and World Order: The Case of Korea," *Asian Survey* 28, no. 11 (November 1978): 1134–51.
72. Reiss, *Without the Bomb*, 94–95.
73. See discussion in Selig Harrison, "A Chance for Detente in Korea?" *World Policy Journal* 8, no. 4 (Fall 1991): 599–632.
74. David E. Sanger, "The Nonproliferation Treaty Bares Its Toothlessness," *New York Times*, Week in Review, 14 March 1993, p. E18.
75. David E. Sanger, "Wary of North Korea, Seoul Debates Building Atomic Bomb," *New York Times*, 19 March 1993, p. A8.
76. Sanger, "Wary of North Korea, Seoul Debates Building Atomic Bomb," A8.
77. Spector, *The Undeclared Bomb*, 75. See Spector's discussion of the Taiwanese case in *The Undeclared Bomb*, 75–79. Cf. Stephen Engelberg and Michael Gordon, "Taipei Halts Work on Secret Plant to Make Nuclear Bomb Ingredient," *New York Times*, 23 March 1988.
78. Spector, *The Undeclared Bomb*, 346, n. 24. See discussion of the early Taiwanese efforts in Edward Schumacher, "Taiwan Seen Reprocessing Nuclear Fuel," *Washington Post*, 29 August 1976; Don Oberdorfer, "Taiwan to Curb A-Role," *Washington Post*, 23 September 1976; and Milton R. Benjamin, "Taiwan's Nuclear Plans Concern U.S. Officials," *Washington Post*, 20 December 1978.
79. Dunn, "Some Reflections on the 'Dove's Dilemma'," 184.
80. Ibid.
81. Spector, *The Undeclared Bomb*, 77.
82. Quoted in Spector, *The Undeclared Bomb*, 346, n. 24.
83. For the role played by U.S. policies in restraining Egypt's and Brazil's nuclear program, see Michael Joe Siler, "Explaining Variations in Nuclear Outcomes among Southern States: Bargaining Analysis of U.S. Non-Proliferation Policies Towards Brazil, Egypt, India, and South Korea" (Ph.D. diss., University of Southern California, May 1992).
84. Dunn, *Controlling the Bomb*, 12. Cf. Beaton and Maddox, *The Spread of Nuclear Weapons*, 98–108.
85. Reiss, *Without the Bomb*, 252.
86. Dunn, *Controlling the Bomb*, 12, 66–67. See also Catherine McArdle Kelleher, *Germany and the Politics of Nuclear Weapons* (New York: Columbia University Press, 1975); Beaton and Maddox, *The Spread of Nuclear Weapons*, 108–20; Horst Menderhausen, "Will West Germany Go Nuclear?" *Orbis* 16, no. 2 (Summer 1972): 411–34; and Betts, "Paranoids, Pygmies, Pariahs & Proliferation," 174.
87. Spector, *The Undeclared Bomb*, 162; Ze'ev Schiff, "Dealing with Syria," *Foreign Policy*, no. 55 (Summer 1984): 94; Shai Feldman, *Israeli Nuclear Deterrence: A Strategy for the 1980s* (New York: Columbia University Press, 1982): 182–84; and Feldman,"Superpower Nonproliferation Policies: The Case of the Middle East," in Steven L. Spiegel, Mark A. Heller, and Jacob Goldberg, eds., *The Soviet-American Competition in the Middle East* (Lexington, Mass.: Lexington, 1988), 107.
88. For contrasting views of the role nuclear weapons played in the 1973 war, see Shlomo Aronson, "The Nuclear Dimension of the Arab-Israeli Conflict: The Case of the Yom Kippur War," *Jerusalem Journal of International Relations* 7, nos. 1–2 (1984): 107–42; and Yair Evron, "The Relevance and Irrelevance of Nuclear Options in Conventional Wars: the 1973 October War," *Jerusalem Journal of International Relations* 7, nos. 1–2 (1984): 143–76.
89. Feldman, "Superpower Nonproliferation Policies: The Case of the Middle East," 107.
90. Robert E. Harkavy, "Pariah States and Nuclear Proliferation," in Quester, ed., *Nuclear Proliferation: Breaking the Chain*, 135–63; Harkavy, "The Pariah State

Syndrome," *Orbis* 21, no. 3 (Fall 1977): 623–49; and Betts, "Paranoids, Pygmies, Pariahs & Proliferation," 166–67. See also Betts, "A Diplomatic Bomb for South Africa?" 91–115; and Pierre Lellouche, "The Garrison State," in William H. Kincade and Christoph Bertram, eds., *Nuclear Proliferation in the 1980s: Perspectives and Proposals* (New York: St. Martin's, 1982), 63–111.

91. The expression is Betts's. "A Diplomatic Bomb for South Africa?" 91.
92. Reiss, *Without the Bomb*, 170 and 312–313, n. 121. According to Art, during 1954–55 Israel asked the United States for a military alliance, but the United States refused. Art, "A Defensible Defense," 29, n. 49. One reason the Kennedy administration increased its commitment to Israel in the early 1960s was to dissuade Israel from building nuclear weapons. These commitments, however, did not amount to a formal agreement. Walt, *The Origins of Alliances*, 95–96.
93. Tom Segev, *The Seventh Million: Israel Confronts the Holocaust* (New York: Hill and Wang, 1993), 11.
94. For discussions of Israel's nuclear history see Seymour M. Hersh, *The Samson Option: Israel's Nuclear Arsenal and American Foreign Policy* (New York: Random House, 1991); Shlomo Aronson with Oded Brosh, *The Politics and Strategy of Nuclear Weapons in the Middle East* (Albany: State University of New York Press, 1992); Spector, *Nuclear Ambitions*, 149–74; Spector, *The Undeclared Bomb*, 164–93; Frank Barnaby, *The Invisible Bomb* (London: I. B. Tauris, 1989); Peter Pry, *Israel's Nuclear Arsenal* (London: Croom Helm, 1984); Fuad Jabber, *Israel and Nuclear Weapons: Present Option and Future Strategies* (London: Chatto and Windus, 1971); Avigdor Haselkorn, "Israel: From an Option to a Bomb in the Basement?" in Robert M. Lawrence and Joel Larus, eds., *Nuclear Proliferation, Phase II* (Lawrence, Kan.: University Press of Kansas, 1974), 149–82; Efraim Inbar, *Israel's Nuclear Policy after 1973* (Los Angeles: Pan Heuristics, 1977); Uri Bar-Joseph,"The Hidden Debate: The Formation of Nuclear Doctrines in the Middle East," *Journal of Strategic Studies* 5, no. 2 (June 1982): 205–27; Yair Evron, *Israel's Nuclear Dilemma* (Hebrew; Tel Aviv: Hakibutz Hameuchad, 1987); Evron, "Israel and the Atom: The Uses and Misuses of Ambiguity, 1957–1967," *Orbis* 17, no. 4 (Winter 1974): 1326–43; Evron, "Opaque Proliferation: The Israeli Case," in Frankel, *Opaque Nuclear Proliferation: Methodological and Policy Implications*, 45–63; Evron, "The Arab Position in the Nuclear Field: A Study of Policies up to 1967," *Cooperation and Conflict*, no. 1 (1973): 19–31; Evron, "A Nuclear Balance of Deterrence in the Middle East," *New Outlook* 18, no. 5 (July/August 1975): 18–19; Shlomo Aronson, *Israel's Nuclear Options*, ACIS Working Paper no. 7 (Los Angeles: University of California, Los Angeles, 1977); Aronson, "Nuclearization of the Middle East: A Dovish View," *Jerusalem Quarterly*, no. 2 (Winter 1977): 27–44; Aronson, "The Nuclear Dimension of the Yom Kippur War"; and the relevant sections in Aronson, *Conflict and Bargaining in the Middle East* (Baltimore: Johns Hopkins University Press, 1978); Alan Dowty, "Nuclear Proliferation: The Israeli Case," *International Studies Quarterly* 22, no. 1 (March 1978): 79–120; Shai Feldman, *Israeli Nuclear Deterrence: A Strategy for the 1980s*; Feldman, "A Nuclear Middle East," *Survival* 23, no. 3 (May/June 1981): 107–15; Feldman, "The Bombing of Osiraq – Revisited," *International Security* 7, no. 2 (Fall 1982): 114–42; Lawrence Freedman, "Israel's Nuclear Policy," *Survival* 17, no. 3 (May/June 1975): 114–20; Paul Jabber, *A Nuclear Middle East: Infrastructure, Likely Military Postures and Prospects for Strategic Stability*, ACIS Working Paper no. 6 (Los Angeles: UCLA, 1977); Robert E. Harkavy, *Specter of a Middle Eastern Holocaust* (Denver: University of Denver 1977); Steven J. Rosen, "Nuclearization and Stability in the Middle East," in Onkar Marwah and Amy Schultz, eds., *Nuclear Proliferation and the Near-Nuclear Countries* (Cambridge, Mass.: Ballinger, 1975), 157–84; Rosen, "A Stable System of Mutual Nuclear Deterrence in

the Arab-Israeli Conflict," *American Political Science Review* 71, no. 4 (December 1977): 1367–83; Rosen, *What a Fifth Arab-Israeli War Might Look Like: An Exercise in Crisis Forecasting*, ACIS Working Paper no. 8 (Los Angeles: UCLA, 1977); Avner Cohen and Marvin Miller, "Nuclear Shadows in the Middle East: Prospects for Arms Control in the Wake of the Gulf Crisis," *Security Studies* 1, no. 1 (Autumn 1991): 54–77. For discussions of the Israeli-French nuclear cooperation see Pieree Pean, *Les deux bombes* (Paris: Fayard, 1981); Bertand Goldschmidt, *Le défi atomique* (Paris: Fayard, 1980); Sylvia K. Crosbie, *The Tacit Alliance: France and Israel from Suez to the Six Day War* (Princeton: Princeton University Press, 1974); Ann Williams, *Britain and France in the Middle East and North Africa, 1914–1967* (London: Macmillan, 1968); and Lawrence Scheinman, *Atomic Energy Policy in France Under the Fourth Republic* (Princeton: Princeton University Press, 1965).

95. Reiss, *Without the Bomb*, 143.
96. Ibid., 151.
97. This is a simplification. For an elaboration of the proposition that Israel's nuclear behavior approximates an ideal-type opaque nuclear behavior, see Avner Cohen and Benjamin Frankel, "Opaque Nuclear Proliferation," in Frankel, *Opaque Nuclear Proliferation: Methodological and Policy Implications*, 14–44. For an account of U.S. attitudes see Hersh, *The Samson Option*.
98. Dunn, "Some Reflections on the 'Dove's Dilemma'," 184; Reiss, *Without the Bomb*, 169, 255–56. Cf. Betts, "Paranoids, Pygmies, Pariahs & Proliferation," 180.
99. See discussion in Feldman, "Superpower Nonproliferation Policies: The Case of the Middle East"; Reiss, *Without the Bomb*, 165–70.
100. For discussion of nuclear proliferation in South Asia see Ashok Kapur, *Pakistan's Nuclear Development* (London: Croom Helm,1987); Brahma Chellaney, "South Asia's Passage to Nuclear Power," *International Security* 16, no. 1 (Summer 1991): 43–72; Stephen P. Cohen, "Nuclear Neighbors," in Stephen P. Cohen, ed., *Nuclear Proliferation in South Asia: The Prospects for Arms Control* (Boulder, Colo.: Westview, 1991), 1–22; Spector, *Nuclear Ambitions*, 63–112; Devin T. Hagerty, "The Power of Suggestion: Opaque Proliferation, Existential Deterrence, and the South Asian Nuclear Arms Competition," *Security Studies* 2, nos. 3/4 (Spring/Summer 1993): 257–84; and Neil Joeck, "Tacit Bargaining and Stable Proliferation in South Asia," in Frankel, *Opaque Nuclear Proliferation*, 77–91.
101. See Zalmay Khalilzad, "Pakistan and the Bomb," *Survival* 21, no. 6 (November/December 1979): 244–50, esp. 246; Dunn, "Some Reflections on the 'Dove's Dilemma'," 183; and Benjamin Frankel, "Explosive Matter: Nuclear Proliferation Policy," *The American Enterprise* 1, no. 2 (March–April 1990): 8–10.
102. Michael R. Gordon, "Nuclear Issue Slows U.S. Aid to Pakistan," *New York Times*, 1 October 1990, p. A3; Paul L. Leventhal, "Cut Off Aid To Pakistan," *Washington Post*, 8 October 1990, p. A23.
103. Betts, "Paranoids, Pygmies, Pariahs & Proliferation," 158. Spector also mentions the Chinese detonation and the 1962 Indian defeat in its border war with China as contributing to the acceleration of the Indian nuclear program. He does not mention the Indian quest for superpowers' guarantees, however. Further, he asserts that the earlier Indian program was motivated by political considerations and that it acquired its military orientation only in the late 1970s in response to the Pakistani nuclear program. Spector, *The Undeclared Bomb*, 82. For good discussion of India's security policies see Raju G. C. Thomas, *Indian Security Policy* (Princeton: Princeton University Press, 1986). See also Siler, "Explaining Variations in Nuclear Outcomes."
104. See discussion in A. G. Noorani, "India's Quest for a Nuclear Guarantee," *Asian*

Survey 7, no. 7 (July 1967): 490–502; M. R. Masani, "The Challenge of the Chinese Bomb – II," *India Quarterly* 21, no. 1 (January–March 1965): 26–27; Raj Krishna, "India and the Bomb," *India Quarterly* 21, no. 2 (April–June 1965): 119–37; Reiss, *Without the Bomb*, 214, 222.

105. Reiss, *Without the Bomb*, 214, 222, 223–224.
106. Ibid., 239.
107. The following is based on Josef Joffe, "Europe's American Pacifier," *Foreign Policy*, no. 54 (Spring 1984): 64–82; Joffe, *The Limited Partnership*, ch. 5; Joffe, "NATO and the Limits of Devolution"; Timothy P. Ireland, *Creating the Entangling Alliance: The Origins of the North Atlantic Treaty Organization* (Westport: Greenwood, 1981); Nicholas Henderson, *The Birth of NATO* (London: Weidenfeld and Nicholson, 1982); Lawrence S. Kaplan, *The United States and NATO: The Formative Years* (Lexington: University Press of Kentucky, 1984); Frank R. Willis, *France, Germany and the New Europe* (Stanford: Stanford University Press, 1965); Robert E. Osgood, *NATO: The Entangling Alliance* (Chicago: University of Chicago Press, 1962); and Sean Greenwood, "Ernest Bevin, France, and 'Western Union': August 1945–February 1946," *European History Quarterly* 14 (July 1984): 319–38.
108. Ronald Steel, "Germany in NATO? Not Important," *New York Times*, 28 June 1990, p. A25.
109. Joffe, *The Limited Partnership*, 183.
110. Ibid., 178–79.
111. Ibid., 178.
112. Ibid., 184.
113. See Stephen M. Walt, "The Case for Finite Containment: Analyzing U.S. Grand Strategy," *International Security* 14, no. 1 (Summer 1989): 5–49. I am grateful to Sean Lynn-Jones for alerting me to this point.
114. Jervis, "The Future of World Politics: Will It Resemble the Past?" 51; cf. 54–55. Jervis goes on to argue that these expectations, the patterns of cooperation they have engendered, and the benefits of such cooperation serve as a strong incentive against the post-bipolar world reverting to the instability that some scholars associate with multipolar arrangements (Jervis also questions whether the world is heading toward a familiar form of multipolarity, and whether previous multipolar orders were indeed more unstable than bipolar ones).
115. Jerry W. Sanders, "Security and Choice," *World Policy Journal* 1, no. 4 (Summer 1984): 698–707; and Jerry W. Sanders and Sherle R. Schwenninger, "Foreign Policy for the Post-Reagan Era," *World Policy Journal* 3, no. 3 (Summer 1986): 375–82.
116. See Joanne Gowa, "Bipolarity, Multipolaroity, and Free Trade," *American Political Science Review* 83, no. 4 (December 1989): 1245–56. For the argument that in the presence of the possibility of war actors worry about both absolute and relative gains, see Joseph M. Grieco, "Anarchy and the Limits of Cooperation: A Realist Critique of the Newest Liberal Institutionalism," *International Organization* 42, no. 3 (Summer1988): 485–507; Arthur Stein, "The Hegemon's Dilemma: Great Britain, the United States, and the International Economic Order," *International Organization* 38, no. 2 (Spring 1984): 355–86; Robert Jervis, "Realism, Game Theory, and Cooperation," *World Politics* 40, no. 3 (April 1988): 334–36; and Michael Mastanduno, "Do Relative Gains Matter? America's Response top Japan's Industrial Policy," *International Security* 16, no. 1 (Summer 1991): 73–113.
117. Walt, "The Case for Finite Containment," 12, n. 17. See also Jan Willem Honig, "The 'Renationalization' of Western European Defense," *Security Studies* 2, no. 1 (Autumn 1992): 122–38. For the argument that the relative/absolute gains distinction does not necessarily lend support to neorealist criticism of inter-

national cooperation, see Duncan Snidal, "Relative Gains and the Pattern of International Cooperation," *American Political Science Review* 85, no. 3 (September 1991): 701–26. For a discussion of the neorealist emphasis on the role of external threat as an incentive for cooperation and alliance, see Gunther Hellmann and Reinhard Wolf, "Neorealism, Neoliberal Institutionalism and the Future of NATO." For an examination of the question of collective goods and alliances, see Mancur Olson and Richard Zeckhauser, "An Economic Theory of Alliances," *Review of Economics and Statistics* 48, no. 3 (August 1966): 266–79; and Wallace J. Thies, "Alliances and Collective Goods: A Reappraisal," *Journal of Conflict Resolution* 31, no. 2 (June 1987): 298–332.

118. "States are alike in the tasks they face, though not in their ability to perform them." Waltz, *Theory of International Politics*, 96.
119. Ibid., 109.
120. Walt, "The Case for Finite Containment," 39. Cf. Keith A. Dunn, "NATO's Enduring Value," *Foreign Policy*, no. 71 (Summer 1988): 171–72.
121. Thomas J. Christensen and Jack Snyder, "Chain Gangs and Passed Bucks: Predicting Alliance Patterns in Multipolarity," *International Organization* 44, no. 2 (Spring 1990): 142.
122. "The superpowers offered security to their Third World clients as well as restraining them. Unless other forces and mechanisms that would serve these functions develop, aggression will be less difficult and, partly for this reason, status quo states in the Third World will worry more about self-protection. Even absent aggressive motives, conflict will often result through the security dilemma: states' efforts to make themselves more secure will threaten others." Jervis, "The Future of World Politics: Will It Resemble the Past?" 60.
123. See discussion in Joffe, "NATO and the Limits of Devolution," 14–17.
124. "Comments by General Guy Méry, March 15, 1976," *Survival* 18, no. 5 (September–October 1976): 226–28. See Philip H. Gordon, *A Certain Idea of France: French Security Policy and the Gaullist Legacy* (Princeton: Princeton University Press, 1993), 85–86.
125. See the discussion in Quester and Utgoff, "U.S. Arms Reductions and Nuclear Proliferation."
126. Waltz challenges this idea. See Waltz, "Nuclear Myths and Political Realities," 739–40.
127. On the effectiveness of a tit-for-tat policy for extended deterrence see Paul K. Huth, *Extended Deterrence and the Prevention of War* (New Haven: Yale University Press, 1988), esp. 49–51.
128. An early exponent of this strategy was Pierre Gallois. See his *The Balance of Terror: Strategy for the Nuclear Age* (Boston: Houghton Mifflin, 1961). See also the discussion in Gordon, *A Certain Idea of France*, 57–60.
129. Echoing this view, two observers write that "The failure of the superpowers to achieve more than a token fulfillment of this commitment [to disarmament] has damaged the treaty's credibility." Smith and Cobban, "A Blind Eye to Nuclear Proliferation," 56.
130. See Joffe's discussion of why middle-powers' nuclear arsenals are not capable of credibly extending deterrence. Joffe, *The Limited Partnership*, 191–95. See the good discussion in Quester and Utgoff, "U.S. Arms Reductions and Nuclear Nonproliferation."
131. Nye, "NPT: The Logic of Inequality," 128.
132. Edward Luttwak, *U.S. Foreign Policy in a Proliferating World*, Discussion Paper no. 68 (Santa Monica, Calif.: California Seminar on Arms Control and Foreign Policy, 1975), 14.
133. Betts, "Paranoids, Pygmies, Pariahs & Proliferation," 164. Spector concurs. "As superpower influence wanes," he writes, "the impact will undoubtedly be felt in

the area of nonproliferation diplomacy, with Washington and Moscow finding it increasingly difficult to exercise the leverage needed to prevent the spread of nuclear arms to additional states. ... regional states that have relied on superpower security guarantees ... may begin to look to nuclear arms as an alternative to weakening commitments from Moscow or Washington." Spector, *Nuclear Ambitions*, 17.

134. Nye, "NPT: The Logic of Inequality," 128.
135. Keeley, "Legitimacy, Capability, Effectiveness and the Future of the NPT," 38, 40; Stephen M. Meyer, *The Dynamics of Nuclear Proliferation* (Chicago: University of Chicago Press,1984), 101–4.
136. Reiss, *Without the Bomb*, 32.
137. Nye, "NPT: The Logic of Inequality," 130.
138. Betts, "Paranoids, Pygmies, Pariahs & Proliferation," 159.
139. "States change or break the rules as their power and interests change." Jervis, "Security Regimes," 189.
140. Jervis, "Security Regimes," 174, 187, 189.
141. Jervis, "Security Regimes," 174.
142. On the failure of the IAEA to prevent Iraq from developing its bomb, see Gary Milhollin, "The Iraqi Bomb," *New Yorker*, 1 February 1993, pp. 47–56; David Kay, "Bomb Shelter," *New Republic*, 15 March 1993, pp. 11–13.
143. Betts, "Paranoids, Pygmies, Pariahs & Proliferation," 163.
144. Jervis, "The Future of World Politics: Will It Resemble the Past?" 63.
145. Art, "Defensible Defense"; Quester and Utgoff, "U.S. Arms Reductions and Nuclear Nonproliferation."

The Realist Nuclear Regime

Zachary S. Davis

NUCLEAR PROLIFERATION provides a testing ground for realist and idealist theories of international relations. Knowledge about the history of the nuclear age since 1945 makes it possible to evaluate the explanatory power of these two dominant schools of thought with respect to proliferation. Such an evaluation finds that classical realism provides a complete explanation for the causes of proliferation and international responses to it – nonproliferation. Realism explains why a few nations have acquired nuclear weapons and why so many countries exercise restraint and cooperate in nonproliferation efforts. It is not necessary to venture beyond the bounds of traditional realism to explain proliferation or its counterpart nonproliferation. Nevertheless, neorealist and liberal theories can contribute to a comprehensive theory of proliferation, although neither eclipses classical realism as the core of such a theory. This study demonstrates the relevance of classical realism for understanding proliferation and examines neorealist and liberal contributions to a theory of proliferation.

Neorealists reach the wrong conclusions about the consequences of proliferation, but for the right reasons. Whereas neorealists correctly emphasize the importance of the systemic factors that drive proliferation, they mistakenly conclude that systemic factors also doom cooperative nonproliferation efforts. By contrast, liberal idealists reach the right conclusions about the resiliency of the nonproliferation regime, but for the wrong reasons. Cooperation against proliferation will continue beyond the cold war, as liberal theorists predict, but it will do so because mutual interests in preserving the nuclear status quo will transcend bipolarity – not as the result of fundamental changes in international behavior.[1] From a classical realist perspective, the continuing threat of unchecked nuclear proliferation makes possible an unusual, even unique, degree of international cooperation. Such extraordinary cooperation constitutes a realist regime.

Classical Realism: Power and Anarchy in a Systemic Theory of Proliferation

Political realism provides a comprehensive framework for analyzing the spread of nuclear weapons.[2] Despite variation within the realist camp, the fundamental concepts of realism are consistent.[3] These concepts are perhaps best articulated in the writings of Hans Morgenthau, who critics and adherents alike credit with reviving the realist legacy after the Second World War. Realists define power as a multifaceted aggregate of physical factors such as natural resources, geography, population, industrial capacity, as well as less tangible (and difficult to quantify) factors such as leadership, form of government, and national character.[4] At the nation-state level, nuclear weapons can contribute to national power, or perceptions of power, in several ways. It is not necessary to recount the destructive potential of nuclear weapons; we can assume they are unique.[5] Beyond the obvious military dimension, psychological factors such as prestige may inspire nations to acquire nuclear weapons. National leaders may view nuclear capabilities as symbols of national achievement. Such symbols of power can be exploited to legitimize government policies, as may be the case in India and Pakistan. It may not even be necessary to actually build nuclear weapons to derive symbolic benefits from nuclear science; it may be enough to demonstrate mastery over some aspect of nuclear technology to evoke perceptions of power. Many countries may be content with implied nuclear capability. Whatever their actual or perceived value for security, nuclear weapons can contribute to national power.

Realists acknowledge the importance of anarchy as an organizing principle of international society. Because of the absence of higher authority, states seek sufficient power to assure their survival and preserve their sovereignty. The self-help system compels nations to acquire potentially useful capabilities – but only those capabilities whose acquisition is not judged by the country's leaders to incur unacceptable risks. Not all forms of power are desirable. The security-power dilemma is a classical realist concept encompassing the systemic implications of anarchy.[6] Thus, classical realism extends well beyond the confines of state-centered theory. Nations accumulate power to reduce insecurity, but they face a dilemma that too much power may cause other states to feel insecure and inspire them to increase their own power. Nations are thus constantly adjusting to the distribution and re-distribution of capabilities throughout the system with an awareness of their relative position within the system. Morgenthau and other realists such as Martin Wight perceived a global balance of power system composed of interrelated subsystems,[7] a system in which anarchy prevails. States generate survival capabilities internally and through alliances. Insecurity may be alleviated or managed by establishing an *appropriate* balance of capabilities among states whose interests could be affected.[8]

The balance of power *system* describes a structure that encompasses subsystems such as the U.S.-Soviet (now Russian) relationship and regional subsystems such as the India-Pakistan relationship. Realist practitioners prescribe policies to acquire enough of the right kinds of power to safeguard a nation's security without causing other nations to respond in ways that ultimately cancel temporary gains. Balance of power politics is, therefore, not equivalent to Hobbes's war of all against all, a blind grab for every potential source of power.[9] Power is not an end in itself; it is a means to serve interests, the primary interest being survival. Realists understand that restraint and cooperation can increase power and advance interests.

Not every nation that views nuclear weapons as a potentially useful form of military and/or political power will make a committed effort to acquire them. The predictable reactions of other countries may make nuclear status self-defeating. Thus, it is advantageous for nations to sign the Non-Proliferation Treaty (NPT) and remain non-nuclear weapon states to avoid nuclear confrontations. Sweden, for example, went to the brink of nuclear status and turned away. South Africa appears to have done the same. More than 150 countries have signed the NPT because their cost-benefit analyses of nuclear status led them to conclude that their interests would be better served by membership than nonmembership in the regime. Membership in the NPT costs very little in the way of forfeited capabilities and has the benefits of avoiding becoming a nonproliferation pariah and of supporting the nuclear status quo. States considering the merits of nuclear status do so with an awareness of how nuclear weapons affect regional and global distributions of capabilities.

Realism predicts that conventions banning nuclear weapons are likely to fail. Just as previous attempts to ban weapons and war itself failed, the NPT alone will not stop proliferation. Neorealists such as Ben Frankel argue that superpower security guarantees, not the NPT or the nonproliferation regime, have controlled proliferation somewhat, but declining U.S. and Soviet/Russian influence in the post-cold war era erodes confidence in cold war security alliances. Declining superpower influence will cause former client nations of Washington and Moscow to compensate for lost security guarantees. For Frankel and other structural realists, arms control only accentuates the growing inability of the United States and the former Soviet Union to sustain the stability of the bipolar system. Thus, "deep reductions in the nuclear arsenals of the two superpowers will lead to the intensification of nuclear proliferation, not its arrest."[10] They argue that the next decade will see numerous defections from the crippled norm of nuclear nonproliferation. Opaque nuclear programs will be launched not only in outlaw states but also in mainstream countries, and threshold states will add to their unacknowledged arsenals.[11] Neorealists expect the nuclear regime to collapse with the end of bipolarity. However, in contrast to Waltz and Mearsheimer who argue that the spread of nuclear weapons may increase stability and security, Frankel views the inevitability of uncontrolled proliferation with alarm.[12]

Classical Realism and Nonproliferation Cooperation

Not all realists subscribe to these predictions. Realists agree that interests and power provide a better guide to international behavior than international organizations, treaties, laws, or the norms that are enshrined in them. Nations that consider nuclear weapons to be essential for their survival will not be deterred by the unenforced norm of nonproliferation. Iraq's secret nuclear weapon program illustrated that determined proliferators can evade national and international nonproliferation barriers, and may even feign membership in the regime as a ruse to hide their real intentions. Not all members of the NPT are beyond suspicion. Nevertheless, the realist emphasis on the primacy of power and interests should not be interpreted to include cynicism towards international norms, conventions, and institutions. Most realists view the United Nations and its affiliated agencies as valuable forums for international problem-solving and would agree that functional co-operation can lead to broader patterns of cooperation.[13] Their criticisms focus on what they consider to be unrealistic expectations for collective security. An assessment that the United Nations has not lived up to the Wilsonian aspirations expressed in its Charter and is inherently incapable of trampling the vital interests of great powers is not dismissive of the positive attributes of the UN or the importance of norms and institutions for controlling the spread of nuclear weapons. Some realists remain optimistic about the prospects for containing proliferation and preserving the nuclear status quo after the cold war.

Most realists subscribe to Hedley Bull's views concerning the merits of international norms, laws, institutions, and other forms of cooperation in moderating the effects of anarchy[14]. Compatible interests may over time engender norms, which may gain legitimacy and ultimately influence international behavior. The norm of nonproliferation developed because it was congenial to the interests of an overwhelming majority of states, including the superpowers, and offensive to only a few. Dissent from the norm, as illustrated by the refusal of France and China to ratify the NPT until 1992, Iraq's nuclear misadventure, and the willingness of major Western countries to sell nuclear and nuclear-related technology to nations known to have covert nuclear programs, did not cause the regime to collapse. Rather, the problems caused by dissent from the norm have inspired efforts to strengthen the regime. The Non-Proliferation Treaty is an expression of aggregate interests in controlling the spread of nuclear weapons. The treaty organizes, institutionalizes, and codifies those interests. The network of treaties, multilateral and bilateral agreements, formal and informal associations, and national laws and policies that constitute the regime evolved as the result of conscious efforts to establish the nonproliferation norm. Co-operation against proliferation has taken many forms, including extraordinary collaboration between Washington and Moscow throughout some of the most polarized periods of the cold war. Although the regime has its origins in the cold war era, mutual interests in maintaining the nuclear status quo make it possible to preserve the regime – perhaps indefinitely. Whereas Waltz and Mearsheimer view proliferation as stabilizing, and Frankel views uncontrolled proliferation as dangerous, another branch of realism remains optimistic about the future of the nonproliferation regime.

Idealism's Contribution to Proliferation Theory and Practice

The main tenants of idealism revolve around the concept of world order. Organization, not anarchy, is (or can be) the central principle of international behavior. Cooperation is seen as an attractive alternative to autonomous action in the pursuit of security. Institutions serve national interests by facilitating mutually beneficial cooperation, which can be administered according to accepted rules. Functional cooperation to achieve shared goals increases the confidence of participants, who are guided by norms of behavior consistent with the preservation of the collective. As norms are reinforced in laws and institutions, political structures are transformed. The security/power dilemma need not condemn nations to eternal conflict. As forms of cooperation evolve, the defining characteristics of international politics, including definitions of sovereignty and security, can be transformed and a more just international order established.

One branch of the idealist tradition has had particular significance for the study of proliferation. The concept of international regimes originated from within the liberal-idealist tradition to explain the emergence of cooperative relations that theorists thought were not accounted for by realism. Regime theorists focused initially on economic relations to identify complex new relationships involving non-state and transnational actors in world politics.[15] Stephen Krasner provided a working definition of a regime as consisting of "principles, norms, rules, and decision-making procedures around which actor expectations converge in a given issue area."[16]

Regime theory postulated that processes leading towards economic interdependence, globalization of culture, declining superpower hegemony, and other transnational phenomena were redefining the concept of interests and developing "new patterns of cooperation."[17] Such new patterns of cooperation could ultimately transform the character of international relations. Regime theorists viewed realism, with its preoccupation with the concepts of power, interests, and sovereignty, as a relic of a bygone era of power politics – its explanatory power exhausted. Regime scholars explored innovative methods to study new forms of communication and commerce which they believed were transforming international politics.

Students of regimes found applications for their theory beyond global economic relations in security and defense studies.[18] The case for applying regime theory to nuclear proliferation is made more credible by the fact that trade-related issues such as international nuclear commerce, export controls, and technology transfer to the Third World are central characteristics of the nuclear proliferation problem. The technology transfer aspect of proliferation, involving multinational procurement networks, transnational corporations, multinational export control organizations, and questions of justice regarding policies aimed at denying technology to developing countries, appears to offer regime theorists an opportunity to apply their hypotheses. Regime theorists cite examples of cooperation to control proliferation, such as the NPT, multilateral export control groups such as the Nuclear Suppliers Group and its equivalents for chemicals (the Australia Group) and missiles (the Missile Technology Control Regime) as evidence for a nonproliferation regime.[19] The conclusion that these instances of cooperation constitute a "security regime," however, is misleading.

Policymakers use the term "regime" in the realist sense to describe the sum of all efforts to impede nuclear proliferation.[20] This common usage is not the same as that which originated in scholarly discourse to describe institutions leading to norm-driven behavior whose motivations are said to transcend parochial interests. What the originators of social science regime theory might call "vulgar regimes" has to a large extent overtaken the original usage. Regimes are in vogue as policy prescriptions for organizing national and international efforts on a variety of issues such as proliferation, arms control, environment, and economy. This parallel usage of the same term to describe two related but conceptually dissimilar phenomena is the source of some confusion. Regime theorists and regime practitioners agree on the goals and methods of the nuclear nonproliferation regime, but part ways on the question of whether the regime provides evidence of movement in a process that is transforming the international system. The overlap between the two uses of the term, however, has the effect of integrating an idealist concept with a realist theory of proliferation. The matrix below compares realist and idealist expectations for regime performance in theory and practice.

	Regime Theory	Regime Practice
Realism	skepticism	descriptive
Idealism	systemic change	evidence of change

Realist and Idealist Expectations for Regimes in Theory and Practice.

Robert Jervis has argued that security regimes are neither necessary nor possible.[21] Patterns of cooperation can be explained adequately by the pursuit of short-term interests. Thus, the norms ascribed to a security regime such as the ones imagined for proliferation will not determine behavior. Cooperative security arrangements are bound to lapse because "States change or break the rules as their power and interests change."[22] According to Jervis, the nonproliferation regime is little more than an alliance against the spread of nuclear weapons, not an expression of shared values. The existence of such an alliance is not by itself evidence that the members of the alliance have accepted particular norms, nor will those norms be a reliable guide to their behavior. And like any alliance, this one will change. Its continuation does not imply progress towards nuclear rollback.

In denying the existence of security regimes, Jervis could not have foreseen the popularization of the regime concept and its use by policymakers to describe the collection of international treaties and institutions, multilateral agreements, and national laws all aimed at stopping proliferation. A non-proliferation regime exists, even if it does not qualify as a genuine regime as defined by regime theory as such. The nonproliferation regime falls short of the requirements of orthodox regime theory, but exhibits characteristics that invite further theoretical scrutiny. Even neorealists such as Ben Frankel recognize the existence of a regime that is "more than an ad hoc constellation of nations motivated by rational self interest to agree on some short-term arrangements."[23] The nuclear regime is more than a conventional alliance yet less than a norm-driven regime. It is an association that is best explained by a realist interpretation of an idealist concept.

Regime theory is not necessarily inconsistent with realism. Scholars from the neoliberal institutionalist camp find common ground between the two schools, much as John Herz did at an earlier stage of the realist-idealist debate. Stephen Krasner and Robert Keohane have elaborated on regime theory so that it is compatible with realist assumptions about the anarchic system. From this perspective, a regime operates against a realist background of anarchy and interests.[24] If we substitute realist/practitioner assumptions about the primacy of power and interests in an anarchic system for liberal/idealist interpretations of the nuclear regime as a form of collective security leading to a progressively more cooperative international environment, we are left with a truncated regime concept. Bereft of expectations of progress, the realist interpretation of the nonproliferation regime does not rule out the possibility that norms exist, or that they influence behavior, but reminds us that they accommodate vital security interests. Interest-driven cooperation still reinforces the mutually beneficial norm of nonproliferation without the expectation that such cooperation is necessarily permanent or irreversible. For example, encouraging developments towards bringing Argentina and Brazil fully into the regime as non-nuclear weapon states could be reversed by a military coup in either country that reinstated military control over nuclear programs. Similarly, it is possible to imagine a failure to extend the

NPT when it comes up for review in 1995. States will never put the integrity of the regime before their own survival. The partial regime has few transformative qualities beyond advancing shared interests in controlling proliferation. Nevertheless, states go to great lengths to accommodate the regime, sometimes at considerable expense and sacrifice. If an orthodox interpretation of regime theory is superfluous for security cooperation, a spare interpretation may still have explanatory value for nonproliferation.

A Realist Regime

Notwithstanding Jervis' critique of security regimes, a nonproliferation regime does exist. The nonproliferation regime is more than an alliance but less than an orthodox regime. In practice, it is an aggregation of multilateral, bilateral, unilateral, formal and informal arrangements aimed at stopping the spread of nuclear weapons. The regime comprises a redundant network of political, legal, technical, economic, military, and normative barriers whose cumulative effect is more than the sum of the parts. It has unusual qualities that make it possible for international inspectors to transgress the sovereignty of weak and strong states alike. Regime members attain an extraordinary level of cooperation in some extremely sensitive areas. Nonproliferation is unusual in the degree to which an international norm and a broad crossection of national security interests are compatible. Such compatibility expands the range of cooperative and collective action beyond that which is normally possible. This alignment of norms and interests explains the existence of a realist regime.

The nonproliferation norm associated with the regime is subject to the inherent limitations of international law and collective security. It is grounded in realist expectations that compliance with norms and enforcement of them will not be uniform. Nevertheless, the alignment of norms and interests improves prospects for enforcement. Hans Morgenthau argued that despite problems of enforcement, "Most rules of international law formulate in legal terms identical or complementary interests. It is for this reason that they generally enforce themselves ..."[25] Interests are still the best guide to behavior, but even vital security interests are often complementary. Such is the case with nuclear nonproliferation. States accommodate the nonproliferation norm because a nonproliferated world serves their interests, and because calculations based on the security/power dilemma suggest that noncompliance with the regime will in nearly all cases cause a net loss of security. States with security problems who might be tempted to exercise their nuclear options can appeal for security assistance from powerful countries as an alternative to seeking security in nuclear weapons. In the past,

U.S. security guarantees tipped the balance in favor of nonproliferation in several countries. The willingness to provide positive and negative security guarantees to potential defectors may define the scope of the proliferation problem in the post cold war era.[26]

The shift away from bipolarity toward a multipolar (or unipolar) distribution of capabilities has not harmed the ability or the motivation to provide security assistance and security guarantees as means to control proliferation. U.S.-Soviet cooperation on nonproliferation was possible in spite of the cold war, not because of it. Washington and Moscow found common cause in creating and maintaining the nonproliferation regime through even the most strained periods of the cold war. The reasons for providing assistance and guarantees have not changed with the passing of bipolarity. Nor are financial constraints so great that the United States, Russia, and other nations cannot provide security assistance and security guarantees to a few potential proliferators. Ukraine, for example, may receive economic and security assistance and security guarantees from a variety of nations, including Russia.[27] Cold war alliance systems clarified which superpower was mainly responsible for providing security guarantees to particular states. Thus, U.S. security assistance and guarantees to allies such as West Germany, Japan, Taiwan, Iran, and South Korea partially satisfied security concerns that could have led those nations to acquire nuclear weapons. Soviet guarantees were probably less significant in preventing specific cases of proliferation.

Contrary to the assertion that deep reductions in U.S. and Russian nuclear arsenals will erode Washington and Moscow's ability to provide extended nuclear deterrence to allies, START II would leave the United States and Russia with thousands of warheads each – plenty to sustain nuclear umbrellas on a global scale.[28] Bipolar structure was not the main determinant of proliferation-related security guarantees or of regime formation. Nor did bipolarity determine the ability of states to offer nonproliferation security guarantees. National interests in maintaining the nuclear status quo, not bipolarity, formed the regime. The same systemic forces that cause proliferation also cause nations to collaborate to stop it. This will not change in a multipolar or unipolar structure.[29]

The passing of bipolarity may actually facilitate efforts to strengthen the regime. The deterioration of old allegiances has created new opportunities for cooperation that would not have been possible under bipolarity. Cooperation by the United States, Japan, China, and Russia to convince North Korea to abort its nuclear weapons program is one example. In the emerging multipolar system, the United States is negotiating security guarantees to prevent proliferation, and is calling on other countries to shoulder part of the burden. Ukraine has demanded security guarantees and economic assistance as its price for giving up nuclear weapons.[30] Other nations may require similar incentives to foreclose their nuclear options and join the NPT. Multipolarity does not hinder efforts to persuade nations to join the regime; it may even aid such efforts. The regime will transcend bipolarity because mutual interests in nonproliferation will endure.

The list of actual and probable defectors is short. Iraq was a rare case of defection, but it illustrates the point that going nuclear does not necessarily increase security. What would have been the reaction of Israel, Iran, or other countries to a surprise announcement by Saddam Hussein that he had nuclear weapons? Threshold states such as India, Israel, and Pakistan did not make non-weapons pledges, but earned pariah status for refusing to join the regime.[31] North Korea's nuclear program counts as a defection because Pyongyang stalled implementation of its safeguard agreement until 1992, seven years after it joined the NPT, and then continued to block inspections. Possible near-term candidates for defection include Iran, Iraq (again), and the nuclear republics of the former Soviet Union, notably Ukraine.[32]

Despite these problems, the regime is growing stronger. In 1991 and 1992, it attracted notable new members, including former threshold states Argentina and Brazil, who implemented a safeguards agreement with the International Atomic Energy Agency (IAEA) and pledged to sign the Treaty of Tlatelolco, and South Africa – another threshold state – which acceded to the NPT and opened its nuclear facilities to inspection.[33]

France and China's accessions to the NPT makes the participation of the declared nuclear weapon states universal. The combination of these additions with the successful enforcement mission in Iraq, and other multilateral and unilateral initiatives to strengthen various aspects of the regime suggest that the regime is increasing its influence on the behavior of nations.

One extraordinary aspect of the nuclear regime is the explicit nuclear bargain on which it is founded. The bargain, which evolved from its origins in Eisenhower's Atoms for Peace proposal, links functional cooperation with arms control. Advanced nuclear nations agree to provide nuclear assistance in exchange for pledges by non-weapons states to use nuclear technology for peaceful uses only and the right to verify that this commitment is being kept. Thus, the appeal of nonproliferation blends good global citizenship with national self-interest. The nuclear bargain links arms control with technology transfer. The regime has its own functional international organization, the IAEA, that administers the nuclear bargain by dispensing technical assistance and inspecting facilities and materials under safeguards.[34]

Another facet of the nuclear bargain is the commitment in Article VI of the NPT for the nuclear-weapons states to "pursue negotiations in good faith on effective measures relating to cessation of the nuclear arms race at an early date and to nuclear disarmament ..." Many non-weapons states interpret this as a commitment to a comprehensive nuclear test ban, but the weapons states have viewed it merely as a statement of broad objectives.[35] While non-weapons states complain about the shortage of technical assistance and the lack of progress towards the disarmament side of the bargain, the weapons states emphasize the need for stricter enforcement of non-weapons pledges. Expectations for nuclear cooperation and for compliance with the non-proliferation norm have exceeded performance on both sides of the nuclear bargain. Nevertheless, the bargain linking cooperation and compliance remains a powerful motivator for states. The nuclear regime's formal link between development and disarmament is unparalleled in security policy.

The IAEA inspection system is a unique element of the realist regime. By its very nature, the IAEA safeguards system infringes on the sovereignty of its members, yet it continues to grow more intrusive. Reforms are now underway to make nuclear programs more transparent, to make inspections more intrusive, to improve access for inspectors, and for members to provide the IAEA with intelligence information about nuclear operations in member countries.[36] The sharing of sensitive intelligence with an international organization – one that includes almost 120 nations – is unprecedented.[37] Why do nations cooperate with the IAEA? Inspections deter cheating, increase the probability that significant cheating will be detected, and maintain confidence that the regime is upholding the nonproliferation norm. The IAEA inspection system gives the regime the capability to police nonproliferation commitments.

Verification is necessary for enforcement. If inspections reveal evidence of noncompliance with nonproliferation commitments, the Director General of the IAEA can report to the United Nations Security Council, which can authorize enforcement actions. The Security Council provides a venue for powerful states to organize collective action to enforce nonproliferation. In the Iraq case, the Security Council passed Resolution 687 and established the United Nations Special Commission (UNSCOM) to oversee its implementation. The Iraq precedent reinforced the Security Council's nonproliferation enforcement potential, particularly if UNSCOM or a similar body were to become a regular nonproliferation enforcement arm of the Security Council. The ability to enforce nonproliferation – even selectively – distinguishes the regime as a unique association. Roger Smith has argued that regime stability depends on the response to non-compliance.[38] The response to Iraq's non-compliance revitalized the regime.

Another aspect of the regime that crosscuts realist and idealist principles is the regime's system of export controls. Regime members forsake economic advantages in the name of nonproliferation. Multilateral supplier control groups such as the Zangger Committee and the Nuclear Suppliers Group (NSG) have succeeded in controlling nuclear commerce, but in so doing have created new opportunities for nonmembers and defectors to offer proliferants alternative sources of supply.[39] This classic "stag hunt" dilemma pits economic interests against security interests. The tension between these competing goals is constant, as the Clinton administration is bound to discover.[40] Supplier control agreements such as the NSG and the Missile Technology Control Regime, national laws such as the Export Administration Act, and federal regulations promote security by preventing easy access to bomb-making technology. Export control violations – when they are detected – broadcast the nuclear intentions and capabilities of suppliers and buyers of nuclear technology.

Export controls are a trigger mechanism for national and international action to block proliferators and to enforce compliance with international nonproliferation standards. United States laws and regulations impose a range of sanctions on individuals, companies, and countries that violate international nonproliferation agreements.[41] Despite their cost and inconvenience, the technology control and sanctions components of the regime are expanding. Controls are expanding both in terms of membership in supplier control groups and in the scope of items that are controlled. New laws in several countries have expanded the use of sanctions as a means to punish a wider array of proliferation-related activities. The ever-widening list of controlled goods started with restrictions on exports of nuclear reactors and fissile material and grew to include enrichment and reprocessing technology, component parts and eventually a wide range of dual-use goods. The latest upgrades of the NSG and MTCR lists include an array of machine tools, raw materials, and scientific equipment.[42] Efforts are underway to expand controls on computers and other dual-use goods. Why do nations participate in this self-denial? Regime members tolerate controls and sanctions despite adverse effects on their trade balances because the nonproliferation regime moderates the tragedy of the commons created by exports of nuclear technology. Export control, like arms control, represents a recognition that not all forms of power are worth the risks they create. This type of cooperation is consistent with the principles of realism. The realist regime organizes the self-denial policies of participating suppliers.

The NPT is the symbolic centerpiece of the regime. The treaty is openly discriminatory in its objective of maintaining the nuclear weapons status quo. Despite regular complaining about this unfair distribution of capabilities in favor of the United States, Russia, China, France, and Great Britain, the NPT continues to attract new members and will probably be renewed when its first 25-year term ends in 1995. Why do all but a few nations flock to a treaty that denies them security and trade options and requires them to submit to inspections by international authorities? How do we explain the durability of the regime despite its many liabilities?

The nuclear regime exceeds the normal boundaries of state-state relations and alliances to establish world nuclear order. The regime harmonizes interests in nonproliferation, economic development, technology transfer, and arms control to moderate world nuclear anarchy. It accommodates obstacles where it must, but maintains steady pressure against challenges to the nuclear status quo. The regime operates at an intermediate level on the continuum between the levels of state and system. It is not necessary for the sake of theory to imagine an untraversable quantum leap separating state and structure. Rigid neorealist notions of structure and idealist optimism about the transformational effects of cooperation lead to incorrect conclusions about the future of the nuclear regime.

Waltz allows that there is two-way interaction between states and the system. The state and system levels need not be hermetically sealed from one another. If it is true that "ambiguity cannot be resolved since structures affect units even as units affect structures,"[44] it is possible that states knowingly affect structure. Since structure defines the arrangement of units within the system, the nuclear regime can be viewed as a structural organization. Under certain conditions, states can be motivated to cooperate within the bounds of the realist domain to establish structural arrangements that enable them to civilize a parcel of the anarchic system. The regime does not revolutionize the ordering principles of the system, but does blur Waltz's definition of structure.[45] For as long as it lasts, the regime calls into question the functional separation of units and the distribution of capabilities among them. As the willful creation of the units, the regime acts as a semi-permanent, anti-hegemonical coalition against concentrations of nuclear capability that threaten the status quo. The regime is a structural arrangement that acts according to realist expectations.

Such "mingling of structural and unit levels" is not inconsistent with the theories discussed here.[46] It does suggest, however, that further study of partial regime behavior is needed to define the types of associations that may exist between state and structure. It may be the case that nuclear anarchy is so far the only threat serious enough to motivate states to form and sustain a security regime.

Operation of a Realist Nonproliferation Regime

A realist-bounded nonproliferation regime operates according to three corollaries:

1) The causes of proliferation are rooted in the security/power dilemma and will persist unless the anarchic system undergoes fundamental change;

2) The security/power dilemma also motivates states to exercise restraint and

cooperate in nonproliferation efforts. The regime organizes interests in preserving the nuclear weapons status quo; and

3) Neoliberal institutionalist principles expand the possibilities for moderating the causes of proliferation. The nuclear bargain uses functional cooperation as an incentive for nations to participate in regime institutions such as the IAEA and the NPT. Such cooperation, and the threat of losing it, is the balm of mutual interests and the key to perpetuating the regime.

These corollaries of an integrated theoretical approach to proliferation and nonproliferation suggest general guidelines for policy. First, we should assume that the underlying systemic pressures that motivate proliferation are chronic but heterogeneous. Proliferation pressures will occasionally erupt in decisions to acquire militarily-relevant nuclear capabilities. Efforts to intervene in such decisions must address the security/power calculus of the states involved. In many cases it may be possible to alter both sides of that equation with positive and negative inducements, both of which can be extremely expensive. The regime pools resources that can be used as inducements.

However, if for certain countries the will to proliferate exceeds efforts to prevent them from doing so, maintaining opaque or threshold status is preferable to open declarations of nuclear capabilities. The regime is fragile. Regime failures should be contained, not rewarded. Moreover, in view of what we now know about the Cuban Missile Crisis, the U.S.-Soviet nuclear deterrent relationship is not a good model for others to emulate.[47] Any furtive nuclear deterrent relationships are more likely than not to encounter similar crises on their way to maturity.[48] Nuclear brinkmanship carries unacceptable risks and should be avoided.

There *is* a nuclear nonproliferation regime, despite its theoretical inequities. The regime aids efforts to dissuade, detect, and interdict proliferation. Nonproliferation is a good investment for nearly all nations, even if proliferators sometimes slip through the regime's net. An active strategy to expand and preserve the key elements of the regime raises the stakes of defection to a point where only the most determined proliferators will challenge the nuclear status quo. Casual proliferators are discouraged. Latent proliferants such as Argentina, Brazil, South Korea, South Africa, Sweden, Taiwan, Libya, and others have been turned back. A few such as North Korea, Iran, and Ukraine appear to be idling while trying to decide whether to challenge the nonproliferation regime. Defections can be minimized by extending the NPT when it comes up for renewal in 1995, expanding political and economic support for the IAEA, expanding cooperation to control exports of nuclear materials and technology, and keeping political and economic pressure on countries of concern. Traditional arms control such as START, START II, and a comprehensive test ban enhance diplomatic efforts to sustain the regime, but have little direct effect on the security calculations of committed proliferators. Regime upkeep and improvement should be a top policy priority.

The regime is composed of capabilities contributed by its nearly universal membership. Those combined capabilities constitute a structural organization that is capable of imposing hardships on those who choose pariah status. The dispensation of costs – like the decision to acquire nuclear capabilities – is guided by national interests, not by adherence to norms, rules, and commitments. Consequently, the performance of the regime is inconsistent. This inconsistency is evident in the examples of Israel, Pakistan, and Iraq, all of which continued to receive preferential treatment from the United States and other countries despite widespread knowledge that they were developing nuclear weapons. Nuclear nonproliferation often conflicts with other strategic, economic, and political interests. Such conflicts reduce opportunities for collective action against proliferators. Even Saddam Hussein would have been allowed to keep his nuclear weapon program if he had not committed aggression that threatened the interests of several great powers. The threat of nuclear proliferation cannot always surmount the inherent limitations of collective action, but it does often bring into alignment an unusually broad array of interests. Regime capabilities are enlisted on a case-by-case basis according to the degree of alignment between the nonproliferation norm and the interests of powerful nations. Alignment is more frequent on less costly measures such as NPT membership and harder to achieve on costly measures such as enforcement. Realist expectations for state behavior also govern regime behavior.

Finally, the fate of the regime and the deployment of its capabilities is largely determined by U.S. leadership. Robert Jervis cites great power interests as the first required condition for forming a security regime.[49] The nuclear regime meets at least this criteria. U.S. leadership, supported by extraordinary cooperation with Moscow and other allies, fostered the development of the regime based on U.S. policies and laws establishing standards for international nuclear commerce.[50] U.S. initiatives have succeeded in expanding controls on nuclear technology and improved safeguards on nuclear facilities and materials. U.S. leadership increased the stability of the nuclear regime by backing the enforcement of United Nations Security Council Resolution 687 by the United Nations Special Commission and the IAEA.[51] The prospects for compliance with and enforcement of nonproliferation improved with the addition of the remaining great power weapons states, France and China, to the NPT and the expansion of the Nuclear Suppliers Group membership and guidelines. French support and Chinese acquiescence may allow for more consistent alignment of interests in the future. U.S. leadership in nonproliferation is essential (but not sufficient) for the regime's success.

The anarchic nature of world politics assures there will be continuing interest in nuclear weapons. Realism predicts states will organize to preserve the nuclear status quo. The regime is a natural expression of vital security interests. However, the future of the nuclear regime is not assured, because nonproliferation policy often conflicts with strategic, economic, and political interests. Nonproliferation requires sacrifices that carry political costs. If the great powers subordinate nonproliferation to other priorities, many countries will do the same, and the regime could collapse. Faced once again with nuclear anarchy, states would begin again to reconstruct the nuclear nonproliferation regime.

Notes

The views expressed are those of the author and do not necessarily reflect policies of the Congressional Research Service or the Library of Congress.

1. I would like to thank Ben Frankel for his assistance in formulating this distinction between neorealist and liberal conclusions about proliferation.
2. The realist tradition is often misunderstood. A good example of common misperception about realism is an editorial by journalist Thomas Friedman of the *New York Times*. Writing about the post-cold war world, Friedman asked, "After all, what does it mean to be a realist when there is no Soviet threat to override human rights considerations?" Thomas Friedman, "It's Harder Now to Figure Out Compelling National Interests," *New York Times*, 31 May 1992, p. E5. By asserting that the collapse of communism will cause the collapse of realism, Friedman falls into the trap of misusing the term realism to refer to policymakers and political scientists who have focused their attentions on the military relationship between the United States and the Soviet Union. Such use builds on the false perception that realism focuses narrowly on calculations of military power. Even historian John Lewis Gaddis – who has written authoritatively and admiringly about no less a realist than George F. Kennan – overlooks fundamental principles of realism when he accuses Hans Morgenthau of being the father of pseudo-scientific, reductionist approaches to the study of politics. "True realism," Gaddis admonished, "is that which acknowledges the power of contingency and therefore how limited our powers of prediction are always going to be." John Lewis Gaddis, "Point of View," *Chronicle of Higher Education*, 22 July 1992, p. A44. Compare Gaddis' statement with this passage on nuclear war from *Politics Among Nations*: "The contingent element in this theory is its utter uncertainty, and this uncertainty is typical of all levels of theoretical analysis and prediction in the field of politics, domestic and international." *Politics Among Nations*, 5th ed. (New York: Knopf, 1978), 21. Morgenthau had deep reservations about the application of scientific methodology to the study of international politics.
3. Recent works on the realist tradition include Michael Joseph Smith, *Realist Thought from Weber to Kissinger* (Baton Rouge: Louisiana State University Press, 1989); Joel Rosenthal, *Righteous Realists: Political Realism, Responsible Power, and American Culture in the Nuclear Age* (Baton Rouge: Louisiana State University Press, 1990); and Greg Russell, *Hans J. Morgenthau and the Ethics of American Statecraft* (Baton Rouge: Louisiana State University Press, 1991). On distinctions

between classical realism and neorealism see Keith Shimko, "Realism, Neorealism, and American Liberalism," *Review of Politics* 54 (Spring 92): 281–301.

4. Hans Morgenthau, *Politics Among Nations*, 5th ed., ch. 9, "Elements of National Power." In response to criticism regarding realism's supposed narrow fixation on military forms of power and its exclusion of other sources of national potency, Robert Gilpin and others have pointed out that realists from Thucydides to Morgenthau have been acutely aware of the importance of relative economic strengths and weaknesses, although it is true that realists do not consider economics to be the central organizing principle of international relations. Robert Gilpin, "The Richness of the Tradition of Political Realism," in Robert Keohane, ed., *Neorealism and its Critics* (New York: Columbia University Press, 1986), 308–10. Gilpin's work exemplifies realist appreciation for the importance of economic factors in international politics. Another example of a realist approach to economic power is Paul Kennedy, *The Rise and Fall of the Great Powers, Economic Change and Military Conflict from 1500 to 2000* (New York: Random House, 1987).
5. Classic works on the destructive capabilities of nuclear weapons include: Samuel Glasstone, ed., *The Effects of Nuclear Weapons* (Washington, D.C.: U.S. Atomic Energy Commission, 1962); and Herman Kahn, *On Thermonuclear War* (Princeton: Princeton University Press, 1960). See also William Daugherty, Barbara Levi, and Frank von Hippel, "The Consequences of 'Limited' Nuclear Attacks on the United States," *International Security* 10, no. 4 (Spring 1986): 15. So far, known chemical, biological, and advanced conventional weapons do not match the destructive potential of nuclear weapons.
6. On the realist roots of the security/power dilemma see John Herz, "Idealist Internationalism and the Security Dilemma," *World Politics* 2, no. 2 (January 1950): 157; Arnold Wolfers, "National Security as an Ambiguous Symbol," *Political Science Quarterly* 67 (December 1952): 481; Wolfers, *Discord and Collaboration* (Baltimore: Johns Hopkins University Press, 1962). Although not a realist, Robert Jervis has elaborated on the security/power dilemma concept. See his "Cooperation Under the Security Dilemma," *World Politics* 30, no. 2 (January 1978): 167. Neorealists, notably Kenneth Waltz, contend that in classical realist theory, power is limited to the state in both its origins and usage. Waltz agrees that variation in the distribution of capabilities among states creates the need for balancing, and posits that structural forces shape national policies. Long before the advent of neorealism, however, the realist concept of the security–power dilemma linked state power with the inescapable universal condition of anarchy.
7. Morgenthau, *Politics Among Nations*, 205.
8. On the structural aspects of classical balance of power theory see Edward Gulick, *Europe's Classical Balance of Power* (Ithaca: Cornell University Press, 1955). Of course, Morgenthau devoted much of his scholarship to explaining the balance of power concept and its enduring relevance to international politics. See *Politics Among Nations*, "Part Four, Limitations of National Power: The Balance of Power." Stephen Walt has elaborated on neorealist balance of power theory with his emphasis on the tendency of states to balance threats rather than "bandwagoning" against unfavorable distributions of capabilities. Stephen M. Walt, *The Origins of Alliances* (Ithaca: Cornell University Press, 1987); and his "Alliance Formations in Southwest Asia: Balancing and Bandwagoning in Cold War Competition," in Robert Jervis and Jack Snyder, eds., *Dominoes and Bandwagons: Strategic Beliefs and Great Power Competition in the Eurasian Rimland* (New York: Oxford University Press, 1991), 51–74.
9. Waltz contends that "Morgenthau thought of the 'rational' statesman as striving ever to accumulate more and more power. Power is seen as an end in itself." This interpretation is out of step with mainstream realist thought, which favors

acquisition of appropriate power in relation to the power and interests of other states. Waltz, "Realist Thought and Neorealist Theory," *Journal of International Affairs* 44 (Spring 1990): 35–36.

10. Benjamin Frankel, "An Anxious Decade: Nuclear Proliferation in the 1990s," in Benjamin Frankel, ed., *Opaque Nuclear Proliferation, Methodological and Policy Implications* (London: Frank Cass, 1991), 7.
11. Frankel suggests that some realists recoil from their own logic to become "lapsed Waltzians" when they conclude that proliferation can be controlled. Other critics of Waltz's theory of proliferation such as Daniel Deudney point out that proliferation is a unit level activity and is therefore disconnected from Waltz's theory of structure. See Frankel and Deudney articles in this volume, and John Mearsheimer, "Back to the Future: Instability in Europe After the Cold War," *International Security* 15, no. 1 (Summer 1990).
12. Kenneth Waltz, *The Spread of Nuclear Weapons: More May Be Better*, Adelphi Paper no. 171 (London: International Institute for Strategic Studies, 1981); Mearsheimer, "Back to the Future."
13. For a realist perspective on the effects of functional cooperation see Kenneth W. Thompson, *Ethics, Functionalism, and Power in International Politics* (Baton Rouge: Louisiana State University Press, 1979).
14. Hedley Bull, *The Anarchical Society: A Study of Order in World Politics* (London: MacMillan, 1977). The crux of Bull's argument is in his chapter "Society and Anarchy in International Relations," in Herbert Butterfield and Martin Wight, eds., *Diplomatic Investigations* (New York: Allen and Unwin, 1966), 40.
15. A major landmark for regime theory is Stephen Krasner, ed., *International Regimes* (Ithaca: Cornell University Press, 1983).
16. Krasner, "Structural Causes and Regime Consequences: Regimes as Intervening Variables," in Krasner, *International Regimes*, 1.
17. Richard Rosecrance, *The Rise of the Trading State: Commerce and Conquest in the Modern World* (New York: Basic Books, 1986).
18. For a good review of the evolution of regime theory see Joseph Nye, Jr., "Neorealism and Neoliberalism," *World Politics* 40, no. 2 (January 1988): 235.
19. On the relevance of regime theory for nuclear nonproliferation see: Joseph Nye, Jr., "Nuclear Learning and U.S.-Soviet Security Regimes," *International Organization* 41, no. 3 (Summer 1987): 371; Roger K. Smith, "Explaining the Non-Proliferation Regime: Anomalies for Contemporary International Relations Theory," *International Organization* 41, no. 2 (Summer 1987): 257; and Roger K. Smith, "Opaque Proliferation and the Fate of the Non-Proliferation Regime," in Frankel, *Opaque Proliferation*.
20. Examples of this usage of the term regime in a policy-related context include: Zachary S. Davis, *Nonproliferation Regimes: A Comparative Analysis of Policies to Control the Spread of Nuclear, Chemical, Biological Weapons and Missiles* (Washington, D.C.: Congressional Research Service, 1991); Janne Nolan, *Trappings of Power: Ballistic Missiles in the Third World* (Washington, D.C.: Brookings, 1991); Joseph Pilat and Robert Pendley, eds., *Beyond 1995, The Future of the NPT Regime* (New York: Plenum, 1990); John Simpson, ed., *Nuclear Non-Proliferation: An Agenda for the 1990s* (Cambridge: Cambridge University Press, 1987).
21. Jervis, "Security Regimes," in Krasner, *International Regimes*, 173.
22. Jervis, "Security Regimes," 189.
23. Ben Frankel and Avner Cohen, "Opaque Nuclear Proliferation," in Frankel, *Opaque Proliferation*, 17.
24. Stephen Krasner has suggested three variations on the regime concept: a realist, a neoliberal, and a reflectionist. The realist variation views regimes as manifestations of the interests of powerful actors. The neoliberal interpretation views

regimes as solutions to the problems of collective security and the prisoner's dilemma. The reflectivist variation considers regimes to be expressions of shared values which can redefine interests. Stephen Krasner, Conference on Social Science Theory and Nuclear Proliferation, Stanford University, 6 February 1993. See also, Robert Keohane, "Neoliberal Institutionalism: A Perspective on World Politics," in Keohane, *International Institutions and State Power* (Boulder, Colo.: Westview, 1989): 7–8.

25. Morgenthau, *Politics Among Nations*, 299.
26. Positive security guarantees consist of military assistance and commitments to aid a nation if its security is threatened. Negative security guarantees consist of commitments not to threaten a particular nation or nations. Positive security guarantees can conflict with efforts to control sales of conventional weapons, since countries with security problems usually request advanced military equipment such as fighter aircraft as inducements not to acquire nuclear capabilities. Security assistance/guarantees to Israel and Pakistan did not end their nuclear weapons programs.
27. Margaret Shapiro, "Ukraine Gets Defense Offer From Yeltsin," *Washington Post*, 16 January 1993, p. A14.
28. START II would eliminate multiple warheads missiles and reduce total warheads to 3,500 for the United States and 3,000 for Russia. On START and START II see Amy Woolf, *START: Central Limits and Key Provisions* (Washington, D.C.: Congressional Research Service, 1992); Woolf, *START II: Central Limits and Force Structure Implications*, CRS, February 1993; Michael Gordon, "A New Amity, A New Treaty," *New York Times*, 30 December 1992, p. A1. On minimum deterrence requirements see: Andrew Goodpaster, *Nuclear Weapons in a New Era*, Atlantic Council Occasional Paper, May 1992; Peter Gizewski, ed., *Toward Minimum Deterrence: How Low Can We Go?*, (Ottawa: Canadian Centre for Arms Control and Disarmament, 1992); *The Future of the U.S.-Soviet Nuclear Relationship* (Washington, D.C.: National Academy of Sciences, 1991).
29. On the distinction between structure and system see Waltz, *Theory of International Politics* (Reading: Addison-Wesley, 1979), 81–83.
30. On Ukraine's nuclear diplomacy see Virginia Foran, "Ukrainian Holdout: The Real Problem with the Treaty," *Washington Post*, 3 January 1993, p. C3.
31. While they have not violated treaty commitments, these countries have violated a number of national nuclear export control laws. Richard Betts defined pariah status with respect to proliferation in his landmark article "Paranoids, Pygmies, Pariahs and Nonproliferation," *Foreign Policy* 26 (Spring 1977): 157.
32. Although all tactical weapons were reportedly returned to Russia, strategic weapons remained in Ukraine, Belarus, and Kazakhstan pending implementation of the START treaty. These three countries agreed to give up all nuclear weapons on their territory and to join the NPT as non-weapon states in protocols to the START treaty signed in Lisbon in May 1992. Failure to implement these commitments would constitute defection from the nonproliferation regime. On Iran and Iraq's nuclear programs see testimony of CIA director Robert Gates before the House Armed Services Committee, 27 March 1992, and before the Senate Governmental Affairs Committee, 15 January 1992. See also Zachary Davis and Warren Donnelly, *Iran's Nuclear Activities* (Washington, D.C.: Congressional Research Service, 1993); and Davis and Donnelly, *Iraq and Nuclear Weapons* (Washington, D.C.: Congressional Research Service, 1992).
33. Leonard Spector, in a recent article, raises concerns about the difficulties of verifying the non-nuclear status of former threshold states. See "Repentant Nuclear Proliferants," *Foreign Policy* 88 (Fall 1992): 21.
34. On the IAEA and its safeguards system see: Lawrence Scheinman, *The International Atomic Energy Agency and World Nuclear Order* (Washington, D.C.:

Resources for the Future, 1987); Zachary Davis and Warren Donnelly, *International Atomic Energy Agency: Strengthen Verification Authority?* (Washington, D.C.: Congressional Research Service, 1993).

35. Article VI, Nuclear Nonproliferation Treaty. On the question of linkage between the NPT and a test ban see George Bunn, Charles Van Doren, and David Fischer, *Options and Opportunities: The NPT Extension Conference of 1995*, Programme for Promoting Nuclear Non-Proliferation, PPNN Study Two, November 1991; Thomas Schmalgerger, *In Pursuit of a Nuclear Test Ban Treaty: Guide to the Debate in the Conference on Disarmament* (New York: United Nations Institute for Disarmament Research, 1991); Warren Donnelly and Katherine Starr, *The UN Conference on a Comprehensive Test Ban: Its Relations to U.S. Nonproliferation Policy* (Washington, D.C.: Congressional Research Service, 1991). For debate in the U.S. Senate on a test ban see the *Congressional Record*, 3 August 1992, s11167-s11212. Following this debate, Congress passed and the president signed test ban legislation.
36. On IAEA reforms see Davis and Donnelly, *IAEA: Strengthen Verification Authority?*; Lawrence Scheinman, *Assuring the Nuclear Non-Proliferation Safeguards System*, Atlantic Council Occasional Paper, October 1992; Hans Blix, "Nonproliferation and Verification in the 1990s: Securing the Future," and J. Jennekens, R. Parsick, and A. von Baeckmann, "Strengthening the International Safeguards System," *IAEA Bulletin* 34, no. 1 (Winter 1992).
37. Intelligence information supplied to the IAEA guided United Nations Special Commission inspections of Iraq's nuclear program. IAEA officials point out that had the agency received such information prior to Operation Desert Storm, its regular inspections might have exposed Iraq's clandestine nuclear program. President Kennedy set a precedent in 1962 when he shared U-2 pictures of Soviet missile emplacements during the Cuban Missile Crisis.
38. Smith, "The Non-Proliferation Regime," in Frankel, *Opaque Proliferation*, 96.
39. William Potter, *International Nuclear Trade and Nonproliferation: The Challenge of the Emerging Suppliers* (Lexington: Lexington, 1990); R. Jones, C. Merlini, J. Pilat, W. Potter, eds., *The Nuclear Suppliers and Nonproliferation* (Lexington: D.C. Heath, 1985).
40. On inherent tensions between support for U.S. high technology exports and nonproliferation policy see R. Jeffrey Smith, "Selling to Scoundrels: Why We Won't Stop," *Washington Post*, 15 November 1992, p. C1.
41. For a review of nonproliferation laws, policies, and sanctions see *Weapons Nonproliferation Policy and Legislation in the 102nd Congress* (Washington, D.C.: Congressional Research Service, 1992). On sanctions see: *Economic Sanctions: Issues Raised by the Sanctions Against Iraq*, CRS, April 1992; *Economic Sanctions: Effectiveness as Tools of Foreign Policy* (Washington, D.C.: General Accounting Office, February 1992).
42. Membership in the Nuclear Suppliers Guidelines expanded to twenty-seven in 1992. The group also expanded its "trigger list" of nuclear dual-use items. See IAEA INFCIRC/254/Rev. 1, July 1992, "Communications Received From Certain Member States Regarding Guidelines for the Export of Nuclear Materials, Equipment and Technology." The MTCR, which now has twenty-two members, has also expanded its controls on missile and missile-related technology. See: "Revision of MTCR Guidelines," U.S. Department of State, statement by spokesman Richard Boucher, 7 January 1992.
43. Waltz insists that it is necessary to maintain the autonomy of the system and unit levels in order to deal with theory intellectually. Waltz, "Realist Thought and Neorealist Theory," *Journal of International Affairs*, 29. I am not convinced that spillover between the two necessarily disrupts the neorealist framework.
44. Waltz, "Realist Thought and Neorealist Theory," 36.

45. Waltz, *Theory of International Politics*, 82.
46. Waltz, "A Response To My Critics," in Keohane, *Neorealism and Its Critics*, 327. Waltz was responding to Ruggie's interest in the concept of "dynamic density," which Waltz contends is a unit-level condition. The composite theory I propose requires no major adjustments in the neorealist concept of unit and structure-level interaction. It would, however, expand the possibility of interaction on the structural axis.
47. On the 30th anniversary of the Cuban Missile Crisis, newly declassified documents confirm that mutual misperceptions brought John F. Kennedy and Nikita Khrushchev close to nuclear war. Declassified documents and testimony of participants confirm that the missiles in Cuba were operational and were not equipped with permissive action links to prevent unauthorized launch by field commanders. Furthermore, Kennedy did not know that in addition to the medium range missiles, the Soviets had placed tactical nuclear weapons in Cuba which were to be used to protect the island during a U.S. invasion. Unlike the missiles, which were to be launched only by a direct order from Moscow, tactical weapons were under the control of Soviet field commanders in Cuba. Against this backdrop, Fidel Castro apparently proposed to Khrushchev "the immediate launching of a nuclear strike on the United States." See Mary S. McAuliffe, ed., *CIA Documents on the Cuban Missile Crisis, 1962* (Washington, D.C.: Central Intelligence Agency, 1992); Fedor Burlatsky, "Castro Wanted a Nuclear Strike," *New York Times*, 23 October 1992; Tad Szulc, "Cuba 62: A Brush With Armageddon," *Washington Post, Book World*, review of recent books on the Cuban Missile Crisis, 15 November 1992, p. 1.
48. The U.S.-Soviet model of deterrence offers other unsettling lessons. After the Second World War, some U.S. diplomats viewed the atomic bomb as a "winning weapon" which could enable them to dictate terms to Russia. Other nations may make similar miscalculations regarding the political-military utility of nuclear diplomacy. Gregg Herken, *The Winning Weapon* (New York: Vintage, 1982). For a review of the debate about U.S. intentions and nuclear diplomacy after Hiroshima see Barton Bernstein, ed., *The Atomic Bomb: The Critical Issues* (Boston: Little, Brown, 1976).
49. Jervis, "Security Regimes," in Krasner, 176.
50. On the evolution of the nuclear regime and an update on its major components see Zachary S. Davis, *Nonproliferation Regimes: Policies to Control the Spread of Nuclear, Chemical and Biological Weapons and Missiles* (Washington, D.C.: Congressional Research Service, 1993).
51. Roger Smith expands on Krasner's general description of regime characteristics to argue that the stability of the nuclear regime depends on "incoherence between norms and rules" and the frequency of non-compliance. Smith, "Opaque Proliferation and the Non-Proliferation Regime," in Frankel, 95.

Paranoids, Pygmies, Pariahs and Nonproliferation Revisited

Richard K. Betts

In the decade and a half since this article was first published in Foreign Policy, *no. 26 (Spring 1977) the cold war was reborn and ended, but these epochal changes had only limited implications for the argument. That argument rested primarily on the* local *security incentives that countries might have to seek nuclear weapons. Most of the reasoning still holds, and with a few exceptions noted in italicized updates below, the reasoning proved a valid guide to the development of the nuclear proliferation question in the years since.*

When I wrote the essay at the end of 1976 I was swimming against the tide of globalist optimism in the first year of the Carter administration.[1] *My main arguments were that there was a contradiction between the aims of U.S. military retrenchment and the prevention of proliferation, and that liberal conventional wisdom about U.S. political intervention and involvement in the Third World and the transfer of conventional weaponry to countries involved in regional conflicts was in many cases inconsistent with liberal opposition to the spread of nuclear weapons. The end of the cold war has validated liberal optimism in many stunning ways, but it does not discredit the logic outlined in the article. The Soviet collapse and burgeoning of democracy on all continents could challenge the logic if we really are witnessing the unfolding of the End of History and the emergence of pacific world order. Much more likely, however, is that events are reinforcing the logic, even though we cannot yet know where the events are taking us.*

The end of the cold war reinforces the logic below in a pessimistic direction[2] *if it serves to justify strategic retrenchment by the only remaining superpower, and a more skeptical attitude about the extent to which Americans should pay for the privilege of policing a world in which the only threat of global proportions has gone away. This would leave more countries around the world to take care of themselves. On the other hand, the end of the cold war reinforces my argument in an optimistic direction if, with the removal of the only potent obstacle to American sway, Washington feels free to interfere and regulate security equations in problematic regions. This linkage of optimism about nuclear proliferation to a dubious rationale for U.S. adventurism underlines the emphasis below on the sad facts that not all good things go together. The priority of other goals may have to mitigate emphasis on nonproliferation as a policy objective.*

This is an updated version of an article which appeared in *Foreign Policy*, no. 26 (Spring 1977). The article is reprinted with the permission of *Foreign Policy* and the Carnegie Endowment for International Peace. The original text is in Roman; the updates are in *italics*.

would have less of a problem, since it is becoming even clearer that recycling of plutonium for use in power reactors – the same process that yields weapons-grade material – will almost certainly not be profitable for developing countries. Were economic rationality a meaningful constraint, Pakistan's ardent quest for a plutonium processing plant would make no sense. Recycling is not economical for civilian uses without at least a 20:1 ratio between power reactors and reprocessing plants – a ratio not within range of less-developed countries in the foreseeable future.[4] Proliferation candidates Pakistan and Argentina plan 100 per cent of their power generating capacity in Heavy Water reactors (HWRS) for which recycling is even less profitable than for Light Water reactors (LWRS). Fuel-conserving alternatives to reprocessing are also emerging, such as the "tandem fuel cycle" option, which would transfer spent fuel from LWRS to HWRS and reburn it.

Beyond second thoughts about the identification of energy economy with plutonium, there also remains the problem of the gap between test detonation of a bomb and the "weaponization" and delivery system costs of meaningful nuclear weapons status. In India, these costs provoked resistance to a full-fledged bomb program in the military establishment, which feared a drain on funds for conventional weapons.[5] At the present level of technology, most nations will not have a reason to acquire weapons-related facilities as a simple by-product of energy programs.

Subsequent developments have reinforced this point with a vengeance. The nuclear power industry neared collapse in the 1980s and virtually nowhere are there plans for any large expansion of electricity generation by nuclear reactors. The countries that had the most energetic programs to develop nuclear weapons in recent years – Pakistan and Iraq – did so with dedicated facilities, not spinoffs from peaceful power programs. There is still no known case of a country that stumbled into interest in getting nuclear weapons because technology acquired to produce energy for civilian consumption made the weapons option more easily available.

The Diseconomic Fallacy

But neither will most nations have compelling impulses not to acquire such facilities for purely economic reasons. Arguments that plutonium reprocessing is economically necessary are still widely accepted. Even if unprofitability is proved, diseconomies of scale remain only a marginal disincentive to many nations, because they pose only a marginal additional expense to the total cost of producing energy. This may be offset by the advantages of greater autonomy and certainty of access to fuel.

The incremental costs of a bomb program, once plutonium is available, are also not dissuasive for a proliferation candidate. Albert Wohlstetter estimates plutonium reprocessing facilities could be built for $50 million if the facilities were only for a bomb program. Using this estimate, investment in reprocessing would be a minor cost when measured as a proportion of defense expenditures of potential proliferators.

Modest delivery system costs are not prohibitive. Most potential proliferators already have nuclear-capable aircraft as part of their conventional force structure. A small nuclear force does not have to be sophisticated enough to be credible against a superpower in order to be useful – or to seem so – against a regional rival. Moreover, for a country dwarfed by a neighboring giant, the goal of sure conventional deterrence is unreachable. Even devoting two or three times the proportional effort to defense, Pakistan will never be able to match India's conventional power. A nuclear force might appear to be a net cheaper way of redressing a gross conventional inferiority (analogous to the Eisenhower administration's "economical" massive retaliation policy).

Finally, if the technological breakthrough of laser isotope separation pans out, as some scientists believe it may within a decade, diseconomies of weapons-related energy programs will evaporate. Laser separation promises to be a very inexpensive source of enriched uranium, which makes as good a bomb as plutonium. In short, energy-related economic incentives or disincentives are important, but they hold the key neither to the problem nor the solution.

The preceeding emphasis on the relatively small burden posed by a nuclear weapons program now appears exaggerated. No one we know of has yet attempted to acquire a "quick and dirty" weapons capability. Pakistan and Iraq invested heavily in sophisticated and diversified components, importing and smuggling numerous materials and technologies for separation of fissionable material and "weaponization" design and fabrication. The fear prevalent in the 1970s that production of at least a crude device would be within the grasp of any highly motivated group that could get some plutonium, some physicists and engineers, and a good machine shop was overdrawn.[7] Countries with incentives serious enough to risk joining the nuclear club do not seem to want childish or risky weapons designs. The arguments that nuclear weapons are not dauntingly expensive in comparison with the conventional military capabilities that a deterrence posture might replace, however, still hold.

The Technicist Fallacy

The reason that nuclear spread is undesirable is that it threatens our security; nonproliferation policy, therefore, has to be a subordinate component of national security policy. Yet there remains a widespread tendency to focus on

Table 1

NUCLEAR WEAPONS DEVELOPMENT AS A PERCENTAGE OF DEFENSE BUDGETS

Selected Candidates	1974 Defense Budget*	Defense Budget +$50 million	% of Total Defense**
Iran	5,280	5,330	0.9
Israel	3,840	3,890	1.3
Egypt	2,100	2,150	2.3
Brazil	1,900	1,950	2.6
Yugoslavia	1,080	1,130	3.8
South Africa	1,010	1,060	4.7
Taiwan	814	864	5.8
South Korea	642	692	7.2
Pakistan	631	681	7.3
North Korea	625	675	7.4
Argentina	573	623	8.0
Libya	412	462	10.8

* in millions of current 1974 dollars
** minimum incremental cost
Source for Defense Budgets: U.S. Arms Control and Disarmament Agency, *World Military Expenditures and Arms Transfers 1965–1974* (Washington, D.C.: Government Printing Office, 1974).

(1974 budgets are the last for which complete and consistent data are available for these countries.)

Table 2

MILITARY INFERIORITY, RESOURCE LIMITS, & INCENTIVES TO ACQUIRE A NUCLEAR DETERRENT

(in constant 1973 dollars)

1974	India	Pakistan
defense expenditures (millions)	2,180	572
defense expenditures as percentage of GNP	3.04	6.68
defense expenditures per capita	3.51	8.37
GNP per capita	116	125
armed forces (millions)	1.62	.5
armed forces per 1,000 people	2.61	7.32

Source: U.S. Arms Control and Disarmament Agency, *World Military Expenditures and Arms Transfers 1965–1974* (Washington, D.C.: Government Printing Office, 1974).

the threat of proliferation as a simple function of the availability of requisite technology. Corollaries are the tendencies to think of candidates for proliferation as an undifferentiated mass of "Nth" countries, and to direct all attention to ways of stopping the diffusion of sensitive components of nuclear technology.

Concern with diffusion of technology is proper and fundamental, but limited in vision. It focuses on the necessary rather than sufficient conditions for proliferation, and hides the significant distinction between capability and the exercise of capability. If capability itself were the driving force behind proliferation, Germany, Japan, Italy, the Netherlands, Belgium, and Sweden would have acquired nuclear weapons long ago. The temptation to conceptualize proliferation apart from the political forces behind it may derive in part from the scientific orientation of many of the most devout arms controllers, or from the tendency in recent years to view the nuclear interaction between the superpowers as something autonomous and self-driving, a numbers-game of reactive deployments, abstracted from political causes. And in some ways, the complexity of stable deterrence has given the U.S.-Soviet nuclear balance a life of its own. It is easy to forget that vertical proliferation was the symptom of East–West conflict rather than the cause.

Proliferation does not have a life of its own; it is a political problem as much as a technical one. Technological mystery, coinciding with international bipolarity, simply gave the United States a long period of grace in which it could afford to pay less attention to the political dimension. Technical barriers coincided with political disincentives. Consciousness of the revolutionary consequences of proliferation should not obscure the traditional nature of incentives behind it: defensive security, and assertive prestige and power. No less than a function of diffusion of technology, proliferation will now be a function of the postcontainment era's diffusion of power. An absolute commitment to nonproliferation unavoidably raises the issue of what we are willing to do to limit this diffusion of power. To seek technological hegemony by denying weapons-related plutonium facilities to developing nations, while simultaneously approving the retrenchment of U.S. commitments and the decline of American military hegemony outside Europe, is necessarily to accept that there are severe limits to what we are willing to do to contain nuclear weapons. The identity of likely proliferators brings this problem into focus.

Candidates

In the early 1960s, with two Berlin crises a recent memory and De Gaulle's distrust of U.S. reliability a live issue, American fears of nuclear proliferation centered on advanced states such as Germany, Japan, and Italy. The multilateral force proposal was in part an attempt to offer an alternative to a larger number of independent European nuclear forces which might set off a downward chain reaction of nuclear spread. It seems less likely now that the nuclear dominoes will fall from the top down in the hierarchy of international power than from the middle sideways (through regional rivalries in the Third World) or middle up (as major powers such as Japan find it harder to abjure the bomb once smaller states such as Taiwan or South Korea have it). Any list of prospective proliferators is debatable. But rather than inferring them simply from technical capabilities it makes more sense to look at those countries that are within shooting distance of the capability and that have significant incentives.

There are two fundamental motives: status and security. Potentially, all nations have the status and prestige incentive. This is likely to be most potent, though, for those states that are emerging as dominant regional power centers with plausible pretensions to being great powers, and without front-line involvement in the East–West superpower confrontation. Current examples are Iran and Brazil (at later dates perhaps Nigeria and Indonesia).

Nonproliferation policy, however, should focus on those incentives that are manipulable more than those that are not. Desire for prestige is difficult if not impossible for an outside power to affect. Trying to coerce or buy off a power-happy state might well backfire; the desire for prestige is the desire not to be in the position of either victim or supplicant. It may be that delaying actions such as obstructing sale of plutonium reprocessing facilities are all that can be done to retard proliferation by regional hegemony-seekers. Or we might try to convince Iran that it has more to gain by remaining the unchallenged conventional military leviathan in its area than by introducing nuclear weapons that might provoke super-power interference. (Similarly, Brazil's net power position might remain better if it does not provoke Argentina into developing a bomb.)

Security incentives, however, are quite susceptible to moderation by outside powers. Security-motivated candidates are thus the ones we should spend the most time worrying about. But here the problem is less what we can do than what we want to do. A vicious circle exists: Those states whose security is threatened seriously enough to provoke acquisition of nuclear weapons are those that the United States has other reasons not to protect unequivocally, which is why they have reason to want a bomb.

The Pygmy States

First are the "pygmies" – nations threatened by much larger neighbors. Pakistan is a prime example, having fought several wars with India, emerging from the last severely battered and even more militarily disadvantaged. Since then, India has detonated its own bomb, demonstrated a penchant for pressure against neighbors such as Sikkim, Nepal, and Bangladesh, and continues to maintain armed forces over three times the size of Pakistan's. India's quasi-alliance with the Soviet Union neutralizes the weight of Pakistan's with China. For the United States to ally with Pakistan, we would have to further damage our already bad relations with a much more important country – India.

Taiwan is another pygmy, as is Yugoslavia. To remove Taiwan's incentive by reaffirming our mutual defense treaty, the United States would have to endanger rapprochement with the mainland. For Yugoslavia to entrust its security to the principal capitalist power, on the other hand, would be not only undignified but foolhardy.

Nuclear weapons may be attractive for a pygmy not only as a deterrent capacity to "tear off an arm" (say, New Delhi, Canton, or Odessa) à la General Gallois' theory, but conceivably for tactical applications against forces concentrated for land breakthroughs or amphibious landings.

The Paranoid States

A second security incentive is paranoia. South Korea's current military capability, on balance, surpasses that of its adversary, North Korea, and barring a catastrophic change in policy, neither China nor the Soviet Union are likely to underwrite any adventurism by the North. Yet South Korea's anxieties remain high, fueled by constant low-level North Korean provocations and sabotage, and it has sought to purchase both a reprocessing plant from France and rocket construction facilities from a U.S. company. And, as the cliché goes, just because you're paranoid doesn't mean they're not out to get you. Few states remain content with military parity or marginal advantage when faced with enemies as bizarre and unpredictable as Kim Il-sung. North Korea, in addition, has neither signed nor ratified the NPT.

The Pariah States

Finally, there are the pariah states, the thorniest problem of all. They combine the disadvantages of pygmies and paranoids along with more visceral and unremitting opposition by their regional enemies and growing isolation from most of the rest of the world. Their capability for producing at least a modest number of crude nuclear weapons also exists already. Israel is universally assumed to have at least ten untested bombs in stock. South Africa has its own facilities for uranium enrichment. Taiwan has been accused of reprocessing spent fuel (this was denied), and has sent military-trained researchers to the Massachusetts Institute of Technology to study ballistic missile-related technology. (MIT terminated their instruction after pressure from the State Department.) Pariahs have the clearest incentives to increase their military power, the least technical distance to go to build a bomb, and the least to lose in doing so.

In the new world disorder following the end of the cold war some of countries with a logical interest in having an independent deterrent capacity have changed. Israel and South Africa are less isolated politically then they used to be. Israel is no less wedded to its nuclear force, but the phasing-out of apartheid is eliminating the status rationale (to get some international "respect" for its independence and power) that may have animated previous nuclear interest in Pretoria.[8] *Other countries, in turn, have become more vulnerable or ostracized than they used to be. Neither Cuba nor North Korea now has a reliable superpower protector against the United States.*

There is no evidence that Cuban leaders see a nuclear capability of their own as desirable; they probably understand that any attempt to acquire one would provoke the United States to do just what they want to deter. North Korea moved toward developing nuclear weapons but found that the initiative brought significant diplomatic grief and barriers to desired foreign investment. With one superpower now in the world, puny nuclear forces aimed at the United States are likely to be more trouble than they are worth, provoking rather than deterring. Pygmy countries are more likely to find very limited deterrents useful against their local enemies or regional hegemonic powers, as long as those enemies are not firmly backed by Washington.

Five Options

In ascending order of activism, the United States has five general options it could exercise toward such insecure nations:

1. Laissez Faire. Recognize the limits of U.S. capacity to prevent proliferation, hope for the best from safeguard inspections by the International Atomic Energy Agency (IAEA), and leave the next generation to worry about the stability of a world with fifteen or twenty nervous nuclear-armed nations.

2. Bribery or "Carrot" Leverage. Induce threatened candidates to forgo the nuclear option in exchange for compensating economic, technical, and other nonmilitary concessions.

3. Alliance Guarantees or Military Assistance. Assuage insecurity by backing vulnerable states with U.S. power or aid in reinforcing their own conventional military capabilities.

4. Denial. Cap the diffusion of weapons-related nuclear technology.

5. "Stick" Leverage or Pre-emption. Threaten to withdraw aid, or to

intervene to prevent development of nuclear weapons, or to change sides and support the candidate's enemies.

Given the attitudes of President Carter and the foreign policy establishment, the first option is unacceptable. The strong forms of the last option are also likely to be rejected as too radical, adventurous, or immoral. Proposals in the arms control community focus predominantly on the fourth option. But more attention is needed to the second and third, and weak forms of the fifth. The problem is that none of the general options are universally applicable – each is undermined by costs and qualifications when applied to concrete cases. A realistic policy will have to combine and hoc variations of several of the options in regard to different candidates. The need to separate the qualifications and costs, and the need to see solutions not as simple absolute preventives but as relative constraints, is suggested by some more fallacies of emphasis.

Cures

The Fatalist Fallacy

One reason that the proper emphasis on security incentives has been lacking in nonproliferation strategy is that many who recognize the salience of these incentives tend to be less absorbed with the immediate threat. One example is the proliferation-is-not-so-bad school which sees nuclear spread as neither irrational, immoral, nor destabilizing. This logic unites many Third World strategists, Gaullist conservatives, and Maoist radicals. To this school, the frenzy of liberal opposition to proliferation is at best hysterical, naive, and doomed, and at worst imperialistic arrogance, reflecting "yellow peril" and "white man's burden" psychology.

Another school consists of analysts who focus on creative and complex projections of alternative proliferation sequences twenty years into the future, and who stress the need to shift emphasis from how to retard the process to how to manage and cope with it. Finally, there are the shoulder-shrugging pessimists who see the issue for candidate states, by now, not as why to get a bomb, but simply why not – which makes universal nuclearization only a matter of time.

The problem with the first school is undue confidence that the Pax Atomica of stable U.S.-Soviet deterrence can be replicated in other regions. The expense involved in creating secure second-strike forces and reliable command and control systems, the incentives for enemy pre-emption in the transitional phase between inauguration and completion of a weapons program, the volatile hostilities in some regions which make use of any available military capability quite likely, the uncertain domestic stability of some candidates,and other imponderables make the theory something that should not blithely be put to the test.

The second approach founders less on grounds of logic than of relevance. Long-range futurology is heuristic and useful, but provides little help in formulating policy at the moment. It also suggests an uncomfortable double entendre in Keynes' hackneyed quip that in the long run we are all dead. The problem with shifting attention to "coping" measures is that it conflicts with preventive efforts by weakening the taboo and by rewarding nations for going nuclear. Stabilizing infant nuclear forces may require aid – technical assistance for intelligence verification, Permissive Action Links and other command controls, targeting advice, hardening delivery systems, bomb design, air defense – much of which would both attract candidates to join the club and weaken our own nuclear posture.

The last group of pessimists actually distracts attention from the importance of incentives by cheapening them, dismissing the significance of disincentives. Luckily there are still powerful reasons not to go nuclear – principally the danger of provocation, making oneself a nuclear target, reducing security instead of enhancing it. Fatalism obscures the difference between the eventual threat of universal proliferation by default and the immediate threat of proliferation by specific states with compelling strategic incentives.

The dam has still not broken. Most observers have consistently overestimated the rate of the spread of nuclear weapons capability since the 1950s. We must be doing something right, or the reasons that seeking nuclear weapons may harm a state's interests rather than promote them must remain as evident and compelling as they ever were. Many countries may make moves toward developing nuclear explosives, but from known indications, only two have successfully accomplished the mission in the years since this article was originally written. Pakistan moved steadily through the late 1970s and 1980s to amass the required components and designs. South Africa was suspected of mounting a nuclear test in the South Atlantic at the end of the 1970s, although the fact that an explosion occurred was never confirmed.

The Legalist Fallacy

The NPT is useful as a symbolic international commitment against proliferation, but it will not stop proliferation. Many candidates, such as Israel, South Africa, Pakistan, Brazil, and Argentina have neither signed nor ratified it. In any case, the treaty itself does as much to lay the groundwork for nuclear spread as it does to contain it. Article IV explicitly commits the nuclear powers to aid the nonnuclear weapons states in development of peaceful nuclear energy capabilities, including "*the fullest possible exchange of equipment*," (emphasis added). Article X provides the right of unilateral abrogation upon only three-months notice.

Thus Libya, in ratifying the treaty, has legally entitled itself to buy all the fuel facilities necessary for weapons capability, while without NPT membership such technology might remain further out of reach. All that would then stand in the way of a decision to build a bomb by that paragon of rectitude, Colonel Qaddafi, is a thin juridical reed. A diplomatic quirk – according to which Taiwan is acknowledged to be part of China, a nuclear weapons power under the treaty – also allows some Chinese on Taiwan to argue captiously that the NPT permits them to have the bomb if they want it. Moreover, in order to gain the widest possible adherence to the treaty, the United States accepted demands for less rigorous IAEA safeguards and bookkeeping systems.[9]

Treaties can only ratify interests. When a consensus develops in a nation that the NPT no longer serves its interests, there is nothing either legally or practically to prevent it from bailing out. There is no reason to believe that the NPT should be any more successful in preserving peace than the greatest international institution: the United Nations. Both are valuable symbols, effective at the margins, but insufficient to their purposes. They should be supported and reinforced, but are hardly to be relied on.

Two of the three most active attempts to develop nuclear weapons capability in recent years were by parties to the NPT, Iraq and North Korea. Enough said.

The Multilateral Fallacy

Other weak comforts have been proposed: multinational nuclear facilities (MNFS) designed to keep reprocessing plants out of candidate governments' hands, and nuclear-free zones. Benefits of MNFS are not clearly greater than their costs. They would increase dispersion of expertise on reprocessing, be least attractive to precisely the states that want a bomb the most and least viable in exactly the most dangerous regions (where neighbors are most antagonistic and cannot cooperate for anything), and would encourage the growth of an international plutonium economy just when antiproliferation energies should be devoted to proving the unprofitability of plutonium recycling.[10]

The nuclear free zone option suffers from the same problems as the NPT: nonadherence by the most relevant parties. The Latin-American zone established by the Treaty for the Prohibition of Nuclear Weapons in Latin America signed in Tlatelolco, is conspicuous by virtue of its incompleteness: neither Brazil nor Argentina are fully bound by it. Pakistan's pleas for a

South Asia nuclear free zone were rebuffed by India. Multilateral solutions are superfluous where they are not needed, and least applicable where they are needed most.

Following domestic political changes in Brazil and Argentina in the 1980s new governments in those countries uncovered secret nuclear weapons programs, dismantled them, and undertook a mutual agreement to refrain from such activities. These refreshing developments, however, owed far more to internal and bilateral politics than to multilateral legal constraints.

The Embargo Fallacy

Barring transfer of enrichment or reprocessing plants would be the most decisive way to hold back proliferation in the short run, but the high political price of a strategy of complete denial makes it justifiable only if it promises absolute effectiveness. An embargo of sensitive fuel facilities would violate Article IV of the NPT, giving the target countries an excuse to violate the treaty themselves. It would enshrine reprocessing capability as a forbidden fruit and promote indigenous development of programs without current safeguards. A minimal step to make such a policy reasonable would be to bear the costs and dangers of buying back and storing all the lethal radioactive spent fuel from foreign reactors (absorbing domestic criticism from environmentalists for doing so) rather than reprocessing it for foreign use or storage.

Embargoing nuclear energy technology is a bald exercise in realpolitik, making American pretensions to supporting equity for the Third World seem disingenuous. It legitimizes counter tactics such as the Organization of Arab Petroleum Exporting Countries' 1973 oil embargo, and further undermines already ailing European nuclear industries. With the profit incentive for breaking an embargo high, could we keep it airtight?[11] Promoting an embargo, at the London nuclear suppliers' conference, has also required strong-arm tactics against France and Germany. After Carter's campaign emphasis on the need for greater sensitivity to our principal allies is this still desirable?

The answer might be yes if there were not so many possibilities that even after all the belligerent arm-twisting, the embargo would fail to prevent proliferation. Despite grudging movement toward the American position at the supplier' conference, neither France nor Germany (as of this writing) are yet willing to apply the policy retroactively and renege on their contracts with Pakistan and Brazil. Even assuming that the NPT, allied harmony, nuclear business profits, and Third World good will are worth sacrificing, an embargo would only cripple the weapons capability of those nations whose potential as proliferators is in the future, rather than those who are already on the threshold. Taiwan, for instance, can reprocess plutonium if it wants to. A U.S. threat to cut off enriched uranium for reactor fuel would not pose an insurmountable obstacle. South Africa is developing its capacity to export enriched uranium – a development which the United States assisted by selling computers necessary for enrichment, out of gratitude for South African uranium used in the U.S. bomb buildup in the 1950s. What is to prevent a consortium of pariah states from combining their already highly developed nuclear capabilities for weapons purposes? Israel, South Africa, and Taiwan have already strengthened their economic, scientific, and military ties as their international ostracism has grown.

The same potential exists for a counter-great power consortium by aspiring regional status seekers. Argentina, which has reprocessing capability, has nuclear cooperation agreements with India and contributes over half the foreign staff of Iran's Atomic Energy Commission.[12] Technology is a moving target. Can a perfected laser enrichment process be kept in bounds as easily as bulky reprocessing plants? Brazil and Iran's technical infrastructures are improving, and they can buy talent if they cannot buy hardware. As there is an inverse relation between security-motivated candidacy and susceptibility to economic and technical *disincentives*, there is an inverse relation between prestige-motive candidacy and economic and technical *barriers*. If after wielding a big stick and alienating recipient countries, the United States only manages to slow proliferation by a few years, will it have been worth it?

Controlling transfer of capabilities is an underkill strategy, because it does nothing to change incentives except to increase them, and it is an overkill strategy because there is no necessary correlation between nuclear capability and candidacy for the exercise of that capability. Germany and Japan are not nuclear weapons powers now not because they cannot be technically, but because they do not want to be politically. As long as their security is enhanced more by association with the United States than by autonomous nuclear forces, their interests make their capability irrelevant. Is an aggressive, upsetting, and probably inadequate effort to hoard the energy-related building blocks for bombs clearly preferable to a more moderate or selective approach to the same end? More importantly, is it a sufficient alternative to more vigorous initiatives to reduce incentives?

Strenuous efforts by the nuclear supplier countries to form a cartel and prevent transfer of sensitive technologies were more effective after the mid-1970s than the preceeding analysis would suggest. The so-called London Suppliers Group did make it extraordinarily difficult for Pakistan and Iraq to get their weapons programs in full swing, and certainly delayed them. Nevertheless, such restraints did not prevent Pakistan from eventually achieving the objective, and would not have prevented Iraq if the 1991 war had not intervened.

The Safeguards Fallacy

Disillusionment with prospects for export control has led some nonproliferationists to put their eggs in the safeguards baskets, hoping to prevent the leap from capability to bomb program by beefing up the IAEA inspection system. But safeguards, while desirable, are no solution.

First, safeguards have no compelling force. India built its bomb with material from a Canadian-built reactor covered by bilateral safeguards, rationalized the breach by denying the "peaceful" explosive was a bomb, and, as Paul Joskow says, "The world just watched."

Second, tightening inspection and accountability procedures to the ultimate degree possible raises objections on grounds of expense, danger of industrial espionage, and effrontery. As Iran's shah said, "you are asking us for safeguards that are incompatible with our sovereignty."[13]

Third, safeguards are hard to rely on in the case of a pariah like Taiwan. Expelled from IAEA membership at Peking's insistence (though inspectors remain, at U.S. behest), with plutonium accumulating from a Canadian-built reactor while Canada has severed diplomatic relations, Taipei would have little reason to continue to adhere to safeguard inspection if the United States were to withdraw support from the regime.

Fourth, safeguards focus on detecting secret diversion of nuclear material, but have nothing to do with the legal accumulation of energy program-generated fissile material in national stockpiles.

The theft or diversion threat has been the focus of the most energetic publicists. But terrorists do not really pose the most imminent danger of proliferation, and the problem is not so much that a country will "cheat in its nuclear bookkeeping," as that it will decide openly to develop nuclear weapons.[14] Focusing on diversion is itself a diversion.

Another justification for centering attention on safeguards has been that while they cannot prevent bomb development, they would serve as an alerting mechanism, providing warning time before a bomb was completed. There are two essential faults in this rationale. One, the early warning function of IAEA is compromised by confidentiality provisions designed to protect industrial secrets; IAEA does not have to report critical information to governments or the public. Two, even if we do get the warning time, so what? What will we do with it? It would be important if we intend to take pre-emptive action to prevent development of the bomb. But in the present climate of U.S. foreign policy such intervention is unlikely, especially if American opposition to protective intervention is what gives the security-motivated candidate the incentive to make the bomb in the first place.

There has still been no case in which legal safeguards or inspections either revealed or derailed an attempt to build nuclear weapons. Iraq, a party to the NPT, was subject to safeguards and inspections, yet managed secretly to marshal a tremendous and nearly successful effort to build weapons. American bombs and extraordinary intrusive inspections imposed on Iraq as a condition for terminating the war, not the normal IAEA *restraints, are what revealed and stopped Baghdad's nascent nuclear force.*

The Umbrella Fallacy

One intriguing proposal frequently offered to alleviate vulnerable states' fear of nuclear attack is for the United States and the Soviet Union to announce in condominium that they will retaliate against any nation that uses nuclear weapons against a nonnuclear country. This solution is simply too sweeping, fanciful, and studded with drawbacks. Paranoid candidates would not trust such a policy. Nor does it solve the problem of pygmy candidates who need a bomb to deter conventional attacks; this policy would leave India free to invade Pakistan, knowing a Pakistani nuclear defense would only invite annihilation by the superpowers. Finally, such a commitment requires Soviet recognition of the legitimacy of all threatened states, while it is precisely the denial of this legitimacy that constitutes the problem for pariahs. Would the Soviets agree to retaliate against a country that uses a nuclear weapon against Pretoria?

An even more novel suggestion is that of Alton Frye: have the United States agree to give a bomb, tit for tat, to any nation attacked with nuclear weapons.[15] This offers too little too late. Who do we give the bomb to after Israel is turned into a cinder? In grandly pretending to guarantee everyone's security, we believably guarantee no one's. And the most fundamental point is that neither of these nuclear guarantees are more desired, plausible, credible, less costly, or provocative than the simple alternative of reducing defensive motivations for a bomb: a conventional bilateral military alliance.

The Two-Wrongs-Don't-Make-a-Right Fallacy

Presidential candidate Carter said, "we cannot be both the world's leading champion of peace and the world's leading supplier of weapons of war." An advisory panel recommended establishing a conventional arms suppliers' conference comparable to the London nuclear supplier's conference. Vice Presidential candidate Mondale also used a great deal of "merchants of death" rhetoric, denouncing the U.S. role in the arms trade. But President Carter now has to face the possibility that we cannot be the world's leading champion of nuclear nonproliferation and at the same time pristinely end arms transfers to vulnerable nations. If anything alleviates objective security incentives for nuclear weapons, an increase in conventional defense capabilities does. Peace is not all of a piece. Rather than companion evils, nuclear weapons proliferation and conventional weapons proliferation may be alternative evils.

Similarly, the moral goal of distancing the United States from repressive or corrupt regimes may be inconsistent with the goal of preventing them from getting the bomb. In the wake of bribery scandals and publicity about KCIA excesses, the prospect of disentangling the United States from support of the South Korean government is increasingly attractive to American liberals. Apartheid makes pressure on South Africa even more desirable. But to demand that a government forgo nuclear weapons is to demand that it compromise its own sovereignty. To expect it to do so without any compensating quid pro quo would not only be condescending but naive.

Jumping from these realizations to the conclusion that the United States should lavish F-14s on all comers, or embrace any reprehensible regime that threatens to build a bomb, would luckily be as myopic as ignoring the trade-offs. Arms transfers are no panacea. They provide delivery capability for a bomb, and may whet rather than satiate a candidate's appetite for grandeur. And for a threatened state, dependent on foreign arms suppliers who retain leverage through the option to stop resupply in a crisis, or embargo spare parts and cripple maintenance, arms aid does not erase the attractiveness of nuclear weapons as an autonomous deterrent. Pakistan learned this in the 1965 war, when the United States cut off arms to both belligerents – leaving India, with an alternative source of supply in the Soviet Union, with a decided advantage. Turkey learned the same thing following the 1974 Cyprus crisis, and Israel avoided the problem in the 1973 war only because U.S. resupply efforts were so energetic.

Nor is toadying to bad regimes clearly worth the price. Besides weakening our moral position, highlighting the nuclear weapons trade-off hands such governments a club with which to threaten us, enabling them to demand progressively greater concessions, turning supplicants into blackmailers. Some weak states are less interested in producing nuclear weapons than in getting something for not producing them.[16] Nevertheless, if nonproliferation is the governing goal it is unrealistic to maintain axiomatically that the United States should never do "wrong" things in the arms trade, or in support of bad governments, in order to bring about the greater good. This problem of lesser evils underlines the need for a discriminating case-by-case policy.

Since this was written the Carter administration tried for its first three years to reduce conventional arms transfers and the Reagan administration loosened restraints, both according to dubious rationales.[17] *In neither case was there any very clear or consistent attempt to link conventional military aid to nuclear restraint by the client. In the case of Pakistan, indeed, both administrations cynically over-*

looked the country's energetic quest, gave certification to Congress that it did not have nuclear weapons, and provided conventional military assistance as long as they needed Islamabad to support the anti-Soviet war in Afghanistan.

With the end of the cold war the United States and European arms suppliers have more potential leverage in the so-called Third World. Local states can no longer play off two superpowers or blocs against each other, and arms suppliers might be able to twist arms successfully on the nuclear issue if they are willing to get dirty in the game of regional balance of power. The end of the cold war, however, has also created new pressures and constituencies to promote solutions modeled on collective security, which cuts against such ad hoc manipulation and discrimination.

The Golden Key Fallacy

Too much frenetic nonproliferation drum beating has been a search for the philosophers' stone. There are no simple solutions that are feasible, no feasible solutions that are simple, and no solutions at all that are applicable across the board. The relative costs of any particular option vary drastically according to who the candidate is. We should not address the problem in terms of "worldwide" proliferation, but rather in terms such as "the balance of power in South Asia" or "alternatives to nuclear weapons in the Middle East." Nonproliferation cannot be segregated from the rest of U.S. policy goals. Establishing the new high-level office for nuclear proliferation in the State Department is a valuable step forward, but it is essential that this not lead to compartmentalizing the problem. Coordination with other policy agencies is crucial. Perhaps "nonproliferation impact statements" for proposed policy changes toward various countries, akin to the arms control impact statements now required by Congress for U.S. military programs, would be a useful exercise for at least forcing overall policy trade-offs into view. We should reverse what Myron Kratzer testified was State Department policy under Kissinger: not "to balance or trade off our nonproliferation objectives against other political objectives."[18] Policy should be developed without illusions about permanent solutions; antiproliferation strategy will have to be a constantly adaptive succession of adjustments and realignments of priorities. But the tendency to see simple solutions remains strong, fed by reticence to make distasteful choices between important goals.

Choices

There is no free lunch in nonproliferation policy; every effective measure has economic, political, or moral price tags. The biggest price would be the reaffirmation or extension of protective alliances which would relieve the insecurities of paranoid, pariah, or pygmy candidates. This is not an agonizing choice for right-wing interventionists or liberal cold warriors, but it is for those liberals – including most arms control enthusiasts – who favor retrenchment of commitments, a lower military profile, and dissociation from objectionable governments.

Yet for those with an absolute opposition to proliferation, alliances have to seem a reasonable price. The United States was able to coerce South Korea into canceling its purchase of a reprocessing plant from France because Seoul still values the American troops stationed south of the 38th Parallel. (The South Korean bribery operation and Park Chong-hui's own authoritarian tightening of his grip on power reportedly began in 1970 at the time the United States took 20,000 troops out of Korea, suggesting a connection between the decline in vigor of the American alliance and objectionable compensating actions by the government in Seoul.) If Carter follows through with plans to withdraw the remainder of those troops, how much leverage will we have left? Indeed, Park himself told two American columnists that if U.S. support falters, South Korea will build its own bomb.[19] And as the lobbying scandal widened, new rumors surfaced that Seoul had tried to reopen negotiations with France for the reprocessing plant. Is keeping South Korean fingers off a nuclear trigger worth keeping a commitment that risks U.S. involvement in another Korean war? Is it worth lending American support to an authoritarian government which tries to subvert our own legislators? And if we have a finite amount of leverage, how much should we expend in keeping the lid on a bomb, and how much on pressing for social or political reforms, or stable oil prices?

The United States has been willing to pay the price for nonproliferation in Germany and Japan. (The value of the Soviet alliance also makes Communist pygmies like Cuba unlikely candidates for nuclear weapons.) Israel is the one country for which a new alliance commitment might be domestically acceptable, but Israel is almost certainly beyond the nuclear candidate stage anyway. How much should the United States pay to keep a covert bomb covert? The intense involvement of both superpowers in the Middle East may be one reason that nuclearization of the whole region has not progressed at a faster pace.

Yet even in 1968, Secretary of State Rusk, testifying in favor of ratification of the NPT, had to assure senators that the treaty would not entail any additional U.S. commitments. Are we any more willing now to extend our power or reassure ornery allies? Should we sacrifice improvement in relations with Peking by maintaining our defense treaty with Taipei? Or would a laissez faire approach, severing the treaty but letting Taiwan develop its own bomb, get us off the hook by ensuring "two Chinas" without U.S. intervention?[20] At the least, it is necessary to calculate priorities, to consider where a new nuclear power might be least costly in net terms, and which candidates we should pay the highest price to keep out of the nuclear club. Proliferation arising from security concerns is only the symptom of a conflict between nations. If the United States does not intend to play a role in regulating that conflict, it can hardly expect to regulate the proliferation.

A different kind of price would be weakening NATO or alienating Japan if the United States tightened the screws too tightly on technology transfers. Senators Ribicoff and Glenn tangled with Kissinger over this issue, criticizing the secretary for failing to ask the Soviet Union to join in pressure on France and Germany against the sale of reprocessing plants. Ribicoff suggested that the United States cold threaten the Europeans with a cutoff off enriched uranium fuel for their own reactors. Kissinger responded that joining an adversary in order to blackmail allies was too much to sacrifice on the altar of nonproliferation.[21]

Other prices may be lower. Opposition to arms transfers can be overcome where a linkage to dampening nuclear spread can be demonstrated. To reverse the economic attractiveness of nuclear power, damaging the commercial vitality of private nuclear industries of the supplier countries (by cutting exports and government subsidies) may be more acceptable to liberals than extending American military power abroad (the reverse may be true for conservatives); but it would still not be costless. As of inauguration day, it was not yet evident that President Carter had faced up to contradictions between his extreme commitment to nonproliferation and his other policy goals, or to the full costs of a strenuous nonproliferation strategy. And the probability that the political give-and-take of the policy-making process will not yield a consensus for paying the total price for any single general option underlines the need for several overlapping limited options.

The relevant mix of nonproliferation policy options is conditioned by two factors. First is the relative price of any option in terms of who the candidate is. Second is the relative price in terms of foreign policy goals not related to proliferation. Particular variations in strategy depend on the administration's overall priorities. A crude, tentative, but illustrative example of a possible distribution of options is the following, with aid in all cases contingent on a quid pro quo not to go nuclear:

Israel: Pay the strategic price. Extend a U.S. alliance, exchanging a trip wire force of American troops and continued arms aid for return to the 1967 borders, a Palestinian state, and ratification of the NPT.

Pakistan: Pay the diplomatic price. Extend military assistance and economic aid, and accept Indian displeasure. Policy as it stood at the end of the Ford administration was not bad: The end of the arms embargo was limited in principle to defensive weapons; sale of A-7 aircraft was conditioned on

Islamabad's canceling its reprocessing deal with Paris; termination of economic aid was threatened if the plant sale went through. These measures were coupled with U.S. pressure on France to back off from the deal and Canadian threats to cut off supply of fuel rods for Pakistani reactors. The one flaw was that the balancing effect of the proposed A-7 sale was vitiated by another U.S. plan to sell A-4s to India.

Taiwan: Fudge the price; maintain the defense pact or laissez faire, whichever China chooses. Continue to avow U.S. desire to improve relations with the mainland, but let Peking decide whether it prefers a non-nuclear Taiwan, still guaranteed against invasion by the United States, or an abandoned Taiwan with a nuclear crash program.

South Africa: Pay the intervention price. Given Pretoria's overwhelming conventional military superiority over black African opponents, the only advantage for South Africa in threatening a nuclear weapons program would be as a bargaining chip to elicit U.S. help. Rather than compromise U.S. support for majority rule or be sucked into diplomatic support for the white regime, Washington could call the bluff, cut Pretoria adrift, hope incentives for the bomb would decline once the gambit failed, and threaten to stomp on the regime if it ever brandishes a bomb against its black neighbors.

In any case, the full cost of preventing a country from going nuclear has to be weighed against contradictory goals. Senator Ribicoff to the contrary, strategy has to accommodate to the fact that the United States has other interests in foreign policy besides prevention of nuclear spread. In some cases, a limited approach, focusing on negative foot-dragging measures to retard technology transfers rather than positive actions to reduce incentives, may be the only reasonable strategy. But arms controllers who plump for this limited approach – rather than incur the costs of military defense commitments, an active role in arms transfers, support for illiberal governments, or alienation of the respective enemies of the potential proliferators – have to admit that the approach is limited, and that our interests in nonproliferation are not absolute. Such a recognition will improve the clarity and realism in nonproliferation policy, make it seem less arrogant and naive to the candidate countries, and make the whole enterprise less of a blustering, blundering, quixotic quest.

The logic in the recommendations penned in 1977 still applies, although it has been overtaken by events in some cases – the normalization of relations with China in the Carter administration, Pakistan's entry into the nuclear club, and Pretoria's moves toward majority rule. For Iraq, the "solution" was provided by war, but one undertaken for other reasons. For North Korea, the solution is not apparent. Both the Israeli preventive raid against the Osirak reactor in 1981, when there was no good evidence ever made available of a serious Iraqi weapons initiative at the time, and which may have spurred Saddam Hussein's nuclear ambitions as much as it constrained them,[22] *and the invasive inspections after the war against Iraq ten years later, showed how unreliable air attack can be for turning countries away from efforts to get the bomb, or rooting out nuclear facilities that have been developed with dedication and secrecy.*

It would be folly to strike North Korea if an attack would not be decisive, but an invasion of the country in order to prevent it from deploying nuclear weapons is hard to imagine. Yet Pyongyang is the perfect example of the violent, uncivilized, risk-prone state about which all anti-proliferationists have nightmares. Enlisting China (North Korea's only ally), Japan (the source of prospective aid and investment for which Pyongyang hopes), and the rest of the international community to pressure the Kim Il Sung to accept unprecedented inspections and guarantees of adherence to NPT *obligations would not assuredly prevent nuclear weapons development, but it might be the least objectionable alternative.*

What new candidates may emerge? Iran, whose nuclear activities have recently begun to recover from the post-Shah shutdown, could be one of the most worrisome. Before the revolution, status appeared to be the main reason that the Shah could develop an interest in nuclear weapons. Now, however, Iran has conceivable security incentives as well. The 1991 war with Iraq distracted attention in the west from the enmity between Washington and Teheran. There have long been sporadic indications that detente could be in the wind, but there are few other countries in the world with which the United States shares as deep an underlying basis of bitterness and resentment. What better way – and what more dangerous way – to show the Great Satan that he does not rule the world and had better not mess with Teheran as he did with Baghdad?

Notes

1. By the time the article appeared in print I was working briefly on the National Security Council staff. Someone higher on the staff told me that my article was given to President Carter in one of his nightly piles of official bedtime reading. Whether or not he ever read it, his policy on the issue showed that he did not agree with it.
2. This assumes that nuclear proliferation is a bad thing, a notion that leading realist scholars in recent years have disputed. For arguments that nuclear weapons enforce caution, deterrence, and stability, and should therefore be welcomed rather than shunned, see Kenneth N. Waltz, *The Spread of Nuclear Weapons: More May Be Better*, Adelphi Paper no. 171 (London: International Institute for Strategic Studies, Autumn 1981); and John J. Mearsheimer, "Back to the Future: Instability in Europe After the Cold War," *International Security* 15, no. 1 (Summer 1990).

3. See, for example, Alva Myrdal, "The High Price of Nuclear Arms Monopoly," *Foreign Policy*, no. 18 (Spring 1975).
4. Richard J. Barber Associates, *LDC Nuclear Power Prospects, 1975–1999*, a report to the Energy Research and Development Administration, 1975; Albert Wohlstetter et al., *Moving Toward Life in a Nuclear Crowd?*, a PAN Heuristics report to the Arms Control and Disarmament Agency, April 1976, pp. 4–5, 74–85, 90–111; Norman Gall, "Atoms for Brazil, Dangers for All," *Foreign Policy*, no. 23 (Summer 1976): 175.
5. George Quester, *The Politics of Nuclear Proliferation* (Baltimore: Johns Hopkins University Press, 1973), 69.
6. Wohlstetter et al., *Life in a Nuclear Crowd*, 40–41.
7. See John McPfee, *The Curve of Binding Energy* (New York: Ballantine, 1975), 96–105.
8. See Richard K. Betts, "A Diplomatic Bomb? South Africa's Nuclear Potential," in Joseph A. Yager, ed., *Nonproliferation and U.S. Foreign Policy* (Washington, D.C.: Brookings, 1980). After this article went to press South African President F.W. de Klerk announced that his country had secretly built six nuclear weapons, but then destroyed them before signing the NPT in 1991. Bill Keller, "South Africa Says It Built 6 Atom Bombs," *New York Times*, 25 March 1993, pp. A1, A12.
9. Michael Nacht, *Nuclear Energy and Nuclear Weapons*, report of the 1975 Aspen Workshop on Arms Control, 2; George Quester, "Taiwan and Nuclear Proliferation," *Orbis* (Spring 1974): 140; Senator Abraham Ribicoff with Paul Leventhal, "A Market-Sharing Approach to the World Nuclear Sales Problem," *Foreign Affairs* (July 1976): 766.
10. Albert Carnesale, testimony in U.S., Congress, Senate, Government Operations Committee, Export Reorganization Act of 1976; Hearings, 94th Congress 2d session, 1976, pp. 182–83; Steven Baker, "Monopoly or Cartel?" *Foreign Policy*, no. 23 (Summer 1976): 218; Wohlstetter et al., *Moving Toward Life in a Nuclear Crowd?*, 93–94.
11. Philip Farley, "Nuclear Proliferation," in Henry Owen and Charles Schultze, eds., *Setting National Priorities* (Washington, D.C.: Brookings, 1976), 154, 158; Paul Joskow, "The International Nuclear Industry Today: The End of American Monopoly," *Foreign Affairs* (July 1976): 788, 791, 800.
12. George Quester, "The Shah and the Bomb," *Policy Sciences* (March 1977).
13. Paul Joskow, "International Nuclear Industry," 802. Shah quoted in Norman Gall, "Atoms for Brazil, Danger for All," 176.
14. Brian Jenkins, "Will Terrorists Go Nuclear?" California Seminar on Arms Control and Foreign Policy, October 1975, pp. 8–12, 17–23; Lincoln Bloomfield, "Nuclear Spread and World Order," *Foreign Affairs* (July 1975): 747.
15. Alton Frye, "How to Ban the Bomb: Sell It," *New York Times Magazine*, 11 January 1976.
16. Michael Handel, "Weak States in the International System" (PH.D. diss., Harvard University, 1974), 219.
17. See the arguments in Richard K. Betts, "The Tragicomedy of Arms Trade Control," *International Security* 5, no. 1 (Summer 1980); and Betts, "Whom Should We Arm?" *Brookings Bulletin* 18, no. 1 (Summer 1981).
18. Kratzer testimony in U.S., Congress, Senate, Government Operations Committee, Export Reorganizations Act of 1976: Hearings, 94th Congress, 2d Session, 1976, p. 412.
19. Richard Halloran, "Korean Chief Linked to Illegal Lobbying," *New York Times*, 1 November 1976, p. 22; Rowland Evans and Robert Novak, "Korea: Park's Inflexibility," *Washington Post*, 12 June 1976, p. A19.
20. George Quester, "Taiwan and Nuclear Proliferation," *Orbis* (Spring 1974): 148.

21. U.S. Congress, Senate, Government Operations Committee, Export Reorganization Act of 1976: Hearings, 94th Congress, 2d Session, 1976, pp. 775–83; Senator Abraham Ribicoff, "Market Sharing Approach," 785.
22. See Richard K. Betts, "Nuclear Proliferation After Osirak," *Arms Control Today* 11, no. 7 (September 1981).

II
State-Oriented Approaches

The End of the Cold War and the Future of Nuclear Proliferation:
An Alternative to the Neorealist Perspective

Glenn Chafetz

NUCLEAR PROLIFERATION has threatened international security since the United States first developed and used atomic weapons almost fifty years ago, but the end of the cold war and the disintegration of the Soviet Union have focused new attention on an old problem. Analysts are now scrambling to understand how the uncertainties associated with the new era in international politics will affect the spread of nuclear weapons: Will proliferation accelerate? Who wants the bomb and why? How can the nuclear-weapons states and some nonnuclear-weapons states prevent proliferation? Is the spread of nuclear weapons throughout the world inevitable? Is proliferation necessarily destabilizing? Analysts cannot agree on the answers to any of these questions, but they do observe that there are a number of states which possess the financial and technological capability to acquire nuclear weapons but choose not to do so.[1] This phenomenon leads to agreement that the future of nuclear proliferation is a political, rather than strictly technical, question. Because of the political nature of proliferation, predictions about its future course depends on one's theoretical conception of the character of international relations as a whole.[2]

In the wake of bipolarity, proponents of structural realism or neorealism suggest that nuclear weapons will proliferate more extensively and rapidly than in the past: beyond control.[3] The neorealists argue that the development and acquisition of nuclear weapons is but one means of many to which states turn in an anarchical international environment. The efforts of states to protect themselves will accelerate in the post-cold war era because the bipolar distribution of power which characterized the cold war was inherently more stable than the multipolar arrangement which will characterize the future. Benjamin Frankel argued, "There is an inherent contradiction between welcoming the end of bipolarity and deploring the spread of nuclear weapons," because, "It is precisely the large size and sophistication of the superpowers' nuclear forces, and the avowed willingness of the superpowers to use them on behalf of their clients, that have prevented a more pervasive weapons proliferation."[4] John Mearsheimer said, "The most likely scenario in the wake of the cold war is further nuclear proliferation."[5]

The neorealists base their predictions for accelerated nuclear proliferation on the assumption that the end of bipolarity means the end of cooperation among the developed, democratic states, or core (the states which constitute the North Atlantic Treaty Organization and more widely, the Organization of Economic Cooperation and Development).[6] They assume that the end of bipolarity will mean that the core states will arm against each other and against states on the periphery as well. Thus, Germany, for example, will desire nuclear weapons to deter French aggression and Japan will acquire nuclear arms to deter North Korea. U.S. promises to shield Japan and Germany will no longer satisfy either state in the absence of a convincing Soviet threat. Moreover, because of the absence of a common threat, the United States will no longer have an interest in providing a nuclear umbrella for other states.

Alternative paradigms, including constructivism and neoliberal institutionalism, challenge the neorealist assumptions and propositions about international relations upon which Frankel and Mearsheimer base their predictions about proliferation.[7] Both constructivism and neoliberal institutionalism emphasize the weakness of the strict dichotomy that neorealism has drawn between domestic and international politics.[8] Both call greater attention to the role played by domestic factors in international relations and international factors in domestic policies.[9] I will show specifically that constructivist and neoliberal institutionalist arguments about the pacifying effects of democracy and complex interdependence mean that the number of states fearful or ambitious enough to seek nuclear weapons will decline rather than increase as a result of the end of the cold war.[10]

There are two parts to this argument. In the first part, I summarize some the problems that many theorists find with the neorealist theory on which pessimistic predictions of increased nuclear proliferation are based. In the second part, I will offer a different model of state behavior which draws from recent research by scholars embracing the neoliberal institutionalist and reflectivist perspectives on world politics. Following the work of Jervis, Keohane, and Goldgeier and McFaul, I will examine the likelihood for proliferation in the world's two subsystems: the core and the periphery. I contend in the first section of Part II that the states which make up the core subsystem of the international system constitute a pluralistic security community (PSC).[11] I argue that proliferation is unlikely within this community because the component states by definition do not regard each other as military threats. The absence of threat derives from shared democratic values and a history of institutionalized cooperation. In the second section, I argue that extended deterrence will continue in the PSC despite the disappearance of the Soviet threat. Thus, members of the community like Japan which face nuclear threats from outside the community will continue to rely on the U.S.

nuclear umbrella rather than develop their own nuclear deterrents, and the United States will continue to maintain an interest in providing that umbrella.

Because few states on the periphery enjoy a history of liberal democratic rule and stable cooperation with each other, I argue in the third section of Part II that optimism regarding proliferation is less justified. Even on the periphery, however, the prognosis is not universally bad. The leaders of the states of the postcommunist world understand that they must choose between a nuclear armed rivalry with one another or integration into the Western community of states; the West will not allow both. So far, indications from the former Soviet states of Russia, Ukraine, Kazakhstan, and Belarus provide reason for cautious optimism. Elsewhere, a small number of anti-status quo states actively seek nuclear weapons. Tense relations among these states – Iran and Iraq, India and Pakistan, China and Taiwan – and between these states and the West – Iraq and Israel, the two Koreas, Libya and the United Nations – continue to resemble the neorealist characterization of international relations.[12] Overall, however, relatively few, less developed, undemocratic states possess strong incentives to acquire or develop nuclear weapons, and even there the possibility exists that the vastly more powerful and integrated states in the developed world can forestall this proliferation. I will conclude with a few observations and general recommendations about specific trouble areas and countries.

Some Problems with Neorealism

The neorealists argue that in an anarchical international environment cooperation on matters of security follows from the distribution of power among the actors in the system. Thus, the cohesiveness of the community of states that resisted the Soviet Union can be explained by the bipolar distribution of power which existed at the end of the Second World War. The end of bipolarity means, therefore, that a multipolar competition for power will replace the cooperation that marked relations among Western states during the cold war.[13] Consequently, as Frankel's comment suggests, the neorealists believe that absent the little order that existed in the international system as a result of bipolarity, no force can prevent the tendency of states to seek greater power through the acquisition of nuclear weapons.[14]

I do not believe that the collapse of the Soviet Union will lead to the disintegration of the pluralistic security community because I do not accept the neorealists' assumptions and propositions about the basis for the anti-Soviet coalition. Unlike the neorealists, I believe that the Soviet Union's expansionist and hostile ideology rather than its power provided the primary basis for the common perception of the Soviet Union as a threat. I appreciate that the cohesion of the anti-Soviet alliance varied in response to different perceptions of the strength and immediacy of the Soviet threat, but all political communities, including those at the subnational and national level, experience variations in cohesion in relation to perceptions of outside threats. The elimination of such threats does not necessarily lead to the disintegration of the community. Recent research on the neorealist assumption of anarchy and conceptions of the balance of power supports my argument.

Anarchy

A number of scholars challenge the neorealist assumption that anarchy characterizes international relations.[15] Drawing on the work of Inis Claude, Milner notes that this assumption overstates the lack of order marking international relations and implicitly understates the lack of order marking domestic politics. First she finds that theorists have inferred two meanings for the term, neither of which accurately characterizes the international system. One meaning for the term is a lack of order, but international relations theorists, among them Hedley Bull, Kenneth Oye, Robert Gilpin, Robert W. Tucker, and Kenneth Waltz, consistently note the orderliness of international relations.[16] Bull, for example, sees order in international society, arguing that it "has always been present in the modern international system because at no stage can it be said that the conception of the common interests of states, of common rules accepted and common institutions worked by them, has ceased to exert an influence."[17] Waltz sees it in the ubiquitous balance of power.[18]

The other meaning of anarchy implied in the literature is the lack of government. This is problematic as well. Defining government as having a monopoly on the legitimate use of force, as Waltz does, presents difficulties because citizens within most countries, including the United States, possess the legitimate right to self-defense. Milner asks, "when the right to use force legitimately (under certain circumstances) is diffused to 240 million people, can a government in Waltz's terms be said to exist?"[19] Thus, such a monopoly probably does not define a government.[20]

Milner notes further that the legitimacy of the uses of force is an international as well as domestic issue. The United Nations coalition action to reverse Iraq's invasion of Kuwait was certainly more legitimate than Iraq's invasion. The former corresponded with accepted norms of international conduct; the latter did not.[21] Milner argues, "This conception of what international politics lacks – a monopoly on the legitimate use of force – is not as clear as it seems, since governments lack such monopolies and since the legitimacy issue arises in international as well as domestic politics."[22]

The third problem with Waltz's definition of government is that "it reduces both international and domestic politics to the use of force."[23] Quoting Weber, on whom Waltz relies, Milner recognizes that this definition is too narrow: "'The threat of force, and in the case of need its actual use, is the method which is specific to political organization and is always the last resort when others have failed.'" It is difficult, if one accepts Weber's definition, to see the distinction between international and domestic politics.[24]

Other scholars base their definitions of government on the existence of laws and institutions.[25] Few scholars would dispute, however, that international institutions and a body of laws, norms, and customs do exist. Instead, they argue that the existing institutions, laws, norms, and customs carry no authority. They are illegitimate. Milner concludes that when Waltz and other theorists argue that anarchy characterizes the international system, they mean that international governing institutions and laws lack legitimacy.[26] To lack legitimacy (defined as some ability to elicit compliance based on the perceived validity of an order rather than the threat of force) is not, however, necessary to lack government.[27] There are a number of governments, including Lebanon, South Africa, Cambodia, Iraq, Turkey, Moldova, and Bosnia, that do not exercise legitimate authority over some of their inhabitants.

Moreover, it is as inaccurate to assert that the international environment is distinguished from the domestic environment by a lack of legitimacy as it is to claim that the difference inheres in the former's lack of government or order. Thomas Franck and others also find that the power of international norms and customs to elicit compliance very often exceeds the power of domestic legitimate arrangements.[28] That over three times more civil than international wars have been fought in this century and that more people have died in the former than the latter provides one indication of the relative legitimacy of the international and domestic environments.[29] The legitimate authority exercised by the Concert of Europe over its member states suggests another.[30] The settlement of the November 1992 U.S.-EC trade dispute, and the United Nations interventions in Iraq, Cambodia, Bosnia, and Somalia provide further evidence that in many cases the international community can act with greater legitimate authority than domestic governments.

The Balance of Power

Despite the preceding examples of post-cold war international institutional problem solving, the neorealists argue that the balance of power explains the relatively greater peace and stability of the international environment. Both Waltz and Mearsheimer argue that the bipolar balance of power, which existed from 1945 until the recent collapse of the Soviet Union, served as the basis for the international order that allowed more domestic than international wars. Frankel argues that the end of the bipolar order will mean more nuclear proliferation. Recent research, however, suggests otherwise.

Balance of power theory provides a useful starting point for understanding the security dilemma that plagues many states, but even neorealist scholars recognize the difficulties presented by Waltz's spare theory. Stephen Walt contends that states "balance against several different kinds of threats, and not just against power alone."[31] According to Walt, the combination of military power, geographic proximity, and intentions compose a threat.[32] In trying to enrich Waltz's spare theory by looking at intentions, Walt implicitly introduces domestic factors.[33]

Intentions are necessarily based on perceptions, and perceptions of the same phenomena vary according to individual psychological, organizational, and societal-cultural factors.[34] Many of these factors are rooted in domestic political structures and ideologies.[35] Moreover, scholars in the field of political psychology observe that decisionmakers are most comfortable with and least threatened by that which is familiar.[36] It is only natural then, that leaders of democratic states will hesitate to attribute hostile intentions to other democratic states and will be quicker to find evidence of hostility in the actions of states governed according to principles avowedly hostile to democracy.[37] Power and capabilities may contribute to decisionmakers' perceptions of the intentions of other actors in the international system, but as Jervis writes:

> Statesmen are often more worried about how the other acts than what it does. That is, the actual gains the other makes – or the losses it inflicts on the state – are often less troublesome than the methods by which they are pursued. A state is likely to be seen as a threat if it displays a willingness to ignore accepted procedure, a disregard of what are usually considered the legitimate rights of others, and an exceptionally high propensity to accept risks in order to improve its position.[38]

American and Soviet policymakers defined and understood the bounds of what was legitimate and acceptable in very different, and ideologically determined ways.[39] Thus, American perceptions of Soviet hostile intentions in the aftermath of the Second World War derived not from Soviet power, but from what the Soviets did with that power; Moscow's repression of democratic movements in Eastern Europe conflicted with the promises Stalin made at Yalta and Potsdam to permit democratization. Western estimates of Soviet strength followed rather than preceded perceptions of Soviet hostility.[40] Had states considered increases in military power as the primary basis for judging other states' intentions, it would have been logical for the United States to target Britain and France upon their acquisition of nuclear weapons and for the latter two to target each other, but no such retargeting occurred. The nuclear powers' perceptions of one another were rooted, therefore, in domestic conceptions of what was legitimate, rather than the other's power.

Ted Hopf's study of international relations in sixteenth century Europe suggests further reason to doubt the explanatory capacities of the balance of power in general and polarity in particular.[41] Looking at multipolar Europe from 1495 to 1521 and bipolar Europe from 1521 to 1559, Hopf finds that changes in polarity cannot account for the constant instability across a polarity change in the system. He shows instead that strategic beliefs, the technical offense-defense balance, and the cumulativity of power resources explain the instability of the period. He concludes that the strategic defense dominance provided by nuclear weapons accounted for the absence of war in Europe during the cold war. Strategic beliefs, that is, ideas about the likely reactions of allies and adversaries to certain events, explain military action on the periphery during the last forty-five years.[42] Because polarity is irrelevant to stability, and the balance of power cannot sufficiently explain it, some other factor(s) must account for the cooperation of the Western, democratic states during the cold war era.

Prospects of Proliferation in the Post Cold-War World

A. Threats from within as an Incentive for Proliferation in the Core

Because of the common values and shared identification within the entire core, the nonnuclear-weapons states which belong to that community do not perceive from other members threats strong enough to induce them to acquire nuclear weapons. Shared democratic values and the experience of the cold war forged a pattern of cooperation within the core that is stronger and more durable than a temporary alliance in response to a common threat, in short, a pluralistic security community. The conditions under which a pluralistic security community will form closely resemble the conditions under which a security regime will form. A regime is the set of "principles, norms, rules, and decision-making procedures around which actor expectations converge in a given issue-area."[43] Jervis argued in 1982 that no regime exists in the security area because he did not believe that four conditions necessary for a regime could be met: "First, the great powers must want to establish it – that is, they must prefer a more regulated environment to one in which all states behave individualistically."[44] In particular, they must all accept the status quo and reject any systemic changes that depend on the use or threat of force. Second, each major actor, defined here as one with enough of a combination of military, economic, and political power to alter the status quo for a significant period of duration, must believe that the other major actors value mutual security and cooperation to the same degree. Third, no major actor can believe that expansion provides the best means for ensuring its security; and finally, all the major powers must view "war and the individualistic pursuit of security as costly.[45]

The democratic developed states meet the four criteria. First, there are five great powers within the core: The United States, Germany, France, Britain and Japan. The United States is a great power because of its military and economic strength. Its political power derives from its military and economy and from its history of postwar leadership. Japan and Germany qualify as great powers because of their great economic power and their alliance with the United States. Britain and France qualify because of their military power, their permanent seats on the United Nations (UN) Security Council, and their alliance with the United States. The continued existence of the North Atlantic Treaty Organization (NATO), the Group of Seven (G-7), the European Community (EC), the General Agreement on Tariffs and Trade (GATT), the Conference on Security and Cooperation in Europe (CSCE), the United Nations, and the U.S.-Japanese security treaty indicates that these states, as well as many others, prefer an international environment regulated by more than the balance of the short-term pursuit of power.

Second, cooperation in the several international institutions cited above also provides evidence of the relatively homogeneous appreciation of mutual security and cooperation among the developed republics. These states often share military technology and participate jointly in military exercises.[46] French and German soldiers serve together in the same combat brigade.[47] Moreover, none of the developed republics prepares for war against any of the others. None deploys troops against any other. Though stresses and strains like the November 1992 U.S.-European Community dispute over agricultural subsidies do occur, they do not lead to war or rupture institutionalized forms of cooperation. Continued political, economic, and military cooperation and the absence of preparations for war also indicate that the core states view war as costly and do not believe expansion provides the best means to ensure their security.

Though meeting the four criteria does not necessarily mean that the actors in the PSC can reach agreement on all the details of a regime, shared liberal democratic values, and the long institutionalized experience of cooperation among the core states have made conflict, and thus proliferation, among them unlikely. Liberal democratic states – those republics which share the characteristics of unrestricted political participation, equality of opportunity, rule of law, and respect for human rights and private property – do not fight each other.[48] This does not mean that liberal democracies do not fight; they have a long history of fighting non-liberal states, but the difficulty involved in finding an example of two democracies which have fought a war suggests the validity of the principle of inter-democratic peacefulness.[49] Theorists have advanced a number of explanations to account for the empirical evidence. Four are particularly compelling. David Lake drew from the microeconomic theory of the state to explain the absence of war among democratic states:

> State rent seeking creates an imperialist bias in a country's foreign policy. This bias is smallest in democracies, where the costs to society of controlling the state are relatively low, and greatest in autocracies, where costs are higher. As a result of this bias, autocracies will be more expansionist and, in turn, war prone. In their relations with each other, where the absence of this imperialist bias is manifest, the relative pacifism of democracies appears.[50]

Michael Doyle followed Immanuel Kant in his three-part explanation of the separate liberal peace. First, he argued that public participation in government ensures that caution will be taken when deciding to go to war because it is the public that must necessarily suffer the consequences of any violent conflict. Thus liberal democracy ensures that the decision to fight must be popular, rather than the result of the caprice of an unaccountable leadership.[51] Second, republics project the tolerance and accommodation of different political viewpoints which characterize their domestic orders on to relations with other republics. As Doyle put it, "Domestically just republics, which rest on consent, then presume foreign republics also to be consensual, just, and therefore deserving of accommodation. The experience of co-operation helps engender further cooperative behavior when the consequences of state policy are unclear but (potentially) mutually beneficial."[52] The claim is supported by the dynamics of the Soviet rejection of cold war policies. Aleksandr Yakovlev, one of Mikhail Gorbachev's closest advisers from 1985-91, responded to calls for intervention to reverse the 'loss' of Eastern Europe by asking, "... if we proclaim principles of freedom and democracy for ourselves can we deny others the same thing?"[53] Finally, economic interdependence exerts a greater pacifying effect on republics than autocracies because liberal states view the market, rather than coercive state policy, to be the source of economic outcomes which affect the republics' fortunes.[54]

B. Peripheral Threats as Incentives for Proliferation within the Core

That the nonnuclear weapons constituents of the PSC do not fear the other members of the PSC enough to want to arm against them does not mean that the former perceive no threats. In terms of predicting proliferation within the

PSC, the important question is whether or not the nonnuclear weapons states of the PSC believe that the nuclear weapons members will continue to provide extended deterrence. Both the deductive and the empirical evidence suggests that they will.

Neither Japan nor Germany, the two countries provoking the greatest fears, have renounced their status as nonnuclear-weapons states. Both countries have positively denied any plan to acquire nuclear weapons.[55] Japan has maintained that it would continue to abjure nuclear weapons despite the existence of threats from North Korea and China.[56] Despite Japanese denials, the North Koreans have accused the Japanese of trying to develop nuclear weapons, and Pyongyang has pointed to Tokyo's decision to import more than a ton of plutonium from France as evidence.[57] There is more evidence, however, to support Tokyo's claim that it will continue to abjure nuclear weapons. First, Japan had the capacity to develop nuclear weapons before; it did not need to purchase the French plutonium to construct a bomb. Second, when Japan decided to pursue a breeder reactor program in 1966, uranium was scarce and expensive, Japan was accumulating a great deal of plutonium waste from its conventional reactors, and scientists had more confidence in the possibilities of breeder reactors. In view of Japan's dependence on foreign oil and uranium for its energy needs, it is reasonable to believe Tokyo's avowal that it wanted the plutonium in order to provide cheap, abundant nuclear power.[58] Third, the density of its population and its status as the only country to suffer an atomic attack have made the Japanese population politically and culturally averse to possessing nuclear weapons. Even having nuclear weapons on U.S. Navy vessels in Japan has created political problems for Japan's leaders.[59]

German Chancellor Kohl responded to the suggestion that Germany should acquire nuclear weapons with an explicit affirmation of his belief in extended deterrence:

> No, under no circumstances. Why should we have them? I am very happy that my French friends have them. I live 40 km from the French border. It does not disturb my piece of mind to know that seven hours flying time from me the U.S. president has the decision-making power over nuclear weapons to protect us Germans and that 40 minutes from my home there is a French president who has the same powers. We must state the facts. We do not need them at all ... .[60]

Because common perceptions of the Soviet threat can no longer account for the continued German and Japanese faith in their nuclear capable allies' provision of extended deterrence, some other factors must account for the former's strategic beliefs. Like James Richter, I contend that "Rapid changes in the international system may force states to reconsider their world roles, but these states cannot completely discard and transform the myths and institutions left over from their previous position."[61] Richter explained how pervasive myths and powerful institutional interests within the Soviet Union and the United States had prevented earlier attempts to end the cold war.[62]

Constructivist scholars support Richter's thesis when they show that the identities and interests of states are not exogenously given, but rather develop as a result of interactions with other states.[63] Just as individuals establish identities according to relationships with other individuals and society – husband, brother, professor – states establish identities according to relationships with other states and international society – leader of the free world, challenger of the status quo, good citizen.[64] States identify their security competitively, individualistically, or cooperatively with other states based on two factors: the values that give the governments of states their legitimacy and the history of interaction with other states.[65] Thus, because the government of the Soviet Union depended on hostility to capitalism for its legitimacy, it was compelled to identify Soviet security interests competitively, that is negatively, with the United States and its allies. As a consequence of the competitive security identification, each side took actions to diminish the other's security. These actions in turn mutually reinforced perceptions of threat and led to the creation of role conceptions and institutions to support the evolving identity of each actor: leader of the free world for the United States and vanguard of communist liberation for the Soviet Union.[66] These role conceptions became so embedded in the domestic politics of each state that neither side could transcend the perception of hostility despite changes in the international system and the best attempts of various leaders to alter the relationship.[67] The domestic actors in each country who resisted changing their respective roles were rewarded while those who attempted transformations – Jimmy Carter and Nikita Khrushchev, for example – were punished. It was not until a crisis of ideological legitimacy, coinciding with a generational shift in Soviet leadership, that the Soviet Union was able to transform its identity and transcend the hostility of the cold war.[68] Even then, many Western leaders were slow to recognize the fundamental nature of the Soviet transformation.[69]

The same theory of identity formation and social interaction that explains continued interstate hostility in the face of demands for conciliation also accounts for continued cooperation in the face of systemic pressures for egoistic competition. Just as the arms race and other mutually perceived aggressive actions confirmed hard-line beliefs in both the Soviet Union and the United States and supported and hardened the myths and institutions of the cold war adversarial roles, the positive interactions among the Western allies during the same period led them to identify their security collectively, rather than individualistically.[70] The cold war rewarded those policymakers and domestic actors whose beliefs supported close cooperation.[71] Of the theories within the core that induce cooperation, one of the most important is the frequently repeated conviction that liberal democracies naturally ally. Indeed, both the happy reaction in the West to the collapse of communism in the Soviet bloc and the desire to encourage democracy in those countries is predicated on the theory of democratic solidarity. Within the PSC, statements of belief in a special democratic community are common within the speeches of its leaders.[72] Evidence that this belief is spreading can be found in the statement of Russian foreign minister Andrey Kozyrev, who said that democratic states "are drawn to other democratic states and see them as friends and potential allies."[73] Though some observers may scoff at these statements as mere rhetoric, the evidence of the power of this rhetoric on mass beliefs and policy is compelling.[74]

The power of shared values and forty-five years of institutionalized cooperation to transform the identities of the Western liberal democracies becomes evident when looking at earlier core attitudes toward nuclear proliferation. As George Quester has pointed out, France, Italy and Germany did not react positively to the nonproliferation regime when it was first proposed but are now among its most fervent supporters. As they began to identify their security more collectively and less individualistically, their interests regarding proliferation have changed.[75] Similarly, the only examples of nuclear proliferation in the evolving pluralistic security community occurred early in its history when not all the members shared the same values. David Schwartz has argued persuasively that the divergence between the United States, on one side, and Britain and France on the other, over imperialism – a clash of values – provided the main impetus for the development of the independent European nuclear forces. Because the Eisenhower administration believed imperialism to be incompatible with liberal democracy, it refused to support the British-French-Israeli adventure against Egypt in 1956, to the point where it applied political and economic pressure on London and Paris at the same time that the Soviet Union threatened them with nuclear weapons. Britain and France derived from the Suez affair the lesson that they could not always depend on U.S. assistance against the Soviet Union and consequently went forward with plans to develop nuclear deterrents independent of Washington.[76] The United States did not, however, see its refusal to support its allies over Suez as a lack of commitment to their security; rather it viewed its action as a sign of its disapproval of imperialism. Thus, the only examples of proliferation within the PSC resulted not from the manifestation of nationalism but rather from a split over ideals that has long since vanished. Increased French and British confidence in the American commitment to their security is evident in that the further the Suez crisis faded in the memories of the allies, the less compelled the British and French have felt to modernize and increase their nuclear forces.

Indeed, as a result of the inertial influence of myths and institutions left over from the cold war, the U.S. lead role in providing for core defense, including extended deterrence for the PSC nonnuclear weapons states, is unlikely to change.[77] Washington has given no indication that it intends to abandon the obligations it assumed during the cold war. Though German and Japanese statements that they will not demand a change in the present nuclear security arrangements are not conclusive, evidence of the truthfulness of their stated intent inheres in the existence of strong domestic impediments in both countries to any attempt to deviate from their non-nuclear-weapons status. Japan, for example, has written into its constitution its desire to be accepted by the international community.[78] Any indication that Tokyo is seeking nuclear weapons will likely involve a constitutional scandal, more so considering the strong pacifist sentiments that continue to exert influence in Japanese politics.[79] Germany operates under many of the same domestic and international conditions that pertain to Japan, including constitutional limitations on the use of force, a strong pacifist movement, and domestic opposition to nuclear weapons.[80] In short, having internalized core norms, both Germany and Japan would suffer negative domestic repercussions from a decision to violate the nonproliferation regime.

C. Proliferation on the Periphery

The states on the periphery have either little or no experience with liberal democracy and thus have not yet established among themselves the norms of peaceful cooperation which govern relations among the states of the core.[81] In this subsystem, therefore, states possess sufficient amounts of fear and ambition – the root causes of any arms program – to make further nuclear proliferation likely.[82] Even within this world, however, three factors can mitigate the tendency of these states to seek security by means of acquiring nuclear explosives. First, the expansion of liberal democracy in the post-communist states and throughout Latin America, Asia, and Africa increased the number of governments that will work to subordinate nationalist fears and ambitions in order to reap the benefits – higher standards of living, greater political stability, and peace – of full integration into the Western economic-political system.[83]

Second, because for the first time in history no state-sponsored ideological movement threatens the security of the core states, they have the relative luxury of being able to devote the bulk of their energies to goals other than providing for their immediate security.[84] Thus, the core will no longer have to make the choice between denying supplies to potential proliferants and losing allies against the Soviet Union. Furthermore, as the United Nations war against Iraq demonstrated, the core can marshal overwhelming military, political, and economic power in defense of its norms, including nonproliferation.[85]

1. *Integration of the Periphery into the Core* Successful integration of the periphery into the core depends on the following: first, the stability and durability of democratic institutions within the peripheral states; second, the willingness of the pluralistic security community to provide political and economic assistance to effect stabilization of those institutions; and third, the effectiveness of various mechanisms to effect peripheral-state compliance with core norms.

As shown above, the democratic states of the PSC cooperate more in the area of security than do nonliberal undemocratic states. As coups in Haiti, Thailand and Peru, and economic instability throughout the former Soviet Union and Eastern Europe demonstrate, the democratic institutions in many of the peripheral states face a number of internal threats. If democracy fails in states that possess the ability to disrupt the nonproliferation norm, their governments will be more likely to disrupt the norm because they will undoubtedly perceive more threats from the core democracies as well as from each other. For this reason, the core states will serve the cause of nonproliferation by providing aid to young democracies that have the power to disrupt the nonproliferation norm. Currently, the European Community and its members, Japan and the United States are funneling some aid to the former communist states, although the effort is small relative to the potential costs of failure and benefits of success.[86]

Shared democratic values and financial support, however, sometimes will not by themselves induce a given state to repudiate nuclear weapons if it is surrounded by hostile neighbors and does not trust the core states to provide for its security, as Israel's predicament shows. Moreover, it may be that even explicit security guarantees from the West could not induce Israel to relinquish its nuclear arsenal.[87] The level of distrust between Israel and its neighbors and the geographical distance between Israel and its potential rescuers is too great. Only the most sincere and credible extended nuclear deterrence – total inclusion into the core – could move Israel to comply with the nonproliferation treaty, and that is unlikely in the foreseeable future.

The core states can, however, induce some peripheral nuclear-capable states to relinquish their weapons. Among the postcommunist states, Ukraine, Kazakhstan, and Belarus have all categorically promised to return to Russia for destruction all of the nuclear warheads which they acquired when the Soviet Union disintegrated.[88] Moreover, all three have already returned the tactical nuclear weapons left on their soil to Russia, and Belarus and Kazakhstan have ratified the Strategic Arms Reduction Treaty (START 1).[89] This treaty calls for Kazakhstan, Belarus, and Ukraine to send the strategic nuclear missiles to Russia for destruction. In return, the West has provided aid, reduced restrictions on the transfer of sensitive technology, and taken steps to include cooperative states more deeply in the PSC security arrangements.[90] Specifically, members of the PSC created two institutions, the North Atlantic Council and the Forum for Cooperation on Export Controls, which will facilitate the eventual integration of the former communist states into NATO and the Coordinating Committee on the transfer of technology (COCOM). In general, the twenty states that emerged from the demise of the Warsaw Treaty Organization and the disintegration of the Soviet Union want to share the benefits of membership in the Western political, economic, and security community. Their interactions with the West illustrate the effectiveness of social proof and membership as mechanisms for encouraging compliance with nonproliferation norms.[91]

First, many of these states are part, or members, of Europe. Most, including Russia, Belarus, Ukraine, Kazakhstan, and the former Eastern Bloc countries are now members of the Conference on Security and Co-operation in Europe, and most explicitly or implicitly accept nonproliferation as a condition of joining the Western community of states. Despite their formal membership in a Western security organization, both Kazakhstan and Ukraine initially petitioned the core states for explicit security guarantees: Kazakhstan wanted a guarantee in return for a promise to abide by the Non-proliferation Treaty; and Ukraine had sought a guarantee from both the United States and Russia in return for compliance with its promise to give up its nuclear arsenal.[92] The West held out the prospect of deeper and wider levels of membership in the core to induce both states to comply without providing explicit guarantees. For Kazakhstan, the anticipated benefits of increased participation in the dominant international society – the core – seemed sufficient to persuade it to eschew immediate behaviors antithetical to the norms of the core.

The records of Ukraine and Russia are mixed. Kiev wants firmer guarantees of its security from the West and more financial compensation for the missiles on its territory.[93] Moscow is selling Russia's overabundance of military hardware for the immediate gain of much needed hard currency. It has made deals with China and sold diesel submarines to Iran. There is no confirmation, however, of any Russian decision to violate the nuclear nonproliferation norm. Russian leaders submitted to very intrusive inspections of its nuclear weapons program and allowed its newly unemployed scientists and technicians to perform research and dismantle weapons under the auspices of an institute formed through the combined efforts of the United States, the EC, and Japan.[94] In addition, Russia signed the Warsaw agreement to limit the supply of materials and expertise which can lead to nuclear proliferation.[95]

In addition to formal membership, social proof has been an effective mechanism for encouraging compliance with the nonproliferation norm for the nuclear weapons-capable ex-Soviet states. According to this principle, many states decide how best to adapt to an environment based on their observations of how other, more successful states adapt.[96] Kenneth Waltz describes the norm of social proof when he argues "each state duplicates the activities of other states, at least to a considerable extent."[97] In a different formulation, the leader-states more actively persuade other states to comply with established norms.[98] The end of the cold war is uncharted territory for all states. The new postcommunist governments, and even other states, will determine their course based in part on how more established, successful states behave, and the more established states will encourage this dynamic. Thus because Germany, Belgium, Sweden, Denmark and Japan continue to refrain from building and possessing nuclear bombs, Belarus will do the same, and the United States will urge that Kazakhstan and Ukraine do as well. As an example of social proof with regard to conventional forces, Leonid Privalov, Deputy Chairman of the Belarussian Parliamentary Commission on National Security, said that Belarus would model its military after Denmark's; because Denmark maintained a force of 27,000 out of a population of five million, Minsk would field a force of "54,000 out of its population of ten million." Privalov added, "Today, Belarus does not see any enemy that we might have to fight – neither at its borders, nor anywhere beyond."[99]

Of the sixteen formerly Soviet and Soviet-bloc states not discussed above, none possesses the combination of incentive and economic and technological capacity which the development of nuclear weapons demands. The war between Armenia and Azerbaijan, based on a territorial conflict and long-simmering ethnic and nationalist hatreds, provides both of those states with the obvious incentives to augment their military power. Barring, however, the transfer of knowledge and materials from Russia, it seems unlikely they will acquire nuclear arms in the foreseeable future. Hungary and Rumania also have long-standing claims on each other's territory, but their desire to integrate into the Western core community exceeds the strength of their dispute. Neither country, therefore, would likely violate norms that would defeat its integrationist aims.

2. *Dominance of the Periphery by the Core* In cases when either the core or peripheral states wish to avoid close political, economic, and military ties – that is, they cannot be socialized, for whatever reason – the core states can punish the peripheral states that refuse to comply with the nonproliferation

norm.[100] Moreover, the end of cold war constraints on core behavior has increased the flexibility of the PSC to act at the same time that it has limited the options for potential violators of core norms. As Prime Minister Mahathir Mohamad of Malaysia observed in a speech to the 1992 gathering of the Non-Aligned Movement, "Where before we had the option to defect to the other side, now we have none ... .Our new option is to submit or resist."[101] He suggested further that movement change its name "to reflect its new role in a unipolar world."[102]

Before resorting to punishment, the evidence suggests that the suppliers of nuclear technology – mostly in the core – will work more carefully now than during the cold war to prevent proliferation to those states outside the PSC. In particular, membership in the group of signatories to the Non-Proliferation Treaty (NPT), in NATO, in the Conference on Security and Cooperation in Europe, the Nuclear Suppliers' Group (NSG), the UN, and other groups in which nonproliferation is a goal, carries with it the obligation to work toward that goal. Membership, as indicated by contracts, treaties and alliances, performs three functions in support of any norm. First, it creates an incentive to enforce the norm because defections are viewed as inconsistent with membership in the group, and are therefore illegitimate. Second, because membership is self-selected, like-thinking members reinforce mutual adherence to the norm. Finally, the very process by which the group forms itself defines and clarifies the expectations of the participants.

Thus, among members of the PSC, membership in the Nuclear Suppliers Group signifies a willingness to observe its guidelines. Discoveries about the extent to which the core states' contributions to Iraq's covert nuclear development program violated the guidelines spurred the NSG to strengthen its guidelines, and by extension, its requirements for membership. On 3 April 1992, twenty-seven nations, twenty-one of which belong to the PSC, signed an agreement in Warsaw which set out, for the first time, detailed rules to limit the international sale of material and machinery that can be used for either peaceful purposes or for constructing nuclear weapons. The restricted products include computerized machine tools, high-strength steel, and industrial robots. In order for any state to receive the goods restricted under the NSG agreement, it must comply with International Atomic Energy Agency's safeguards on the use of all fissionable material.[104] By adhering to the voluntary agreement and membership in the NSG, the signatories indicate their willingness to cooperate with the nonproliferation norm and must suffer the injury to their reputations, self-perceptions, or membership in the international community if they violate the norm.

The Warsaw agreement provides a further example of how the end of the cold war will strengthen the ability of the core constituents to police themselves in regard to compliance with nonproliferation norms because none of the compliant states need worry that enforcement will detract from the formerly primary task of containing the Soviet Union. During the cold war such an agreement would have been difficult or impossible. Secretary of State Henry Kissinger exemplified this problem in 1976 when he refused to allow the United States to join the Soviet Union in placing pressure on France and Germany for selling reprocessing plants capable of producing weapons-grade plutonium. He said that joining an adversary, with the consequent alienation of allies, demanded too much for the benefit of nonproliferation.[105] Though the dilemmas created by the cold war have disappeared, the prospect of further nuclear proliferation remains. Where integration and the denial of necessary supplies fail to dissuade potential proliferants, the PSC has sufficient power and flexibility to punish violators.

The power and cohesion of the core states makes them almost strong enough to constitute a global, hegemonic security regime.[106] It falls short because one of the major powers in the system – China – does not accept the international status quo. Beijing perceives itself to be ideologically isolated and fears that external democratic actors are trying to foment anticommunist revolution.[107] Nevertheless, China formally accepts the nonproliferation norm – it signed the NPT in March 1992 – because it desires the benefits of trade and economic assistance from the West, and because it maintains an interest in denying nuclear status to Kazakhstan, Taiwan and India.[108]

Moreover, despite the existence of a revolutionary major power in the global system, the ability and propensity of the core to act in defense of its norms will increase, rather than decrease in the aftermath of the cold war. During the cold war, the overriding concern of both superpowers with containing each other distracted them from preventing proliferation in two ways. First, the Soviet threat preceded nonproliferation in the hierarchy of NATO (and NATO-aligned) interests. The United States refused to pressure both supplier-states and consumers of weapons material and technology if their role as ally was deemed important enough. Thus, the United States overlooked Islamabad's obvious progress toward acquisition of nuclear weapons because of Pakistan's service as a front-line state in the battle against the USSR in Afghanistan. The precedence of the Soviet threat precluded punishment. Following the end of the cold war the United States imposed economic sanctions on Pakistan.

The cold war also inhibited Washington and Moscow from trying to prevent proliferation to each others' allies for fear of provoking each other. Once the cold war ended, however, the United States began to pay greater attention to the dangers of nuclear proliferation. In March 1991, President Bush cited Iraq's progress toward acquisition of nuclear capability as a motivating factor in the continued U.S. pressure on Iraq.[109] In March 1992, according to U.S. Secretary of State James Baker, nonproliferation ranked among United States' highest foreign policy priorities.[110] President Clinton's nominee to oversee policy toward the former Soviet Union also said that nuclear nonproliferation was a top priority of the Clinton administration.[111] Furthermore, the establishment of the U.S. Central Intelligence Agency's (CIA) 100-person Non-Proliferation Center reflects the increased Western commitment to the nonproliferation goal.[112] The core-dominated United Nations continues to punish Iraq for its violations of the Non-Proliferation Treaty, though it is not clear that the punishment would have been as severe if Iraq had not invaded Kuwait. Nevertheless, for the first time in its history, all five permanent members of the Security Council are signatories of the NPT, and, reflecting the core consensus that nuclear proliferation represents a threat to world peace, the Security Council has become much more active in its efforts to halt proliferation.[113]

North Korea serves as a better test of Western commitment. Pyongyang appears to be very close to the development of nuclear weapons.[114] The United States and South Korea have taken steps to elicit North Korea's compliance with the NPT, including removing all U.S. nuclear weapons from the Korean peninsula. North Korea responded positively, but Pyongyang remains ideologically isolated and hostile to its southern neighbor and the West.[115] Moreover, there are indications that Pyongyang is trying to deceive the IAEA.[116] Thus, North Korea's continued violation of the NPT in the form of a covert nuclear weapons' program may require the PSC to resort to punishment to ensure compliance. Punishment succeeded against Kazakhstan when the United States refused to release aid to Alma Ata in April 1992 because of its contradictory statements regarding its nuclear intentions.[117] North Korea, however, stands less susceptible to Western inducements and sanctions than Kazakhstan: Kim Il Sung can withstand Western threats because he has already intentionally isolated his state from the international community. Thus, if Pyongyang remains obstinate, the core states will have to resort to military means to destroy its nuclear capacity, and, although the United States is considering military action, it is not clear that the either the United States, alone, or as part of the PSC, has the will to pursue that option.[118] Still, the NPT can survive if North Korea acquires a nuclear capability as long as the United States can convince South Korea and Japan that it continues to guarantees their security.

Conclusions and Recommendations

The end of the cold war has not driven violent international conflict into obsolescence, nor has it eliminated the incentive of many states to acquire more powerful and frightening weapons, but it has produced a situation in which the liberal democratic, developed states can dampen the urges to conflict and the danger of proliferation in much of the world. Within the core, further proliferation will occur only if democracy in one or more of the core countries succumbs to an internal challenge. This is very unlikely.

On the periphery, the status quo actors from the core can keep deviants "small in number, isolated, and manageable, the three conditions which James Keeley describes as necessary for an effective and legitimate nonproliferation regime.[119] In Latin America and sub-Saharan Africa, because of the dominance of the United States in the former and because of the weakness of states and illegitimacy of almost every state's borders throughout the latter, incentive for international conflict, and thus proliferation, is unlikely.[120] South Africa's transition to black majority rule has eliminated the source of its conflict with its neighbors and led it embrace the nonproliferation regime as a nonnuclear state.[121] Moreover, Brazil and Argentina have also retreated from aggressive nuclear weapons-development programs and appear ready to accept the intrusive safeguards and inspections necessary to build mutual confidence.[122]

The threat of proliferation remains high, however, in an arc running from North Korea through the Middle East. In these areas, ideological, historical, and nationalist incentives for conflict abound. Violations, especially unpunished violations, of the nonproliferation regime in these areas will not destroy the regime, but they will erode it.[123] While Kenneth Waltz and other theorists correctly argue that limited proliferation will not necessarily destabilize international security, many statesmen do not contemplate the prospect of increased nuclear proliferation with equanimity, and if decision-makers believe that proliferation presents dangers and are prepared to act on those beliefs, then indeed the spread of nuclear weapons does destabilize international security. The Israeli raid on Osirak, Iraq in 1981 demonstrates the potential of proliferation to increase regional tensions, and the discovery of Iraq's progress since then has heightened tensions worldwide.[124]

The core states should pursue a mix of five policies. First, in cases where mutual interests and values exist, the core states should facilitate integration of the peripheral states into the core system as it is already doing, albeit with insufficient effort, in Eastern Europe and the former Soviet states. Second, where long-standing disputes within the periphery continue to fester, and there is a danger of proliferation among the aggrieved parties, the core states should redouble their efforts to effect settlements. Washington's good offices in the current Middle East peace talks exemplify this track. Though it is probably not possible to convince Israel to relinquish its nuclear arsenal, a mutually satisfactory settlement may persuade Syria and other Arab nations to renounce atomic weapons.

Third and fourth, if a settlement proves to be impossible, the core states should be prepared to pursue two options: either to insist on the most intrusive inspections and prepare to enforce cooperation with the Non-Proliferation regime or to abandon hope of settling the conflict. The first option will be necessary in the event that the core states believe that the revolutionary aspirations of one or more of the aggrieved actors – that is, Iraq, Iran, or North Korea – exceed the confines of the periphery and threaten the core. The second option, while regrettable, would be more appropriate in cases where nationalist aspirations were limited, spillover unlikely, and interference rather than inaction posed a greater threat to core security. The Pakistani-Indian conflict falls into this category because neither state has provided the international community with any indication that their security dilemma can be mitigated by mediation or force.[125] New Delhi believes that Pakistan actively foments the separatist tendencies of India's Muslim population in Jammu and Kashmir in order to strip the territory from India, while Islamabad fears that India's attacks on the Indian Muslim population may lead to attacks on Pakistan.

Finally, because democratic states are far less likely to question the dominant norms of the international system, the core states should not hesitate to continue promoting democracy in the nondemocratic world. It is true that in the short run, promoting human rights and democratic reform in states like China and North Korea risks their further alienation and gives them incentive to disrupt the status quo. Thus there will probably be cases in which the core states must trade a vigorous human rights policy for cooperation on issues of nonproliferation. In the long term, however, nonproliferation and vigorous human rights advocacy will not contradict each other: even without core pressure on their human rights policies, revolutionary actors already possess enough antipathy to the status quo to want to disrupt it. Ultimately, however, not even they will benefit from the dangers inherent in the spread of nuclear weapons.

Notes

I would like to thank John Duffield, David Shumaker, David Mason, Jim Doyle, Benjamin Frankel, Mike Desch, Avery Goldstein, Jim Morrow, George Quester, Etel Solingen, Scott Sagan, Paul Chrzanowski, Sarah Mendelson, Stephen Flank, Devin Hagerty, Daniel Deudney, and an anonymous reviewer for their comments and suggestions on earlier drafts of this paper.

1. See Benjamin Frankel, "An Anxious Decade: Nuclear Proliferation in the 1990s," *Journal of Strategic Studies* 13, no. 3 (September 1990): 4.
2. See, for example, Frankel, "An Anxious Decade," 3; Richard K. Betts, "Paranoids, Pygmies, Pariahs and Nonproliferation," *Foreign Policy*, no. 26 (Spring

1977), pp. 157, 163; James F. Keeley, "Legitimacy, Capability, Effectiveness and the Future of the Non-Proliferation Treaty," in David Dewitt, ed., *Nuclear Nonproliferation and Global Security* (London: Croon Helm, 1987), 37; John J. Mearsheimer, "Back to the Future: Instability in Europe After the Cold War," *International Security* 15, no. 1 (Summer 1990): 5–56; and Roger K. Smith, "Explaining the Non-Proliferation Regime: Anomalies for Contemporary International Relations Theory," *International Organization* 41, no. 2 (Spring 1987): 253–81.

3. Mearsheimer, "Back to the Future," 12–29; and Frankel, "An Anxious Decade." Unlike the classical realists E. H. Carr, Reinhold Niebuhr and Hans Morgenthau, who found that the human need for security and power drove state behavior in the international environment, neorealists like Kenneth Waltz and John Mearsheimer believe that the anarchical structure of the international system impels states to seek power. See Kenneth N. Waltz, *Theory of International Politics* (Reading, Mass.: Addison-Wesley, 1979); and "The Origins of War in Neorealist Theory," in Robert Rotberg and Theodore Rabb, eds., *The Origin and Prevention of Major Wars* (New York: Cambridge University Press, 1989), 41.
4. Frankel, "An Anxious Decade," 7.
5. Mearsheimer, "Back to the Future," 12, 37–39. Mearsheimer confines his discussion to Europe (including the United States and Canada), but I think it is fair to assume that the extension of the parameters of the system does not alter his theories or conclusions.
6. This argument follows those of Robert Jervis in "The Future of World Politics," *International Security* 16, no. 3 (Winter 1991/92): 31–69; James M. Goldgeier and Michael McFaul, "Core and Periphery in the Post-Cold War Era, *International Organization* 46, no. 2 (Spring 1992): 467–91; and Robert O. Keohane, *International Institutions and State Power: Essays in International Relations Theory* (Boulder, Colo.: Westview, 1989), 9. All three authors posited a higher level of cooperation among the core states than the peripheral states. I include in the core the following states: the NATO countries (excluding Turkey), Japan, Sweden, Switzerland, Austria, Australia, South Korea, and New Zealand. Though democratic, Israel is excluded in the discussion of proliferation because the governments of the states that surround it depend in large measure on antipathy to Israel for their legitimation. Moreover, given the circumstances of its origins, its small size, and its history, Israel has more reason than any other core state to believe that the other members of the core might not intervene in time to protect it from attack.
7. For a concise summary of the major themes of neoliberal institutionalism, see Keohane, *International Institutions and State Power,* 1–20. On reflectivism, sometimes called constructivism, see Alexander Wendt, "Anarchy is What States Make of It: The Social Construction of Identity," *International Organization* 42, no. 2 (Spring 1992): 391–425; Anne-Marie Burley, "Liberal States: A Zone of Law," (Paper presented at the Annual Meeting of the American Political Science Association, Chicago, 3–6 September 1992); Nicholas Onuf, *World of Our Making: Rules and Rule in Social Theory and International Relations* (Columbia, S.C.: University of South Carolina Press, 1989); Thomas M. Franck, *The Power of Legitimacy Among Nations* (New York: Oxford University Press, 1990); and John Gerard Ruggie, "Human Rights and the Future International Relations Community," *Daedalus* 112, no. 4 (Fall 1983): 93–110.
8. In addition to the books and articles listed in n. 7, see also see Helen Milner, "The Assumption of Anarchy in International Relations Theory: A Critique," *Review of International Studies* 17, no. 1 (January 1991): 67–85. Milner draws her main theme from points made by Inis L. Claude, Jr., *Power and International*

Relations (New York: Random House, 1962). See also Robert Putnam, "Diplomacy and Domestic Politics: The Logic of Two-Level Games," *International Organization* 42, no. 3 (Summer 1988): 427–60.

9. See for example, Milner, "International Theories of Cooperation Among Nations," *World Politics* 44, no. 3 (April 1992): 488–95; and Jack Levy, "Domestic Politics and War," in Rotberg and Rabb, eds., *The Origin and Prevention of Major Wars*, 79–99; and James G. Richter, "Perpetuating the Cold War," *Political Science Quarterly* 107, no. 2. (Summer 1992): 271–301.
10. Nonoccurrences provide the empirical evidence for the authors who investigate the absence of war of among democratic states. Here, I try to provide as many concrete and significant example of positive cooperation among democratic states as possible, but I also rely on the absence of war and the absence of military preparations among those states as evidence of their separate peace. On the evidentiary value of nonoccurrences, see Alexander L. George, *Presidential Decisionmaking in Foreign Policy: The Effective Use of Information and Advice* (Boulder, Colo.: Westview, 1980), 60. See also Ernest R. May, *"Lessons" of the Past: The Use and Misuse of History in American Foreign Policy* (New York: Oxford University Press, 1973), 44–45.
11. See Jervis "The Future of World Politics"; Goldgeier and Michael McFaul, "Core and Periphery in the Post-Cold War Era"; Keohane, *International Institutions and State Power: Essays in International Relations Theory* (Boulder, Colo.: Westview, 1989), 9. See also Karl Deutsch et al., *Political Community and the North Atlantic Area: International Organization in the Light of Historical Experience* (Westport, Conn.: Greenwood, 1969; reprint, Princeton: Princeton University Press, 1957).
12. I will use the term "core" from Goldgeier and McFaul, as a synonym for pluralistic security community and will use the terms interchangeably. I base my taxonomy on the proposition that the core states are in general economically more developed, militarily more powerful, and politically more stable than the states in the periphery. Moreover, the core states are more interconnected and interdependent than states within the periphery or between the core and the periphery.
13. For the best example of this argument, see Mearsheimer, "Back to the Future."
14. Frankel, "An Anxious Decade," 7.
15. Milner, "The Assumption of Anarchy," 68.
16. Milner, "The Assumption of Anarchy," 70. See Hedley Bull, *The Anarchical Society: A Study of Order in World Politics* (New York: Oxford University Press, 1977), 42; Robert Gilpin, *War and Change in World Politics* (New York: Cambridge University Press, 1981), 28; Kenneth Oye, "Cooperation Under Anarchy," *World Politics* 38, no. 1 (October 1985): 226; and Waltz, *Theory of International Politics*, 102.
17. Milner, "The Assumption of Anarchy," 69. See Hedley Bull, *The Anarchical Society*, 42.
18. Milner, "The Assumption of Anarchy," 70. She derived the observations of Gilpin and Waltz from Gilpin, *War and Change in World Politics* , 28; and Waltz, *Theory of International Politics*, 102.
19. Milner, "The Assumption of Anarchy," 72. See Waltz, *Theory of International Politics*, 103–4. Even in Hobbes's Leviathan, individuals possessed the legitimate right to defend themselves against the government.
20. Waltz, *Theory of International Politics*, 103–4.
21. It should be noted that I used legitimacy in a relative sense. For a discussion of the relativism of legitimacy and the irrelevance of that relativism to legitimacy's power, see Franck, *The Power of Legitimacy Among Nations*, 41–49.
22. Milner, "The Assumption of Anarchy," 72.

23. Ibid.
24. Ibid. See Max Weber, *Economy and Society*, ed. Guenther Roth and Claus Wittich (Berkeley: University of California Press, 1978), 54.
25. Milner, "The Assumption of Anarchy," 71. See, for example, Martin Wight, *Power Politics* (Harmondsworth: Penguin, 1978), 102; and Harry Eckstein, "Authority Patterns: A Structural Basis for Political Inquiry," *American Political Science Review* 67, no. 4 (December 1973): 1142.
26. Milner, "The Assumption of Anarchy," 71. See also Oye, "Cooperation Under Anarchy," 1–2.
27. Milner, "The Assumption of Anarchy," 74.
28. Franck, *The Power of Legitimacy*.
29. See Melvin Small and J. David Singer, *Resort to Arms: International and Civil Wars, 1816–1980* (Beverly Hills: Sage, 1982), 59–60, 79–80, and 221–32.
30. Jervis, "Security Regimes," in Stephen D. Krasner, ed., *International Regimes* (Ithaca: Cornell University Press, 1983): 177.
31. Stephen M. Walt, "Alliance Formation in Southwest Asia: Balancing and Bandwagoning in Cold War Competition, in Robert Jervis and Jack Snyder, eds., *Dominoes and Bandwagons: Strategic Beliefs and Great Power Competition in the Eurasian Rimland* (New York: Oxford University Press, 1991), 52.
32. Ibid.
33. Walt, "Balancing and Bandwagoning in Cold War Competition," 51–74. For the view that domestic factors are the most important determinants of state action, see Robert Cox, *Production, Power, and World Order* (New York: Columbia, 1987).
34. Yaacov Y. I. Vertzberger, *The World In Their Minds: Information Processing, Cognition, and Perception in Foreign Policy Decisionmaking* (Stanford: Stanford University Press, 1990), 18. See also Alexander L. George, *Presidential Decisionmaking*; John Odell, *U.S. International Monetary Policy* (Princeton: Princeton University Press, 1982); and G. John Ikenberry, ed., *The State in American Foreign Economic Policy* (Ithaca: Cornell University Press, 1990).
35. See George, *Presidential Decisionmaking*, 39–47; Jervis, "Perceiving and Coping with Threat," in Robert Jervis, Richard Ned Lebow, and Janet Gross Stein, *Psychology and Deterrence* (Baltimore: Johns Hopkins University Press, 1985), 4; and Vertzberger, *The World In Their Minds*, ch. 5.
36. Jervis, "Perceiving and Coping with Threat," 4.
37. On the tendency of policy makers to seek the familiar, see Vertzberger, *The World In Their Minds*, 160–63; Jervis, "Perceiving and Coping with Threat," 19, 22; and Robert Putnam, *Beliefs of Politicians* (New Haven: Yale University Press, 1973). For the argument that democracies ally with other democracies, see Randolph M. Silverson and Juliann Emmons, "Birds of a Feather: Democratic Political Systems and Alliance Choices in the Twentieth Century," *Journal of Conflict Resolution* 35, no. 2 (June 1991): 285–306. For more general explanations of peaceful relations among democratic states, see Michael W. Doyle, "Liberalism and World Politics," *American Political Science Review* 80, no. 4 (December 1986): 1151–69; Doyle, "Kant, Liberal Legacies, and Foreign Affairs, pt. 1," *Philosophy and Public Affairs* 12, no. 3 (Summer 1983): 205–35; Doyle, "Kant, Liberal Legacies, and Foreign Affairs, pt. 2," *Philosophy and Public Affairs* 12, no. 4 (Fall 1983): 323–53; David A. Lake, "Powerful Pacifists and War," *American Political Science Review* 86, no. 1 (March 1992) : 24–37; Randall L. Schweller, "Domestic Structure and Preventive War: Are Democracies More Pacific?" *World Politics* 44, no. 2 (January 1992): 235–69; Rudolph Rummel, "Libertarianism and International Violence," *Journal of Conflict Resolution* 27, no. 1 (March 1983): 27–71; Zeev Maoz and Nasrin Abdolali, "Regime Types and International Conflict," *Journal of Conflict Resolution,* 33,

no. 1 (March 1989): 3–36; Stephen Van Evera, "Primed for Peace: Europe After the Cold War," *International Security* 15, no. 3 (Winter 1990/91): 26–28; and A. A. Yazkova, "Natsional'nye Protsessy i Mezhnatsionalnye Protivorechiya v SSSR i Vostochnoy Evrope," *Mirovaia Ekonomika i Mezhdunarodnye Otnosheniya* no. 1, 1992, p. 87. This point is discussed in greater detail below.

38. Jervis, "Perceiving and Coping with Threat," 14–15. See also Raymond Cohen, "Threat Perception in International Crisis," *Political Science Quarterly* 93, no. 3 (Spring 1978): 93–107.
39. For discussions of the Soviet perspective, see William Taubman, *Stalin's American Policy: From Entente to Detente to Cold War* (New York: Norton, 1982); and Vojtech Mastny, *Russia's Road to the Cold War* (New York: Columbia University Press, 1979). For the American perspective, see John Lewis Gaddis, *Strategies of Containment: A Critical Appraisal of Postwar American National Security Policy* (New York: Oxford University Press, 1982); and Deborah Larson, *Origins of Containment: A Psychological Explanation* (Princeton: Princeton University Press, 1985).
40. In addition to the books listed in n. 39, see also see Jervis, "Perceiving and Coping with Threat," 14; and Matthew Evangelista, "Stalin's Post-War Army Reappraised," *International Security* 7, no. 3 (Winter 1982/83): 110–38. For the opposing view, see Mearsheimer, "Back to the Future," 26–29; and Josef Joffe, *The Limited Partnership: Europe, the United States, and the Burdens of Alliance* (Cambridge, Mass: Ballinger, 1987), 177–214.
41. Ted Hopf, "Polarity, the Offense-Defense Balance, and War," *American Political Science Review* 85, no. 1 (June 1991): 475–93. See also Woosang Kim, "Power, Alliance, and Major Wars, 1816–1975," *Journal of Conflict Resolution,* 33, no. 2 (June 1989): 255–73. Kenneth Waltz argued in *Theory of International Politics*, p. 135, that bipolarity produced greater stability. George Kennan argued the opposite in *Memoirs: 1925–1950* (Boston: Atlantic-Little Brown, 1967), 463; and *Russia, the Atom and the West: The BBC Reith Lectures, 1957* (London: Oxford University Press, 1958), ch. 3.
42. Hopf, "Polarity, the Offense-Defense balance, and War," 489–90.
43. Krasner, "Structural Causes and Regime Consequences: Regimes as Intervening Variables," in Krasner, *International Regimes*, 1.
44. Jervis, "Security Regimes," in Krasner, *International Regimes*, 177.
45. Ibid., 177–78.
46. On international cooperation among defense industries, see James B. Steinberg, *The Transformation of the European Defense Industry: Emerging Trends and Prospects for Future U.S.-European Competition and Collaboration* (Santa Monica, Calif.: RAND, 1992).
47. For a discussion of both the Franco-German brigade and the larger issue of European military integration, see Edward Foster, "The Franco-German Corps: A 'Theological' Debate?" *RUSI Journal* 137, no. 4 (August 1992): 63–67.
48. I rely on the definition of liberalism in Rummel, "Libertarianism and International Violence," 30–32; and Doyle, "Liberalism and World Politics," 1151–69. Doyle bases this argument principally on the works of Immanuel Kant and the observations of Clarence Streit, *Union Now: A Proposal for a Federal Union of the Leading Democracies* (New York: Harpers, 1938). See also Francis Fukuyama, "The End of History," *The National Interest* no. 16, (Summer 1989): 3–18; and Maoz and Abdolali, "Regime Types and International Conflict," 3–36.
49. Lake argues that only twice in history have democratic states fought each other or on opposite sides of wartime coalitions: Spain and the United States in 1898, and Finland, on the side of the Axis powers during the Second World War. See Lake, "Powerful Pacifists and War," 33; Schweller and Doyle do not grant even

those examples. See Schweller, "Domestic Structure and Preventive War," 235–69; and Doyle, "Liberalism and World Politics," 1151–69.

50. Lake, "Powerful Pacifists and War," 24.
51. Doyle, "Liberalism and World Politics," 1160.
52. Ibid., 1161.
53. Aleksandr N. Yakovlev, in *Pravda*, 8 February 1990, p. 5. For a relatively recent American articulation of this belief, see Robert B. Zoellick, "The New Europe in a New Age: Insular, Itinerant, or International? Prospects for an Alliance of Values," Address by Counselor of the State Department before the American-European Community Association International's Conference on U.S./EC Relations and Europe's New Architecture, Annapolis, Maryland, 21 September 1990, *Dispatch of the United States Department of State Bureau of Public Affairs* 1, no. 4 (September 24, 1990): 121. Yakovlev's comment suggests that the Soviet Union was determining its interests in terms of its emerging liberal democratic identity, rather than simply its power vis-à-vis other states as realist theory would predict. For a critique of this facet of realism, see Michael Joseph Smith, *Realist Thought From Weber to Kissinger* (Baton Rouge: Louisiana State University Press, 1986), 223; Robert O. Keohane, ed., *Neorealism and its Critics* (New York: Columbia University Press, 1986); and Alexander Wendt, "Anarchy is What States Make of It," *International Organization* 42, no. 2 (Spring 1992): 391–425. For the role of ideology in redefining Soviet interests in Eastern Europe, see Glenn Chafetz, "Soviet Ideological Revision and the Collapse of Communism in Eastern Europe," *International Relations* 11, no. 2 (August 1992).
54. Doyle, "Liberalism and World Politics," 1161. An interesting way to test this last thesis would be to poll Americans on the level of their antipathy to Japan given two different premises: (1) the U.S. trade deficit with Japan results only from Americans' preference for Japanese goods; or (2) the U.S. trade deficit with Japan results from Japanese government policy.
55. For Japanese denials, see the statements of foreign ministry spokesman Masamichi Hanabusa, KYODO, 21 February 1992, cited in *Foreign Broadcast Information Service* (FBIS), FBIS-EAS-92-035, 21 February 1992, p. 4; and *KYODO,* 8 April 1992, cited in FBIS-EAS-92-068, 8 April 1992, p. 1. From Germany, see the speeches of German defense minister Gerhard Stoltenberg in *Frankfurter Allgemeine,* 31 January 1992, p. 4, cited in FBIS-WEU-92-022, 3 February 1992, p. 17; Chancellor Helmut Kohl, *Berlin ADN*, 15 July 1992, cited in FBIS-WEU-92-136, 15 July 1992, p. 15; and *Le Figaro,* 20 May 1992, pp. 6–7, cited in FBIS-WEU-92-100, 22 May 1992, p. 18.
56. See Takashi Inoguchi, "Japan's Role in World Affairs," *Survival* 34, no. 2 (Summer 1992): 71–87.
57. See the statements of Japanese foreign ministry spokesman Masamichi Hanabusa, *KYODO* 21 February 1992, cited in FBIS-EAS-92-035, 21 February 1992, p. 4; and *KYODO*, 8 April 1992, cited in FBIS-EAS-92-068, 8 April 1992, p. 1. See also David E. Sanger, "Japan May Cut Atomic Shipments," *New York Times,* 14 November 1992, pp. 1, 5.
58. Japan requires plutonium, rather than the cheaper and less dangerous uranium, for use in developing and running fast-breeder nuclear reactors. See Karen Lowry Miller, "Japan's Nuclear Gamble Ignites a White-Hot Debate," *Business Week*, 22 June 1992, pp. 76–77. See also Sanger, "Japan May Cut Atomic Shipments," "Japan Says Technical Problems Will Force Storage of Plutonium," *New York Times,* 28 November 1992, p. 3; and "Japan's Nuclear Fiasco," *New York Times,* 20 December 1992, Section 3, pp. 1, 6.
59. See Ezra F. Vogel, "Pax Nipponica?" *Foreign Affairs* 64, no. 4 (Spring 1986): 752–67; and Masamichi Hanabusa, *KYODO*, 8 April 1992, cited in FBIS-EAS-92-068, 8 April 1992, p. 1.

60. Kohl, in *Le Figaro*, 20 May 1992, pp. 6–7, cited in FBIS-WEU-92-100, 22 May 1992, p. 18.
61. James G. Richter, "Perpetuating the Cold War," 274. See also Allen Lynch, *The Cold War is Over – Again* (Boulder, Colo.: Westview, 1992); Alexander L. George, "Domestic Constraints on Regime Change in U.S. Foreign Policy: the Need for Policy Legitimacy," in Ole R. Holsti, Randolph R. Silverman, and Alexander L. George, eds. *Change in the International System* (Boulder, Colo: Westview Press, 1980), 233–62. Kenneth Waltz makes a similar argument with regard to the effects of nuclear weapons when he asserts that concerns about military doctrine and numerical parity in the nuclear age were "a hangover from conventional days," in "Nuclear Myths and Poltical Realities," *American Political Science Review* 84, no. 3 (September 1990): 737. I am indebted to Daniel Deudney for this observation.
62. For earlier works on how domestic politics responds to the effects of the international environment, see Peter Gourevitch, "The Second Image Reversed: the International Sources of Domestic Politics," *International Organization* 32, no. 4 (Autumn, 1978): 881–911; Barrington Moore, Jr., *Social Origins of Dictatorship and Democracy* (Boston: Little, Brown, 1966): 413–44; and Otto Hintze, "Military Organization and the Organization of the State, in Felix Gilbert, ed., *The Historical Essays of Otto Hintze* (New York: Oxford University Press, 1975): 178–215.
63. See Onuf, *World of Our Making*; and Wendt, "Anarchy is What States Make of it," 391–425.
64. Wendt, "Anarchy is What States Make of it," 398.
65. The typology comes from Wendt, "Anarchy," 400, but he did not include the ideological component of identity.
66. On the role conception and national identity, see K. J. Holsti, "National Role Conception in the Study of Foreign Policy, *International Studies Quarterly* 14, no. 3 (September 1970): 233–309; Martha L. Cottam, *Foreign Policy Decision Making: The Influence of Cognition* (Boulder, Colo.: Westview, 1986); and Chih-yu Shih, "A Cognitive Approach to International Organization: Perspective and Application," *Behavioral Science* 34, no. 3 (July 1989): 176–98.
67. On embeddedness, see John Gerard Ruggie, "International Regimes, Transactions, and Change: Embedded Liberalism in a Postwar Economic Order," in Krasner, *International Regimes*, 195–232. The research from studies of political psychology and on norms of international behavior supports this logic. Policy makers dealing with states categorized as adversarial interpret ambiguous information from those states to confirm the original image, while leaders dealing with states previously understood to be friendly will interpret ambiguous information to confirm the friendly image. See Robert Putnam, *Beliefs of Politicians*; and Jervis, "Perceiving and Coping with Threat," 19, 22. On threats to identity, see Terrell Northrup, "The Dynamic of Identity in Personal and Social Conflict," in Louis Kreisberg et al., eds., *Intractable Conflicts and Their Transformation* (Syracuse: Syracuse University Press, 1989), 55–82. On resistance to change, see Timur Kuran, "The Tenacious Past: Theories of Personal and Collective Conservatism," *Journal of Economic Behavior and Organization* 10, no. 2 (September 1988): 143–71. On mass constraints on elite attempts to change foreign policy, see Mark Peffley and Jon Hurwitz, "A Hierarchical Model of Attitude Constraint," *American Journal of Political Science* 29, no. 4 (November 1985): 871–90; Hurwitz and Peffley, "How Are Foreign Policy Attitudes Structured? A Hierarchical Model," *American Political Science Review,* 81, no. 4 (December 1987): 1099–1120; and Hurwitz and Peffley, "Public Images of the Soviet Union: The Impact on Foreign Policy Attitudes," *Journal of Politics,* 52, no. 1 (February 1990): 3–28. On norms, see Robert

Axelrod, "An Evolutionary Approach to Norms," *American Political Science Review* 80, no. 4 (December 1986): 1097–1108.

68. See Chafetz, *Gorbachev, Reform, and the Brezhnev Doctrine* (Westport, Conn: Praeger, 1993).
69. See Michael R. Beschloss and Strobe Talbott, *At the Highest Levels: The Inside Story of the End of the Cold War* (Boston: Little, Brown, 1993).
70. See Wendt, "Anarchy," 405; and Ruggie, "International Regimes, Transactions, and Change," 195–232. The collective identification of security was a strategic belief. In addition to the literature on strategic beliefs cited above, see John Odell, *U.S. Monetary Policy* (Princeton: Princeton University Press, 1982); and Jervis, "The Future of International Politics, 52. For the view that cooperation continues when the interest groups that benefit most from cooperation dominate the ones that benefit least in the policy process, see Peter Gourevitch, *Politics in Hard Times* (Ithaca: Cornell University Press, 1987); and Milner, *Resisting Protectionism* (Princeton: Princeton University Press, 1988). See also Richard Rosecrance, *The Rise of the Trading State*, (New York: Basic Books, 1986), ch. 7. For the argument that bureaucratic institutional interests are more important, see G. John Ikenberry, ed. *The State in American Foreign Economic Policy* (Princeton: Princeton University Press, 1990); See also Peter Haas, *Saving the Mediterranean* (New York: Columbia University Press, 1990).
71. On the dynamics of rewarding, see Thomas W. Milburn and Daniel J. Christie, "Rewarding in International Politics," *Political Psychology* 10, no. 4 (December 1989): 625–45. See also Harvey Starr, "Choice, Learning, and Security Communities," *Journal of Peace Research* 29, no. 2 (May 1992): 207–13. Starr argues that learning characterizes the interaction that led the liberal democracies to the development of the security community. Our arguments are consistent.
72. See the speeches by President George Bush at the United Nations Security Council Summit in New York, 31 January 1992, printed in *Vital Speeches of the Day* 18, no. 10 (1 March 1992): 290–91; French president François Mitterand at the United Nations Security Council Summit in New York, 31 January 1992, printed in *Vital Speeches of the Day* 18, no. 10 (1 March 1992): 292; British prime minister John Major at the United Nations Security Council in New York, 31 January 1992, printed in *Vital Speeches of the Day* 18, no. 10 (1 March 1992): 295; German chancellor Helmut Kohl at the 106th Annual Convention of the American Newspaper Publishers Association in New York, 5 May 1992, printed in *Vital Speeches of the Day* 18, no. 16 (1 June 1992): 485–86; Canadian prime minister Brian Mulroney at Johns Hopkins University, Baltimore, Maryland, 21 May 1992, printed in *Vital Speeches of the Day*, 18, no. 20 (1 August 1992): 615. See also the speech of Bill Clinton (as a candidate) in Milwaukee, Wisconsin, 1 October 1992, printed (excerpts) in *New York Times*, 2 October 1992, p. A21.
73. See Vladimir Abarinov and Vitaly Tretyakov, "Andrey Kozyrev: 'Soyuz Ostavil Rossii Polkhoe Vneshnepoliticheskoe Nasledstvo'," *Nezavisimaya Gazeta*, 1 April 1992, p. 3. Leaders of nondemocratic states tend to operate according to different beliefs. In states where democracy is tenuous, conflicts have emerged over the theoretical validity of democratic solidarity. Kozyrev's authoritarian opponents have argued that such conceptions of international relations are "romantic" and inaccurate. See, for example, Yuri Glukhov, "Lyubov bez Vzaimnosti," *Pravda*, 26 May 1992, pp. 1, 3.
74. For demonstrations of the impact on policy of this kind of rhetoric, see B. Thomas Trout, "Rhetoric Revisited: Political Legitimation and the Cold War," *International Studies Quarterly* 19, no. 3 (September 1975): 251–84; and Richter, "Perpetuating the Cold War," 271–301. On the constraints exercised by mass beliefs, see Peffley and Hurwitz, "Attitude Constraint," 871–90; Hurwitz and

Peffley, "How Are Foreign Policy Attitudes Structured?" 1099–1120; and Hurwitz and Peffley, "The Impact on Foreign Policy Attitudes," *Journal of Politics* 52, no. 1 (February 1990): 3–28.

75. Comment of George Quester at the conference on Social Science Theory and Nuclear Weapons Proliferation, at the Center for International Security and Arms Control, Stanford University, 6 February 1993.
76. David N. Schwartz, *NATO's Nuclear Dilemmas* (Washington, D.C.: Brookings, 1983), 60–61. See also Charles De Gaulle, *Memoirs of Hope: Renewal and Endeavor,* Terence Kilmartin, trans. (New York: Simon and Schuster, 1971), 201–2, 204, 209, 213.
77. For a discussion of extended deterrence and how it is likely to function in the post-Cold War world, see Jervis, *The Meaning of the Nuclear Revolution*, ch. 3; *The Illogic of American Nuclear Strategy* (Ithaca: Cornell University Press, 1984), chaps. 5–6; and Jervis, "The Future of World Politics," 48.
78. Inoguchi, "Japan's Role in World Affairs," 74.
79. Ibid., 77–78. See also Franck's argument that deviations from international norms produce domestic consequences, *The Power of Legitimacy.*
80. The idea that a state would forego immediate external advantages in order to avoid the domestic consequences of violating a norm indicates that the state has internalized the norm. Thus, Richter's notion that international roles and behaviors become embedded in the domestic politics of states is consistent with the description of internalization as a means of eliciting compliance with norms. See Axelrod, "An Evolutionary Approach to Norms," 1104. For an overview of Japan's incentives and disincentives to develop nuclear weapons, see James E. Doyle and Sarah S. Doyle, "Japan and Nuclear Weapons: Considerations for U.S. Japanese Security Cooperation," *Strategic Review* 20, no. 3 (Summer 1992): 75–80. For a brief explanation of the reasoning behind Germany's determination to abjure the bomb, see Stanley Hoffman's letter in "Correspondence: Back to the Future," Part II: International Relations Theory and the Future of Europe, *International Security* 15, no. 2 (Fall 1990): 192.
81. See especially A. A. Yazkova "Natsional'nye Protsessy i Mezhnatsionalnye Protivorechiya v SSSR i Vostochnoy Evrope," *MEiMO*, no. 1, 1992, p. 87. She contends that war is more likely among states in which there are no liberal protections for minorities. Thus, Serbia, for example, attacked its neighbors because of fears that the Bosnians, Croatians, and Slovenians would not protect the Serbs left in those countries as a result of the dissolution of Yugoslavia. It should be noted that given the tendencies of decision-makers to project their experiences on to adversaries, it is not probable that Serbia was prepared to protect the rights of other Yugoslavs left in Serbia. For evidence of this view, see the comments of Serbian political commentator, Aleksandar Nenadovic, who implied that the Yugoslav civil would never have occurred if Germany had insisted that Croatia and Slovenia ensure the rights of ethnic minorities on their territories; see "As Serbs See War: Foreign Plots and Ingratitude," *New York Times,* 2 September 1992, p. A2.
82. See, for example, Betts, "Nonproliferation," 157; and Jervis, "The Future of World Politics," 60.
83. See Fukuyama, "The End of History," 3–18. I do not argue that democratization is irreversible.
84. On the hegemony of ideological consent, see Enrico Augelli and Craig Murphy, *America's Quest for Supremacy and the Third World: A Gramscian Analysis* (London: Pinter, 1988).
85. See Jervis, "The Future of World Politics," 62–69; "Security Regimes," 177–78; and Smith, "Explaining the Non-Proliferation Regime," 253–81.

86. Paul Lewis, "Kohl Urges Japan to Increase Help for East Europe: A Blunt German Request," *New York Times*, 6 May 1992, p. A1.
87. Betts, "Nonproliferation," 182.
88. Barbara Crossette, "Bush is Told of Threat to Soviet Aid Bill," *New York Times*, 7 May 1992, p. A14, and "4 Ex-Soviet States and U.S. in Accord on 1991 Arms Pact," *New York Times*, 24 May 1992, pp. 1, 12; and *Krasnaya Zvezda*, 28 January 1992, p. 3.
89. "Belarus Approves First Arms-Limitation Pact," *New York Times*, 5 February 1993, p. A6.
90. See Yury Kovalenko, "Uchenym-Yadershchikam SNG Naidetsya Rabota na Rodine: V Moskve Budem Mezhdunarodnyy Tsentr Nauki i Tekhnologii," *Izvestia*, 13 March 1992, p. 6; and David Hoffman, "Aid Set for Nuclear Scientists: Center to Employ Ex-Soviet Experts; Yeltsin, Baker Meet," *Washington Post*, 18 February 1992, pp. A1, A10; Kovalenko, "Obrashchenie Rossii Zastalo NATO Vrasplokh," *Izvestia*, 23 December 1991, p. 6; Yelizaveta Leontyeva, "Kogda My Ushli ot Messianstva," *Rossiyskaya Gazeta*, 21 January 1992, p. 5; and "Plowshares from Russia," *New York Times*, 2 September
91. The effects of the lure of the West on the behavior of Hungary, Czechoslovakia, and Poland are discussed in David Shumaker, "The Origins and Development of Central European Cooperation 1989–1992," *East European Quarterly* (June 1993, forthcoming).
1992, p. A18.
92. Margaret Shapiro, "Ukraine Gets Defense Offer from Yeltsin," *New York Times*, 16 January 1993, pp. A1, A18; Crossette, "Bush is Told of Threat," p. A14; and Chrystia Freeland, "Ukraine Seeks Western Protection from Russia," *Washington Post*, 29 April 1992, pp. A28-A29.
93. Shapiro, "Ukraine Gets Defense Offer from Yeltsin," pp. A1, A18; and "Belarus Approves First Arms-Limitation Pact," p. A6.
94. See Kovalenko, "Uchenym-Yadershchikam SNG," 6; and Hoffman, "Aid Set for Nuclear Scientists," A1, A10.
95. See International Atomic Energy Agency, Information Circular, 254/Revision 1/parts 1 and 2, July 1992.
96. Axelrod, "An Evolutionary Approach to Norms," 1105.
97. Waltz, *Theory of International Politics*, 96.
98. Ikenberry and Kupchan posit a more active persuasion of the followers by the leaders in "Socialization and Hegemonic Power," 253–81.
99. "Iz Minska: Pervaya Partiya Yadernogo Oruzhiya Vyvezen Belarus v Rossiyu," *Krasnaya Zvezda*, 28 January 1992, p. 3.
100. Dominance is self-explanatory. For a discussion, see Axelrod, "An Evolutionary Approach to Norms," 1103.
101. Quoted by Philip Shenon, "Non-Aligned Bloc Holds Talks on Finding a Reason for Being," *New York Times*, 2 September 1992, p. A4.
102. Ibid.
103. See Axelrod, "An Evolutionary Approach to Norms," 1105. See also Wendt, "Anarchy is What States Make of It," 391–425.
104. See International Atomic Energy Agency, Information Circular, 254/Revision 1/parts 1 and 2, July 1992. See also "Suppliers Agree on Dual-Use Controls, Full-Scope Safeguards Export Policy," *Nuclear Fuel*, 13 April 1992, 4; and Roland Timerbaev, "A Major Milestone in Controlling Nuclear Exports," *Eye on Supply*, no. 6 (Spring 1992): 58–65. The members of both the NSG and the PSC include: Australia, Austria, Belgium, Canada, Denmark, Finland, France, Germany, Greece, Ireland, Italy, Japan, Luxembourg, Netherlands, Norway, Portugal, Spain, Sweden, Switzerland, Great Britain, and the United States. The other members of the NSG are: Bulgaria, The Czech and Slovak Federal

Republic, Hungary, Poland, Romania, and Russia. Emerging supplier which did not subscribe to the new NSG guidelines are: China, India, South Korea, Argentina, and Brazil.

105. United States Congress, Senate, Government Operations Committee, *Export Reorganization Act of 1976: Hearings*, 94th Congress, 2nd Session, 1976, pp. 775–83; and Senator Abraham Ribicoff with Paul Leventhal, "A Market-Sharing Approach to the World Nuclear Problem," *Foreign Affairs* 54, no. 4 (July 1976): 766; both cited in Betts, "Nonproliferation," 171.
106. See the discussion of the literature on hegemonic regime theory in Smith, "Explaining the Non-Proliferation Regime," 262–64.
107. I define China as a major power because its size, military potential, and permanent seat on the United Nations Security Council give Beijing the ability to disrupt the international order. See Nicholas D. Kristof, "As China Looks at World Order, It Detects New Struggles Emerging," *New York Times*, 21 April 1992, pp. A1, A10. See also David Shambaugh, "China's Security Policy in the Post-Cold War Era," *Survival* 34, no. 2 (Summer 1992): 88–106.
108. See Shambaugh, "China's Security Policy," esp. 96–106.
109. President Bush, "The World After the Persian Gulf War," Address before a joint session of Congress, Washington, D.C., 6 March 1991, *Dispatch of the United States Department of State Bureau of Public Affairs* 2, no. 10 (11 March 1991): 161.
110. Secretary of State James A. Baker, III, "Statement Before the House Appropriations Subcommittee on Commerce, Justice, State, Judiciary, and Related Agencies, Washington, D.C., 3 March 1992," *Dispatch of the United States Department of State Bureau of Public Affairs* 1, no. 10 (4 March 1992): 190. Also, John Courtney, the U.S. Ambassador to Kazakhstan reported to an American visitor that his highest priority was the removal of nuclear weapons to Russia for destruction.
111. Elaine Sciolino, "U.S. Will Draw up a Strategy to Aid Ex-Soviet States," *New York Times*, 8 February 1993, pp. A1, A7.
112. Leonard S. Spector, "Repentant Nuclear Proliferants," *Foreign Policy* no. 88 (Fall 1992): 23.
113. France became the final permanent member of the Security Council to sign the NPT in August 1992. On the increasing activism of the UN on proliferation, see Spector, "Repentant Nuclear Proliferants," 23.
114. Michael R. Gordon, "North Korea Rebuffs Nuclear Inspectors, Reviving U.S. Nervousness," *New York Times*, 1 February 1993, p. A1; David E. Sanger, "North Korea Reveals Nuclear Sites to Atomic Agency," *New York Times*, 7 May 1992, p. A8; and Don Oberdorfer, "North Korea Releases Extensive Data on Nuclear Effort: In Preparation for UN Inspections Pyongyang Reveals It Has Produced Plutonium," *Washington Post*, 6 May 1992, p. A11.
115. See Spector, "Repentant Nuclear Proliferants," 27–29; and Oberdorfer, "U.S. Welcomes Koreans' Nuclear Accord," *Washington Post*, 1 January 1992, p. A1.
116. Gordon, "North Korea Rebuffs Nuclear Inspectors," p. A1.
117. R. Jeffrey Smith, "3 Former Soviet Republics Meet US Arms Terms," *Washington Post*, 27 April 1992, p. A1.
118. See Jim Hoagland, "Will Kim Blink," *Washington Post*, 21 April 1992, p. A19.
119. Keeley, "Legitimacy," 29.
120. Jervis, "The Future of World Politics," 58.
121. See Spector, "Repentant Nuclear Proliferants," 21.
122. See Spector, "Repentant Nuclear Proliferants," 26–27, 31; Spector and Jacqueline R. Smith, *Nuclear Ambitions: The Spread of Nuclear Weapons, 1989–1990* (Boulder, Colo.: Westview, 1990): 221; and John R. Redick, *Argentina and Brazil: An Evolving Nuclear Relationship,* Program for Promoting Nuclear Non-

Proliferation, Occasional Paper Seven (Southampton, England: The Center for International Policy Studies, 1990).

123. See Keeley, "Future of the NPT," 33; see also Steven R. David, "Why the Third World Matters," *International Security* 14, no. 1 (Summer 1989): 50–85.

124. Although the danger from the point of view of most governments is apparent, it is a controversial question among analysts. For arguments on the stabilizing effects of nuclear proliferation, see Waltz, *The Spread of Nuclear Weapons: More May Be Better*, Adelphi Paper no. 171 (London: IISS, 1981); R. Robert Sandoval, "Consider the Porcupine: Another View of Nuclear Proliferation," *Bulletin of the Atomic Scientist* (May 1976): 17–19; Shai Feldman, "A Nuclear Middle East," *Survival* no. 23 (May–June 1981), 111–15; S. Neil MacFarlane, "Don't Worry, Be Happy: The Collapse of the Soviet Union and its Implications for the Atlantic Security Community" (Paper presented at the Conference on European Security, Queens University, Kingston, Ontario, 14–15 May 1992); and Mearsheimer, who argues in "Back to the Future," 37–40, that a limited proliferation to Germany will increase stability. For the more common, traditional viewpoint, see Lewis A. Dunn, *Controlling the Bomb* (New Haven: Yale University Press, 1982), 69–94; Betts, "Paranoids," 168–69; and Mearsheimer's argument in "Back to the Future," that poorly managed proliferation will reduce stability and security, 37–40.

125. See Richard N. Haass, *Conflicts Unending: The United States and Regional Disputes* (London and New Haven: Yale University Press, 1990), 78–98.

Proliferation Optimism and Theories of Nuclear Operations

Peter D. Feaver

SOCIAL SCIENCE theory ought to shed light on the behavior of proliferating states after they develop nuclear weapons. This article proposes factors such a theory of nuclear proliferation must incorporate in order to explain one important component of behavior: nuclear operations. The result is admittedly tentative; I do not develop a completely specified theory, nor do I conclude with specific predictions about individual proliferators. But I do identify important relationships and suggest some plausible hypotheses of cause and effect. At a minimum, I hope to indicate the possibility for a more systematic analysis of nuclear operations – and therefore of nuclear behavior – than analysts of nuclear proliferation have hitherto attempted.

A theory of nuclear behavior is especially important as the focus of proliferation studies shifts from simply determining which states are developing nuclear weapons (and, by extension, how we can stop them from doing so), to understanding the political and strategic effects of regional proliferation (and so what can be done to mitigate or manage the negative consequences).[1] Analysts are attending more and more to the problem of dealing with those countries which successfully defy the nonproliferation regime.[2] Managing nuclear proliferation, however, begs a deeper understanding of nuclear behavior. How will the new states handle their new nuclear forces? Will the new nuclear nations control their weapons as the United States and the Soviet Union controlled theirs? Will the proliferators adopt similar strategic preferences, for example favoring secure second-strike capabilities designed to maximize stable deterrent postures? In short, how will nuclear behavior vary over the field of nuclear proliferators?

Some scholars, notably Kenneth Waltz and John Mearsheimer, have advanced a controversially optimistic answer to the question of nuclear behavior.[3] As explained in greater detail below, these iconoclasts argue that nuclear weapons have a pacifying affect on states and so the spread of nuclear weapons will result in the spread of stability and peace. In their starkest form, optimists view nuclear proliferation as a *solution* to concerns about state behavior, not a factor exacerbating those concerns. This line of argumentation makes the quest for a theory of nuclear behavior moot. Explaining the behavior of individual states is not a particularly interesting theoretical puzzle if we already know that nuclear proliferation will spread peace. Thus, before advancing my proposal for developing a theory of nuclear behavior, I must address the argument that this entire exercise is unnecessary, given what we know about the pacifying effects of nuclear weapons. Along the way, I will show how some optimistic views of nuclear proliferation depend implicitly on the very elements I argue should form the basis of a theory of nuclear behavior: namely, state-level factors.

My argument proceeds in four stages. In section one, I briefly explain the derivation of proliferation optimism from basic rational deterrence theory. I also examine reasons for remaining pessimistic about nuclear proliferation, and so for remaining interested in explaining variance in the behavior of individual proliferators. In section two, I survey the three sets of factors shaping nuclear operations in proliferating countries: strategic systems, strategic environment, and strategic culture. In section three, I consider how the special features of "opaque" proliferation might modify nuclear operations. In section four, I conclude with a brief discussion of the research needed to resolve uncertainties still attending my argument about nuclear operations.

What, Me Worry?

Optimism about proliferation rests on two pillars, one theoretical and one empirical. In theoretical terms, the central logic of rational deterrence theory (RDT) gives good ground for being confident about nuclear proliferation: rational states will not go to war if their enemies can convince them that the expected costs of doing so will far exceed the expected benefits.[4] Deterrence success thus depends upon the defender's ability to make credible threats of such magnitude that the potential aggressor decides to abandon any plans for mischief. Deterrence can still fail, but it is unlikely to do so when the threats are sufficiently terrifying.

This logic, which holds generally across all forms of conflict, applies a fortiori to nuclear war where the costs are likely to be astronomical. Nuclear weapons make deterrence relatively easy because nuclear deterrence threatens certain and unacceptably horrible devastation. If both sides in a contest have nuclear weapons, they will reasonably conclude that escalating to nuclear use will result in the complete destruction of both societies. They will determine, accordingly, that nuclear weapons are inherently unusable and so will never initiate a nuclear strike. Nuclear nations will also become cautious about engaging in conventional war, calculating that any conflict has the potential to escalate to the nuclear level. They will, as a consequence, avoid getting on any slippery slope; the outcome is general international stability.

Rational deterrence theory (RDT) is not, of course, without its detractors. RDT enthusiasts admit the theory is still under development; critics argue it is deeply flawed.[6] While an extensive review of the rational deterrence theory debate is obviously beyond the scope of this paper,[7] one criticism emerges as particularly relevant to the nuclear proliferation question: whether RDT pays inadequate attention to the role of key factors, psychological and otherwise, which complicate deterrence calculations. Rational leaders are nevertheless subject to miscalculation and misperception and so may misread otherwise credible deterrent threats. Moreover, not every leader will have the same degree of risk-acceptance. Thus, the same deterrent threat may be sufficient or insufficient to deter war, depending on whether the leader to be deterred has miscalculated or is willing to take a gamble.[8] RDT may only outline the conditions for deterrence success and failure under ideal circumstances rarely achieved in the real world.

But even if these criticisms have merit when applied to rational deterrence theory generally, they may not, in and of themselves, refute optimism about nuclear proliferation. Indeed, Waltz's fundamental claim is that the problems of deterrence are far more tractable under conditions of nuclear proliferation. Much the same way a hanging is said to concentrate the mind, the awesome destructive power of nuclear weapons compels countries to conform their behavior to the expectations of rational deterrence theory and so overcome petty psychological problems.[9] Nuclear weapons, McGeorge Bundy argues, can offer "existential" deterrence: the mere existence of nuclear weapons is enough to deter an adversary.[10]

The empirical pillar of proliferation optimism lends apparently confirmatory support to the claims of rational deterrence theory, at least with respect to nuclear weapons. The world has, after all, experienced many decades of nuclear peace. The records of leaders as singular as John F. Kennedy, Richard Nixon, Josef Stalin, Nikita Krushchev, Mao Tse-tung, and Chou En Lai tell a common tale: all brandished the nuclear sword without ever stumbling into a nuclear war. The world survived periods of nuclear monopoly, nuclear superiority, and nuclear parity, without a single nuclear calamity. Reasonably secure nations embarked on a nuclear development program (Brazil), as did rather insecure ones (Israel); no one besides the United States has actually exploded nuclear weapons in anger. It is tempting, therefore, to concede that nuclear proliferation conforms most closely with the expectations of rational deterrence theory.

If this logic is true, why worry about nuclear proliferation? Optimists say we should not, and yet we most certainly do. With the end of the cold war, nuclear proliferation has re-emerged as a priority topic in policy circles.

Political scientists and policy analysts are turning and returning to proliferation studies with renewed energy and purpose, as evidenced by this volume. If anything, there is a glut of proliferation studies which take as their point of departure the diametrically opposite assumption that nuclear proliferation is very worrying indeed. If nuclear proliferation will not increase the risk of nuclear war why worry about it?[11] What justifies this obstinate fretting about nuclear proliferation in the face of RDT's crystalline logic?[12]

One reason is disarmingly simple: nuclear deterrence *can* fail, so proliferation *could* lead to nuclear war. At best, rational deterrence theory can predict that nuclear deterrence should assure peace most of the time. Most is not all. Indeed, two of its more zealous defenders, Christopher Achen and Duncan Snidal, point out that if RDT could successfully predict peace 99.5 percent of the time it would therefore miss .5 percent of the time.[13] This would qualify RDT for the social science theory hall of fame, but it would not make nuclear proliferation trivial. In nuclear studies, as in seismology, it is axiomatic that extremely rare events are nevertheless extremely interesting and important. Given the stakes involved, it is reasonable to worry about nuclear proliferation – even if RDT is correct that most proliferators will avoid war most of the time. The "rare events are interesting" axiom makes proliferation an arresting policy problem, of course, but it is not by itself enough to make nuclear behavior an intriguing problem for theory.

The second justification for pessimism speaks more directly to the question of whether we need a theory of nuclear operations. Forty-five years without a superpower war is not ultimately conclusive evidence of the pacifying effects of nuclear weapons.[14] As all the participants in the lively debate acknowledge, many factors contributed to the long peace: the bipolar system structure, the absence of irredentist claims between the superpowers, the development and observance of "rules of the game," and so on. Many of these factors may be absent in the regions of nuclear proliferation. It remains an open question whether nuclear weapons will keep the peace under conditions that are decidedly *not* ceteris paribus. Indeed, Mearsheimer casts his broader argument in terms of a experiment in which unfolding events will serve as "laboratory" test of the reliability of the various theories' predictions.[15] Mearsheimer intends to test theories of war and peace, but ancillary theories about whether nuclear operations affect nuclear behavior and about how operations vary are also worthy candidates for the experiment – provided we can identify competing explanations of nuclear operations.

Finally, in championing their theory's policy relevance,[16] RDT enthusiasts indirectly provide the grounds for developing a theory of operations. RDT prescribes steps states should take to make deterrence more robust, and identifies pitfalls states should avoid lest they weaken their deterrent threats.

Significantly, the policy relevance extends to both conventional and nuclear deterrence. This implies, however, that not all nuclear postures are equal; some postures strengthen deterrence while others undermine it. RDT argues that rational states ought to, and will, adopt nuclear behavior which bolsters deterrence – but positing this statement as a prescription underscores the variance in possible nuclear behaviors, and the possibility that some countries may fail to adopt appropriate conduct. Indeed, proliferation optimists are careful to qualify their confidence; in their caveats lies the fertile ground for a theory of nuclear behavior. Waltz bases his conclusion that "the slow spread of nuclear weapons will promote peace and reinforce international stability" on the provisos that states can and surely will (since "they have every reason to do so") adopt secure, second-strike force postures.[17] Mearsheimer similarly concludes that "well-managed" proliferation in Europe may reduce prospects for war, but " ... mismanaged proliferation could produce disaster ... " He identifies four conditions of well-managed proliferation: the transition is peaceful, states adopt second-strike force postures, states adopt proper doctrines, and the arsenals are secure and safe.[18] Nuclear optimism, then, is predicated on the assumption that states will adopt proper postures and appropriate nuclear behavior; they might, but this begs a theory explaining how and why states choose among different patterns of deployment and behavior.

For Waltz, and presumably for other RDT-proliferation optimists, structural realism provides the answer: systemic variables, in this case the presence of other nuclear states, will constrain nuclear behavior. But in *Theory of International Politics*, Waltz is careful to point out that he does not offer a theory of foreign policy (state behavior); his theory tells us what a state will have to react to, but not exactly how a particular state will act.[19] At most, then, proliferation optimists can argue that rational states ought to exhibit certain nuclear behavior. As Waltz acknowledges, non-systemic factors play an important role in the formation of foreign policy and, by extension, nuclear behavior. To understand variations in nuclear behavior a theory must thus incorporate a wider array of components, including systemic and subsystemic elements.

Do variations matter? After all, despite big differences in the kinds of nuclear systems and behavioral patterns among the various nuclear powers, the outcome – if measured in numbers of nuclear wars – has been happily uniform: none. But, as noted above, even optimists concede that not all nuclear behavior is equally desirable. This concession is a tacit admittance of the importance of nuclear behavior, if only at the margins. Waltz cites as grounds for optimism the fact that the superpowers lived through the early cold war period, a time of relatively irresponsible nuclear behavior.[20] While this is evidence that survival is possible, surely it cannot serve as proof that survival was predetermined. Scott Sagan, in reviewing the previously classified record of certain key nuclear operations during the cold war, documents several near-disasters that have hitherto escaped public notice.[21] Counterfactual arguments are never conclusive, but Sagan's findings corroborate the relevance of the very operational patterns we would expect to vary across different proliferating countries.

In sum, optimists and pessimists alike must rely on a theory of nuclear behavior, a theory of how states will deploy the new nuclear forces at their disposal. To explain variations in behavior, a theory must include more variables than structural realism allows; that is, a new theory is needed.[22] The next section offers a first cut at this project by identifying both important factors which the theory should incorporate and, where possible, the causal relationships between those factors.[23] To make the task manageable, I consider only one (albeit crucial) component of nuclear behavior: operations.

Explaining Nuclear Operations: Three Domains

Nuclear operations refers to the day to day management of nuclear weapons and the doctrinal concepts by which the military use nuclear weapons in peacetime and combat.[24] It is distinct from nuclear strategy, which rationalizes political goals with nuclear means, and also from nuclear force structure, which describes different mixes of weapons. There is considerable overlap and synergy between these three components of defense policy and a complete treatment of nuclear proliferation must incorporate each. I focus only on operations for three reasons. First, operations place an important constraint on nuclear behavior by defining (a) what is *possible* for a state to do, and (b) how the nuclear system is *likely* to perform in a crisis.[25] Second, toward the end of the cold war nuclear operations emerged as perhaps the liveliest sub-field of nuclear studies – especially with the declassification of key documents long held back from scholars – and so it is a particularly fruitful area for analysis. Third, it is the least-studied aspect of nuclear proliferation and thus a high research priority.

The operational angle has been slighted in proliferation studies, and the explanation goes to the heart of the problem of building useful theories: there are virtually no data available on operations in emerging nuclear nations. With no data, analysts are reluctant to make claims about operations in current or prospective proliferating countries.[26] However, as more is known about the history of nuclear operations during the cold war, it is possible to identify the general factors that drove superpower nuclear operations. The superpower experience, then, is a useful point of departure for a more general theory of nuclear operations; this experience suggests that factors from three domains shape nuclear operations: the strategic environment, strategic culture and strategic systems. Particular aspects of these domains will vary across different countries and this will necessarily modify the way countries will respond to the "nuclear revolution," but the general classification holds.[27]

Strategic Systems

Strategic systems refers to the characteristics of the arsenal: its size, the type of weapons in it, their mode of delivery, and the character of the command and control system. Each of these features has an independent effect on political behavior generated by the nuclear revolution. The importance of these factors has not been lost on students of nuclear proliferation; even primarily inductive studies rely implicitly on theories about how differences in arsenal characteristics alter assessments about the effects of nuclear proliferation.[28]

Despite the implications of the title of Brodie's seminal work, *The Absolute Weapon*, politicians quickly decided that not all nuclear weapons were created equal. Late in the 1940s the United States countered the burgeoning Soviet nuclear program with two divergent engineering initiatives: one producing extremely high-yield thermonuclear weapons and the other producing low-yield weapons optimized for battlefield use. While both sets are immensely destructive in comparison to conventional weapons, their effects were sufficiently dissimilar that analysts theorized about distinctive political-military strategies based on different mixes of weapons.[29]

Just as different yields potentially recommend different strategies, so also do differences in the size and complexity of the overall arsenal suggest different patterns of state behavior. A country with a few clumsy weapons clustered in one or two depots is apt to consider its arsenal differently from a country that has tens of thousands of sophisticated weapons widely dispersed throughout its zone of strategic activity. The latter arrangement suggests more flexibility and more strategic options for state leaders. It also complicates command and control arrangements, as I discuss below.

Delivery systems vary according to five crucial features, each of which can influence state behavior by changing the capability of the arsenal in dramatic ways: *range*, *promptness*, *accuracy*, *pre-launch survivability*, and *post-launch survivability*. The shorter the range of the delivery system, the more constrained the options for the political leadership. Very short range systems allow only for a doctrine of deterrence by denial, while longer range systems allow for deterrence by punishment. Similarly, proliferators will vary in the promptness with which they can execute nuclear strikes. Traditional strategic theory argues that prompt delivery systems increase the relative power of the country but greatly decrease the stability of strategic relationships by shrinking the time leaders will have to manage crises.[30]

The accuracy of the available delivery systems affects the kinds of enemy targets proliferators can hold at risk. In general, more accurate delivery systems permit counterforce strategies while less accurate delivery systems constrain the proliferator into countervalue targeting.

The pre- and post-launch survivability of the delivery vehicles can also constrain proliferators in important ways. If a delivery system is vulnerable to attack before it is launched, both sides will have a powerful incentive to strike first in any conflict. Similarly, if the post-launch survivability of a delivery system is considerably higher than its pre-launch survivability – that is, an enemy stands some chance of destroying the weapon on the ground but little chance in flight – the enemy may also have a strong incentive to strike first.

The command and control system is another broad component of the strategic system affecting behavior.[31] Command and control as used here refers to the social system designed to balance the always/never dilemma inherent in nuclear operations. Leaders want the nuclear weapon always available for use, but never detonated by accident or without authorization, and they rely on the command and control system to achieve these sometimes conflicting goals. The command system is comprised of three types of components: *hardware*, *procedures* for using the weapons, and the *people* in the chain of command. Hardware refers to engineering devices, often built into the weapon, which provide some physical assurance that the weapons cannot be easily misused. For instance, most modern weapons have Environmental Sensing Devices (ESDs) which block detonation until the weapon undergoes a prescribed environment (for example, achieves free-fall or reaches a certain altitude). This protects against an unwanted detonation triggered by an airplane crash or a weapon falling off a loading dock. Likewise, most modern weapons have Permissive Action Links (PALs) which block detonation until a code is inserted, effectively preventing the holder of the weapon from using it without authorization.

Administrative rules and procedures reinforce the engineering measures. Thus, both the Americans and the Soviets instituted a "two-man rule" which directed that every step in the delivery, maintenance, and launch sequence involve at least two people. Also, senior commanders often kept weapons components or launch codes apart from the delivery systems, thereby frustrating attempts at unauthorized use.

These engineering and administrative procedures are embedded in a social system comprised of fallible human beings. Consequently, human reliability received a lot of attention throughout the cold war in both the United States and the USSR. The Americans adopted the Personnel Reliability Program, an elaborate oversight and review system which monitored the behavior of all individuals with nuclear responsibility in an effort to weed out irresponsible people and to make sure that the military operators would in fact obey an authorized order to execute a nuclear strike.

All these command and control components work together to insure that the nuclear weapons are *reliable*, that is, unlikely to fail at the moment when leaders want to use them, *safe*, that is, unlikely to detonate accidentally, and *secure*, that is, resistant to efforts by unauthorized people to detonate them. These goals are partially contradictory, however, and so every command system must cut some sort of balance among them. For instance, hardware and administrative configurations that minimize the chance of an unauthorized use of nuclear weapons also introduce vulnerabilities into the arsenal that a well-crafted and well-executed enemy strike could exploit.[32] Systems which are optimized to assure the always side are *delegative*; they are designed to minimize vulnerabilities to a disarming enemy first strike, but they necessarily raise the risk of an accidental or unauthorized use of nuclear weapons. Conversely, systems which are optimized to assure the never side are *assertive*; they minimize the chance of an accidental or unauthorized use but they increase one's vulnerability to an enemy first strike.

The strategic systems domain – consisting of arsenal size, weapon mix, and character of the command and control system – greatly influences the political-military behavior of a proliferating country. A proliferator with a small, slow, and assertively controlled arsenal would have a very different range of military options from a proliferator with a large, rapid, and delegatively controlled arsenal. An analysis of nuclear proliferation that glossed over these differences would likely misstate the range of nuclear behavior one could expect from a proliferating state. The relationships can be summarized in a series of propositions. Ceteris paribus, the smaller and cruder the arsenal, the more vulnerable it will be. The more vulnerable the arsenal, the greater the temptation for the proliferator to rely on delegative command systems. The more delegative the command system, the greater the chance for undesirable nuclear operations. This domain is not in itself determinative of all nuclear behavior, however; rather the features of the strategic system interact closely with the second general domain: strategic environment.

Strategic Environment

Strategic environment refers here to those aspects of a new proliferator's geostrategic situation which bound nuclear behavior. The strategic environment alters nuclear behavior by influencing the likelihood a state would feel the need to use nuclear weapons. A state facing no external threat would likely manage its arsenal differently from one threatened by aggressive regional competitors. Two features of the strategic environment are of special importance: geopolitical stability and crisis stability.

Geopolitical stability is a function of the proliferator's position in the regional balance of power and the war-proneness of the region due to other factors, for example contested borders or incompatible demands for justice.[33] At one end of the continuum, one could imagine a local hegemon in a region where inter-state borders enjoy wide legitimacy (Brazil). At the other end of the continuum is a proliferator facing an overwhelming coalition of enemies each of which has a pressing irredentist claim against the new nuclear nation (Israel).

Obviously, the strategic environment places different strains on the nuclear arsenal at different points along the continuum. One would expect that the less secure a state's geostrategic position, the more pressure it would feel to wield its nuclear capability for immediate political benefit, either by adopting deployments and strategies that allow for ready use of nuclear weapons or by brandishing the weapon in exercises of nuclear diplomacy.[34] The pressure would grow even more acute as the threatened state exhausted other diplomatic and conventional military options.

Crisis stability refers to the special pressures a nuclear proliferator will feel if its relations with hostile neighbors escalate to the point of crisis. All other things being equal, a stable situation is one in which there are few or no military imperatives for resorting to the use of military force early in a crisis even as the political situation degenerates. In an unstable situation, the opposite obtains: as the political circumstances degenerate, pressures to use force increase dramatically. The greatest crisis instability is when one or both sides in a conflict have the capacity to deliver a knockout blow, *provided they strike first*. Under such a condition both sides have a strong incentive to escalate quickly to war lest their opponent seizes the advantage.

Crisis stability thus depends heavily on several of the factors noted earlier in the discussion on strategic systems. If a state's arsenal is small and assertively controlled, it is more susceptible to a knockout blow and so its strategic environment will be that much more crisis-unstable.[35] Crisis stability is also a function of the capabilities and disposition of a proliferator's enemy. A small, easily targeted nuclear force is not vulnerable (in a way that would affect crisis stability) if none of the proliferator's competitors has the military wherewithal to execute a credible first-strike.

Geopolitical stability and crisis stability collectively structure the strategic incentives facing a new nuclear proliferator in times of crisis and, in so doing, are important determinants of the state's behavior. The strategic environment sets the parameters for the proliferator's choice of nuclear doctrine, that is, the plans detailing how to use nuclear weapons to achieve military goals. One of the more important features of doctrine is "time-urgency," whether the arsenal must be ready for immediate use early in a conflict or whether the arsenal can be maintained at a relatively low state of readiness. The more hostile the strategic environment the more likely the new proliferator will feel compelled to adopt a time-urgent nuclear doctrine.[36] Such a move, however, can feed back and exacerbate the strategic environment. First, adopting a time-urgent posture puts pressure on your opponents to do the same; if your opponent fails to counter this move, he may be vulnerable to a knockout first strike. Second, a time-urgent posture often involves a delegative command and control system. But a delegative command system is itself more prone to accidents and unauthorized use in a crisis thereby making any emergency that much more dangerous.

Strategic Culture

The third domain, strategic culture, is embedded in the domestic political level of analysis and thus has been slighted by rational deterrence theorists. When a state crosses the nuclear threshold, its new capability is necessarily grafted onto an existing military and political culture. Even if the country invents new agencies or governmental organs to manage the new weapons these new actors must operate within the old social/political/military environment. And each country's distinctive environment, which I call strategic culture, shapes nuclear behavior in meaningful ways.

Strategic culture is an ambiguous term in the literature but here refers collectively to a wide variety of domestic factors including three crucial elements:[37] (1) the pattern of civil-military relations; (2) the character of and relations among other governmental institutions; and (3) the long shadow of military and political history which influences attitudes about the usefulness of military force (including nuclear weapons) as an instrument of state policy.

The way a new proliferator wields its nuclear power will depend, in part, on the way civilian and military leaders interact. Traditional civil-military relations theory, from Clausewitz to Huntington, posits a grand social contract wherein the military promise to submit to civilian rule in exchange for a civilian promise to respect military autonomy on operational matters.[38] The problem, of course, is that few countries match this ideal arrangement. Some polities face severe challenges from would-be military dictators; others have secure civilian control over the military in political terms, but face a conflict over the appropriate degree of military autonomy on expressly military concerns. Operations like the functioning of command and control are central to nuclear behavior and are precisely the issues over which civilian and military leaders are most likely to disagree. Thus, different patterns of civil-military relations will suggest different day-to-day management styles and, as a consequence, different nuclear behaviors.

Patterns of civil-military relations vary according to the assertive-delegative continuum I outlined above in the discussion on command and control.[39] Assertive civil-military relations are those in which the civilians exercise close and direct control over military affairs. Delegative civil-military relations, in contrast, are characterized by a high degree of civilian trust and military autonomy. The preferences of the actors follow the traditional pattern of central control in large bureaucracies. Insiders naturally resist outsiders' interference in cherished operations, jealously guarding their autonomy of activity. The military tend to prefer the operational autonomy afforded by delegative control. In contrast, central decisionmakers in a large organization seek mechanisms to enhance their control over the behavior of critical elements of an organization. Thus, civilians tend to prefer assertive control over nuclear operations.[40]

In the United States, this distribution of preferences has motivated a rich history of civil-military struggle over the control of nuclear operations.[41] Although they favor assertive control, however, American civilian leaders have actually tolerated a relatively high degree of delegation in the nuclear command and control system, thus making the system more survivable and the deterrent more credible. This phenomenon is explained partly by the evolution of the strategic environment. But the nature of civil-military relations also seems relevant. The United States tolerated more delegative control than the Soviet Union did, even though the Soviet Union's strategic environment dictated that *its* arsenal should be the one more delegatively controlled because Soviet forces were more vulnerable to a first strike.[42] Since the United States has a long tradition of military professionalism and voluntary military subordination to civilian control, American policymakers have not confronted stark challenges to civilian rule. Military disobedience, though a concern, has not dominated all other worries about the reliability of the nuclear arsenal.

It is reasonable to expect a nation with a different experience in civil-military relations to respond to the inherent conflict over operational policy in a different fashion. Totalitarian and single-party states where political leaders have used extraordinary control measures are more likely to have assertive patterns of civil-military relations. States in this camp are more likely to be ruled by leaders who value direct and even personal control; thus, they will employ extraordinary measures such as para-military organizations and separate chains of command – as the Soviet Union did with the KGB,[43] and Iraq does with the secret police.[44]

Such assertive control is likely to extend to the nuclear realm as well, because the mechanisms for assertion already exist and because the leaders will naturally expect this kind of control.[45] A checkered history of civil-military conflict – for instance a history of coups – would have a similar effect.

In such cases, the national leader may be a former military officer who exploited his position to achieve political power, or the leader may be a civilian who faces some political competition from the military. In this volatile situation, nuclear weapons could be considered important symbols of political power and so could themselves become stakes in a domestic power struggle. Nuclear weapons have played such a role in attempted coups, especially during the revolt of the French generals in Algeria in 1961.[46] The best way to prevent a domestic rival from exploiting the political power of nuclear weapons is to retain as severe an assertive grip on the arsenal as possible. Hence, the greater the fear of coups, the more likely the weapons are under assertive control.

Brazil provides an interesting outlying case. Brazil has an unhappy tradition of coups and other military interference in domestic politics. Some reports indicate, however, that the military controlled Brazil's nuclear program and did so secretly even during periods of nominal civilian control of the government.[47] Brazil's case may indicate yet another variant: in countries where the military is a more stable institution than the civilian government, a determined military may be tempted to direct the program itself, start to finish. It is unclear, however, what effect this would have on the command and control system, and so, by extension, on nuclear behavior generally. Possibly, whoever was leader of the military program would recognize that his position was precarious and so seek guarantees that the arsenal could not be exploited by his political competitors.

In sum, the command and control system is likely to reflect the underlying pattern of civil-military relations and so nuclear operations in a proliferating country cannot be reliably explained without reference to the very domestic political factors systemic theories ignore. Nuclear weapons are grafted on to an existing power structure and must be wielded by the existing actors in the political spectrum. Accordingly, the prevailing pattern of civil-military relations, which sets the broad parameters for the brokering of power in the state, will influence the command and control system.[48]

The foregoing merits an important caveat. We would expect that, given a volatile pattern of civil-military relations, the top leadership will want assertive control over nuclear weapons. If civil-military relations are too volatile, however, the top leadership may be incapable of asserting control. Countries with pathological civil-military relations are characterized by persistent and pervasive military insubordination. In such cases, the de facto command and control system might be highly delegative, in spite of the strong incentives for tough assertive control. Because of such deductive uncertainty, this caveat must be applied on a case by case basis to countries with extremely volatile civil-military traditions.

The character of and relations among other governmental institutions will also affect nuclear behavior. In the U.S. case, the evolution of nuclear operations was steered not only by civil-military relations but by patterns of relations within successive presidential administrations and between the executive and legislative branches. For instance, different presidents held different expectations for the nuclear command and control system. President Eisenhower accepted a delegative pattern of operations while President Kennedy, facing roughly the same systemic environment, demanded more assertive control.[49] Moreover, the powerful Joint Committee on Atomic Energy (JCAE) played a dominant role in shaping the development of American nuclear weapons during the first two decades of the cold war. Crucial turning points like the decision simultaneously to develop both high-yield and low-yield weapons, or the decision to deploy Permissive Action Links on NATO nuclear forces, cannot be explained without reference to the unique role of the JCAE and its often adversarial relationship with successive administrations.[50]

The argument extends even further to capture the effects of bureaucratic politics at lower levels. For instance, the development of safety and security devices on U.S. weapons partly grew out of the competition between the various weapons laboratories for claims on budget resources.[51] Similarly, it is virtually impossible to explain the dramatic growth in the size and complexity of the U.S. nuclear arsenal without showing how inter-service rivalry resulted in separate Army, Navy, and Air Force strategic nuclear programs. Even the character of command and control systems varies across the services, influenced heavily by individual service cultures and traditions; during the cold war, civilians continued to tolerate delegative control over naval nuclear weapons even while asserting control over army and air force weapons.[52]

These twists and turns in the American nuclear program are more than mere historical detail. As I argued earlier, command and control is an important determinant of nuclear behavior; similarly, the capabilities of the arsenal, driven as they are by the vicissitudes of bureaucratic politics, are themselves important drivers of state behavior in the international system. A complete theory of nuclear behavior should offer explanations for why nuclear programs take different paths. Civil-military relations and other bureaucratic factors contribute to that explanation.

It is possible, however, that these factors will be less important for proliferating countries, especially if the proliferating countries are small. Domestic political factors may be more consequential in large states like the United States than they are in small ones.[53] It is at least plausible that while bureaucratic politics did have an important effect on nuclear behavior in the United States, the impact would be sufficiently slight in smaller countries to be safely ignored in a theory about proliferation. This remains an important question to be resolved as more data about proliferating countries become available.

The resolution of this question is made more vexing by the difficulty of formulating the role bureaucratic factors play in terms of testable hypotheses.[54] Whereas the civil-military factor suggested relatively straightforward preferences for assertive or delegative nuclear operations, these secondary bureaucratic factors are more indeterminate in their effects on nuclear behavior. We know that bureaucratic conflict changes nuclear behavior, but the precise causal relationship is ambiguous. Does interservice rivalry always breed nuclear overexpansion? Does inter-agency competition always result in assertive command and control? It is not very satisfying to say that the factor is important if we cannot specify what its influence will be. But it is also not very convincing to explain the nuclear behavior of the superpowers while ignoring entirely how these factors influenced the course of events at crucial junctures. Perhaps better information about the role of bureaucratic politics in nuclear countries will suggest more testable hypotheses. In the meantime, the hypotheses do serve a useful purpose by directing the attention of researchers to this question.

One last component of strategic culture deserves mention: the long shadow of military and political history which generates attitudes about the usefulness of military force (and particularly nuclear weapons) as an instrument of state policy.[55] Bernard Brodie was the first American to write that nuclear weapons could not be used for traditional military ends.[56] Forty years later, President Reagan seemingly echoed this sentiment with his famous claim that "a nuclear war could not be won and must never be fought."[57] In the intervening years, of course, hundreds of billions of dollars and the considered energies of generations of military leaders, policymakers, and analysts were expended in pursuit of strategies for fighting just such a nuclear war. Nevertheless, the conventional wisdom fastened on the non-usability of nuclear weapons; or rather, on the idea that nuclear weapons were only good for making deterrent threats which themselves increasingly resembled bluffs.[58]

This nuclear maxim necessarily affected the evolution of nuclear doctrine and nuclear behavior. Nuclear weapons which had a primary war-fighting role, for example, battlefield nuclear weapons, were held in disrepute and their longevity owed more to bureaucratic factors than to any conviction that they were vital components of the national security posture.[59] Nuclear brinkmanship enjoyed only the briefest of runs during the 1950s and it was used very sparingly after the shock of the Cuban missile crisis.[60] The point is not that nuclear weapons were inconsequential; rather, the point is that the American strategic culture generally embraced the idea that nuclear weapons were not very usable for narrow military goals.[61] This directly affected the way the United States wielded its nuclear sword, especially once the Soviets achieved some sort of parity in the late 1960s.

It is an open question whether the Soviet Union's strategic culture cultivated a similar understanding of the limited useability of nuclear weapons. Colin Gray, in his critique of the American style, argues implicitly that the Soviet Union was much more willing in the 1970s to envision military uses for nuclear weapons and, consequently, adopted very different nuclear doctrines.[62] The different Soviet strategic culture potentially augured very different behavior if a crisis ever escalated to the level of armed hostilities.[63] Thankfully, Gray's argument was never proven in combat. But it is suggestive of how different strategic cultures can approach nuclear weapons in different ways.

Stephen Rosen similarly argues that appreciating the role of strategic culture is an essential step in understanding the development of nuclear weapons on the Asian sub-continent. Rosen traces Indian strategic canon back hundreds of years and shows how it differs in marked ways from both the American and the Soviet doctrines.[64] Obviously, pending a test no one wants to see, the question of whether Indian nuclear behavior in a crisis will conform to the expectations of deterrence theorists or to the expectations derived from strategic culture must remain open.[65]

Again, it is difficult to identify causal propositions based on the long shadow of military and political history. Waltz already has rejected sharply the more simplistic and ethnocentric version of this argument, that is, that new nuclear states in the Third World are by nature more irresponsible because of cultural backwardness.[66] But Waltz rebuts a straw-person, ignoring a more reasonable and potentially more testable formulation: proliferating states with strategic cultures that accept nuclear weapons as useful will adopt deterrent postures based on war-fighting doctrines. This hypothesis may place an almost intolerable burden on subjective assessment. Absent independent evidence – such as speeches, leaked confidential memos, or interviews with senior policymakers – how do we know that the proliferator has a particular attitude about the usefulness of nuclear weapons without reference to their nuclear behavior or their nuclear doctrine?

The ambiguity of the strategic culture factor is ample reason for being cautious about making dogmatic claims for its precise influence in proliferating countries. But a nagging fact prevents us from dismissing it altogether. The U.S. and the Soviet Union did develop different strategic doctrines and different ways of managing their nuclear arsenals. These differences cannot be explained solely with reference to system-level factors. Some of the differences, however, can be traced to the influence of civil-military relations, relations among other governmental institutions, and cultural attitudes about the usefulness of nuclear weapons. We should expect, therefore, that these factors have leverage in explaining the behavior of proliferating states as well.

The Special Case of Opaque Nuclear Proliferation

The argument advanced above applies with greatest force to countries that openly cross the nuclear threshold. This type of proliferation remains relevant – witness the possible spread of nuclear weapons in the former Soviet Union. Open proliferation, however, is not the only important model for future proliferation theories. Recently, scholars have argued that the current wave of nuclear proliferation is likely to be opaque; proliferating countries will deliberately obscure whether or not they have in fact developed nuclear weapons.[67] The effect of opacity on nuclear operations merits its own study and I can only briefly sketch out some suggestive hypotheses here.

Opaque nuclear proliferation refers to the gray area between no nuclear development program and an openly declared arsenal. The classic opaque proliferator is Israel.[68] The existence of Israel's nuclear weapons program is widely recognized yet Israel has never officially admitted it owns or is seeking to own a nuclear arsenal. Seven basic features characterize opaque proliferation: (1) the proliferator does not conduct a nuclear test; (2) the proliferator denies that it possesses nuclear weapons; (3) the proliferator does not face a direct nuclear threat; (4) the proliferator does not declare a military doctrine involving nuclear weapon; (5) the proliferator does not openly deploy nuclear weapons in a military mode (although it might do so covertly); (6) elites within the proliferating country do not openly debate the role of nuclear weapons; and (7) the nuclear program is rigidly "insulated" from routine security policy activities.[69]

Opacity may be a desirable feature for many would-be proliferators. An opaque proliferator can continue to develop nuclear weapons without incurring world opprobrium as a de jure violator of the nuclear nonproliferation regime. Moreover, if the opacity lifts enough to persuade potential enemies the weapons' program is real, then the opaque proliferator can enjoy the deterrent benefits of possession at minimal cost. From the perspective of nuclear operations, however, opacity introduces important conditions, some of which may ameliorate operational dilemmas and some of which may exacerbate them. The decisive characteristic in this respect is weaponization, the degree to which an opaque proliferator turns a nuclear development program into a deployed military weapon integrated with that country's armed forces.

Since operational behavior changes with different degrees of weaponization, it makes sense to distinguish between four levels of opaque proliferation. The first level of opacity is characterized by virtually no weaponization of the arsenal. Here proliferation is little more than a research program, the development of knowledge which the country someday could exploit for military purposes. Such a condition approximates the deterrent posture Jonathan Schell recommends in *The Abolition*: countries deter each other with the knowledge that if threatened a victim could choose to develop nuclear weapons.[70] Mutual deterrence inhibits both sides and so neither weaponizes, ever. By definition there are no nuclear operations without nuclear weapons so most of the discussion of operational behavior is moot. Under Schell-esque opacity there is no always-never dilemma and command and control reduces to controlling the decision to weaponize. This latter task is not inconsiderable, as France and perhaps Brazil discovered,[71] but it is qualitatively different from the challenge of nuclear operations as defined here.

The second level of opacity is characterized by minimal weaponization, an arsenal barely more embodied than technical knowledge and far less robust than a nuclear military posture. Nuclear operations at this level reduce to Bundy's existential deterrence: the proliferator need not concern itself with the always-never dilemma since any minute degree of uncertainty suffices to deter an adversary.[72] In this case, opacity reinforces deterrence; because the contours of the nuclear program are so shrouded, the enemy of a proliferator is even less confident about any preemptive strike. Opacity breeds uncertainty, and uncertainty breeds existential deterrence. Provided that the proliferator wholly buys Bundy's argument, operational dilemmas are again moot. Command and control means, in this case, controlling the commitment to existential deterrence and eschewing further weaponization.

Arguably, this form of opacity is widespread. India and Pakistan have so far refrained from fully weaponizing their nuclear arsenal. Even Israel might be thought of as an existential opaque proliferator, if the awkwardness of the Yom Kippur War is disregarded (on which more below). I am skeptical, however, of the longevity of these countries' respective commitments to existential deterrence. As noted above, neither the United States nor the Soviet Union ran their nuclear arsenals as if existential deterrence mattered. On the contrary, each side devoted hundreds of billions of dollars (or rubles) in pursuit of marginal returns from a carefully calculated deterrence. Proponents of existential deterrence focus on the enemy, asking whether the enemy will ever truly trust its preemption option. But the question can be turned on its head: will a proliferator ever truly believe that its arsenal is invulnerable? Neither the United States nor the Soviet Union acted like they did, and they each possessed the most survivable weapons of all, submarine-launched ballistic missiles. Even if new proliferators begin with a higher degree of trust in the absolute weapon, any country's overall commitment to existential opaque proliferation probably depends on the likelihood of that country being involved in a major military conflict.[73] When war is unlikely, existential deterrence is cheap. As war becomes more likely, the pressures to assure retaliation will mount. The faith of some members of the policymaking elite (probably the military) will waver as conventional war looms. In a shooting conflict, particularly one that is going badly, the pressures to weaponize further may be irresistible. If I am right, level two opacity is at best a temporary condition, decaying sooner or later into level three.

The third level of opacity is delayed weaponization. During peacetime, the opaque proliferator avoids weaponizing the nuclear program. The military do not develop a use doctrine, nor do they practice with the weapons which are themselves segregated from the rest of the force structure. The elites do not debate the merits of this or that strategic program, and public consideration of strategic dilemmas is virtually nil. Because of the extremely compartmentalized nature of the nuclear program, these strategic issues are scarcely attended to even within government. The always/never dilemma and the other features of nuclear operations are non-issues – until the country faces a crisis that forces a rapid weaponization. This form of opacity is the most dangerous of all; the factors which shape nuclear operations (discussed in section two above) will all come into play in a moment of extreme national emergency. The risks of aberrant behavior are greatest precisely because the opacity has inhibited preparing the national leadership for weighing the trade-offs wisely. Accounts of the 1973 Yom Kippur War suggest that Israel confronted this problem of opacity after the failure of the Israeli armored counter-offensive in the Sinai on 8 October day of the war.[74] Seymour Hersh describes a panicky group of advisors meeting at the last possible moment to discuss nuclear options – the implication being for the first time.[75] If these accounts are credible, Israel confronted serious nuclear operational dilemmas and had very little time to resolve them. The reversal of fortunes in the Golan and the Sinai saved Israel from a potential disaster, but the incident is instructive of the risks inherent in this kind of opacity. And it suggests that the crucial element of control of nuclear operations is controlling the timing and circumstances of weaponization.

The fourth level of opacity is covert weaponization. Here the arsenal has many of the features of an open nuclear capability: the nuclear program is weaponized and senior leadership have confronted the operational dilemmas inherent in a nuclear posture. But this kind of opaque proliferation differs in one crucial respect: the refusal to acknowledge the nuclear capability openly constrains strategic discourse on and even awareness of operational issues. The need to maintain the articificial status of a non-nuclear country chokes off public discussion and prevents a thorough vetting of operational tradeoffs within the governmental elite. This is a difference in degree not kind, for all countries restrict information, and hence discussion, about nuclear operations. Yet the opaque proliferator is that much more restrictive; hence debate in an opaque country is that much less informed, making the management of operational tradeoffs that much more difficult. Since the public is rarely involved in operational policymaking anyway, the effect of opacity is significant chiefly to the extent that it hampers management within the politico-military elite. In particular, nuclear operations are hampered by the inability of the appropriate military forces to practice their nuclear mission regularly and openly; as any military officer knows, operational problems can only be overcome through thorough field tests and regular training.[76]

With this caveat, nuclear operations in this kind of opaque proliferator are likely to follow the pattern described in section two above: the relevant leaders (a smaller and possibly less well-informed group this time) will confront the same set of factors from strategic systems, strategic environment, and strategic culture and will balance the same operational dilemmas. The Israeli arsenal may be moving to this level of opacity. The scanty information available on Israeli nuclear operations suggests a fairly advanced degree of weaponization; at the very least, Israel appears to have some forces dedicated to the nuclear mission (the missiles at Hirbat Zachariah) and so at least some units must have trained for the nuclear mission.[77]

Opacity, therefore, may require some modifications to a theory of nuclear operations. The more weaponized the opaque arsenal becomes, the more its operations will face the constraints derived from the three domains; the less weaponized, the less relevant are operational considerations. Delayed weaponization and covert weaponization may actually exacerbate operational challenges because opacity could hinder strategic discourse and effective management of nuclear operations. While the nuclear behavior of proliferators under the first two forms of opacity may conform to the expectations of proliferation optimists, behavior is likely to vary in the second two forms. This brief discussion, then, underscores the need for a more reliable theory of nuclear behavior – one that can account for variation across different kinds of nuclear proliferation.

Toward a Richer Theory of Nuclear Behavior

Rational deterrence theory has an undeniable advantage over other candidate theories of nuclear behavior. It is parsimonious and embedded in a sophisticated and generalizable understanding of international relations. But its very parsimony limits its application, at least in the realm of nuclear policy. Because it only considers systemic factors, it cannot explain adequately problems couched at a different level of analysis: for example, variations in U.S. nuclear operations over time. Its usefulness as a reliable predictor of nuclear behavior in proliferating states is limited precisely because it concludes that variances in such behavior are inconsequential. Yet its own sanguine conclusions about proliferation depend, in part, on the proliferators adopting the appropriate nuclear operations.

A more useful theory should incorporate systemic and sub-systemic factors. While such a theory is less parsimonious, it offers the prospects of more nuanced explanation and, perhaps, more useful prediction. Where parsimony yields uncertain bias, there are powerful reasons for including more relevant factors, regardless of their origin.

This analysis is an early cut at this project. I identified the three domains which mediate the causal relationship between nuclear weapons and political behavior, and I specified how key features from each domain influence nuclear behavior. The strategic systems domain affects the manageability of the arsenal because not all nuclear weapons are equal. A small nuclear arsenal consisting of unwieldy devices more resembling engineering experiments than weapons of war grants a proliferator a very different capability than does one based on a modern arsenal consisting of prompt, survivable, and accurate missiles. The former offers at best an existential deterrent option; the latter could provide more flexible, and potentially more useable, options. Similarly, a delegative command and control system augurs very different nuclear behavior from an assertive one; in a crisis, a country with delegative control faces sharper risks of an accidental or unauthorized use while a country with assertive control may be unable to execute a nuclear strike at all.

These factors help determine whether proliferation is likely to constitute a major threat to regional or global stability. The same small vulnerable arsenal that leads to crisis instability in a hostile strategic environment can be a minimal threat in a pacific strategic environment. A hard-pressed new proliferator may feel compelled to adopt a time-urgent doctrine which itself will constrain its options in a crisis. A more secure proliferator may feel free to tolerate a nuclear posture that, though vulnerable, makes for moderate political behavior (for example, developing the capability to build nuclear weapons but not actually crossing the threshold into doing so).

The third domain, strategic culture, is at once the most ambiguous and the most intriguing in its effects. If strategic systems and the strategic environment set broad parameters for choice, strategic culture influences the way the nuclear proliferator will frame those choices. A long history of military obedience will make a state more disposed to respond to environmental incentives for delegating control over nuclear operations. Domestic political pathologies may inhibit the way a proliferator manages vulnerabilities in its arsenal. It is at least plausible that new nuclear nations will see their arsenals as more militarily useful than did the superpowers near the end of the cold war, particularly if the strategic systems and the strategic environment domains are similarly conducive to a war-fighting strategy. If their strategic culture supports such a posture, nuclear proliferation may not be nearly as stabilizing as deterrence optimists expect.

Opaque proliferation can either eliminate or complicate nuclear operations. If opaque proliferation means never weaponizing an arsenal then operational considerations never materialize. But if it means weaponizing in a crisis or covertly, then the basic operational dilemmas obtain. A comprehensive theory of nuclear operations, therefore, can incorporate the opacity factor; doing so, however, dilutes prediction due to the difficulty of determining levels of opacity. Almost by definition, opaque proliferation defies robust analysis. A theory which depends upon knowing explicit levels of weaponization is doomed to imprecise forecasts so long as the crucial information is unavailable because the countries involved never lift the shroud of secrecy. Perhaps theories of opaque proliferation will necessarily be somewhat opaque themselves, at least until the database expands.

This analysis illuminates how different nuclear proliferators may behave in very different ways even while undergoing the same nuclear revolution. Examining the factors from the three domains and determining the degree of weaponization will enable experts to make more plausible predictions about the strategic consequences of open and opaque nuclear proliferation in their region of interest.

At least at this stage, however, two hurdles remain before the hypotheses presented here can be considered a complete theory of nuclear behavior. First, the theory should be able to specify outcomes when the factors make for contrary predictions. For instance, how will a new proliferator behave if it has a large arsenal (which is therefore somewhat difficult to manage), is surrounded by capable enemies (and so faces pressures to adopt delegative operations), but has a volatile pattern of civil-military relations (and so feels compelled to adopt assertive operations)? Deterrence theory shares this problem of indeterminacy. Because it dismisses nuclear behavior as inconsequential, it cannot predict how a state will respond to strategic vulnerability: that is, whether to seek survivable strategic systems or delegative command and control measures.

A second hurdle is the need to specify the relative weights for the influence of each domain. Is the strategic culture factor of volatile civil-military relations more important than the strategic environment factor of vulnerability? Here, deterrence theory has an edge (at least in explicitness), by stating categorically that the first two domains dominate the third. I argue that this gravely understates the importance of strategic culture, but I cannot identify unequivocally what its appropriate weight should be.

The best way to resolve these problems is by testing these various hypotheses against the historical record. A comparison of the expectations with the behavior of a large sample of nuclear countries should produce one of three possible outcomes. First, if vastly different nuclear proliferators do behave in roughly similar ways the usefulness of rational deterrence theory is confirmed. Second, if differences in strategic cultures do correlate with different nuclear behaviors then the desire for parsimony is weakened. Third, even if the results are not decisive, the exercise will suggest refinements to specific propositions advanced above. In particular, a more detailed picture of nuclear operations in proliferating states is certain to improve our understanding of the general link (if any) between bureaucratic politics and nuclear operations.

Unfortunately, such a theory test is not possible at this time for the simple reason that the data on nuclear proliferators do not exist. We have a fairly sophisticated understanding of U.S. nuclear programs and our knowledge of the Soviet program, at least as it was during the cold war, is rapidly improving. But we know comparatively little about even mature arsenals like the British, French and Chinese. Despite ample attention over the years, our understanding of third generation proliferators like Israel and India is very unreliable on crucial details about nuclear operations. And for those countries yet to cross the threshold, the information is scarce to nonexistent.

But it is not merely a problem of data. Even if the information were available, a robust theory test would be difficult because the competitor theory to rational deterrence theory presented above needs to be fleshed out and refined so that the propositions are more direct and less equivocal. I believe I have demonstrated the need to open the black box of nuclear behavior theory; I do not claim to have specified or organized all its contents.

The foregoing suggests a research agenda for nuclear proliferation studies. More scholars need to mine the growing mound of declassified information on cold war era nuclear programs and behavior in the U.S. and the former Soviet Union. Even if the results are primarily descriptive, the potential contributions to theory are significant. As information accumulates, scholars should be able to model U.S. and Soviet behavior more accurately, possibly identifying causal relationships that went unnoticed in our earlier superficial accounts. A rich description of our nuclear history should, through process-tracing, help identify crucial interactions at the domestic level that can become hypotheses about nuclear cause and political effect.

This emphasis on the United States and the Soviet Union makes a virtue out of necessity. Theory about nuclear proliferation is bound to be informed by the U.S. and Soviet cases simply because we know those two best. Therefore, it makes sense to fashion as theoretically sophisticated *and* descriptively accurate an understanding as possible. The agenda does not end there, however. New insights about the influence of domestic institutions derived inductively from the U.S. and Soviet cases should be refashioned into generalizable deductive propositions that can guide research directly on nuclear proliferation. This paper identifies important factors that should cue research on individual countries: for example, patterns of civil-military relations, strategic rationales about the useability of nuclear weapons, and the institutional frameworks governing the production of nuclear weapons. Other studies will add to this list. Ultimately, the circle will close when insights derived from studies of proliferating countries can be turned to refine our understanding of the U.S. and Soviet cases as well.

This last point suggests yet another fruitful line of inquiry for proliferation studies. A research project that focuses on the influence of domestic factors on system behavior can profitably be turned to explain better the influence of system factors on domestic behavior. This is the question outlined by Peter Gourevitch in "The Second Image Reversed" and recently addressed in some fashion by a variety of studies.[78] It makes sense to study the domestic institutions in proliferating countries not only to predict how those institutions will affect nuclear behavior, but also to understand how undergoing the nuclear revolution changes the domestic institutions themselves.

With the close of the cold war, the focus of strategic studies has shifted inevitably to post-cold war problems. Nuclear proliferation has emerged at the top of policymaking agendas and so the renaissance of proliferation studies is predictable. The timing is fortuitous in one respect. This time our study of other nuclear countries benefits from recent studies of cold war nuclear operations. And that understanding tells us that domestic factors are too important to be left out of the equation. A generalizable theory of nuclear behavior ought to incorporate all the significant relationships between nuclear cause and political effect, including those operating in all three domains: strategic systems, strategic environment, and strategic culture.

Notes

For their careful reading of earlier drafts and useful suggestions, I thank Stephen Biddle, Paul Chrzanowski, Michael Desch, Lynn Eden, Benjamin Frankel, Joseph Grieco, Ole Holsti, Peter Lavoy, Steven Miller, Emerson Niou, Scott Sagan, David Welch, and the participants in the Stanford Conference on Social Science Theory and Nuclear Weapons Proliferation, 5–6 February 1993. I also thank Roger Cooper for his diligent and energetic research assistance.

1. Thomas Graham has criticized the management approach as a needless distraction from the traditional, and for him, achievable goal of stopping proliferation. See Thomas W. Graham, "Winning the Nonproliferation Battle," *Arms Control Today* (September 1991): 8–13.
2. The topic is not entirely new. One of the first systematic treatments was Joel Larus, *Nuclear Weapons Safety and the Common Defense* (Columbus: The Ohio State University Press, 1967). Another early expression is W. B. Wentz, *Nuclear Proliferation* (Washington, D.C.: Public Affairs Press, 1968). Lewis Dunn has also long championed a balance between prevention and management; see, for example, Lewis A. Dunn and Hermann Kahn, *Trends in Nuclear Proliferation, 1975–1995: Projections, Problems and Policy Options* (Croton-on-Hudson, N.Y.: Hudson Institute, 1976). The question of managing nuclear proliferation has

enjoyed something of a renaissance, however, beginning in the early 1980s and then accelerating with the collapse of the Soviet Union. Michael D. Intriligator and Dagobert L. Brito, "Nuclear Proliferation and the Probability of Nuclear War," *Public Choice* 37, no. 2 (1981): 247–59; John J. Weltman, "Managing Nuclear Multipolarity," *International Security* 6, no. 3 (Winter 1981/82); Bruce Bueno de Mesquita and William Riker, "An Assessment of the Merits of Selective Nuclear Proliferation," *Journal of Conflict Resolution* 26, no. 2 (June 1982): 283–306; Dagobert L. Brito, Michael D. Intriligator, and Adele E. Wick, eds. *Strategies for Managing Nuclear Proliferation* (Lexington, Mass.: Lexington Books, 1983); Shai Feldman, "Managing Nuclear Proliferation," in Jed C. Snyder and Samuel F. Wells, eds., *Limiting Nuclear Proliferation* (Cambridge, Mass.: Ballinger, 1985); Daniel Caldwell, "Permissive Action Links: A Description and Proposal," *Survival* 29, no. 3 (May/June 1987): 224–38; Lewis A. Dunn, *Containing Nuclear Proliferation*, Adelphi Paper no. 263 (London: IISS, 1991); Lewis A. Dunn and Gregory F. Giles, *Nuclear Proliferation Contingency Planning: Defining the Issues* (McLean: Center for National Security Negotiations, 1991); Mark D. Mandeles, "Between a Rock and a Hard Place: Implications for the U.S. of Third World Nuclear Weapon and Ballistic Missile Proliferation," *Security Studies* 1, no. 2 (Winter 1991): 235–69; Gregory Giles, "Safeguarding Undeclared Nuclear Arsenals," *Washington Quarterly* 16, no. 2 (Spring 1993): 173–86; and Robert D. Blackwill and Albert Carnesale, eds., *Coping with New Nuclear Nations* (forthcoming).

3. Waltz's formulation of proliferation optimism and Mearsheimer's recapitulation have garnered the most notoriety, although neither originated the idea that nuclear proliferation was good. The French strategic thinkers, especially Pierre Gallois, were probably the first to draw the conclusion which itself derives from Brodie's analysis in *The Absolute Weapon*. See Pierre Gallois, *The Balance of Terror* (Boston: Houghton Mifflin, 1961); Bernard Brodie, ed., *The Absolute Weapon: Atomic Power and World Order* (New York: Harcourt Brace, 1946); and see citations in n. 2. I treat Waltz and Mearsheimer as my foils precisely because their elegant treatments are now standard and because of their obvious links to neorealist theory. Waltz and Mearsheimer take different paths to their common conclusion about nuclear proliferation. Peter Lavoy notes that Waltz is optimistic about the potential of nuclear deterrence, while Mearsheimer is pessimistic about the inevitability of proliferation (and so seeks to make a virtue out of necessity, at least where Germany is concerned). Waltz's formulation poses a higher hurdle for developing a theory of nuclear behavior and so I consider it more closely in the text. Peter Lavoy, "Learning to Live with the Bomb: The Politics of Nuclear Proliferation in South Asia" (Ph.D. diss., University of California–Berkeley, forthcoming). Kenneth N. Waltz, *The Spread of Nuclear Weapons: More May be Better*, Adelphi Paper no. 171 (London: IISS, 1981), passim; Kenneth N. Waltz, "Nuclear Myths and Political Realities," *American Political Science Review* 84, no. 3 (September 1990): 731–45; and John J. Mearsheimer, "Back to the Future: Instability in Europe After the Cold War," *International Security* 15, no. 1 (Summer 1990): 38. Stephen Van Evera challenges much of Mearsheimer's argument but nevertheless concurs that a German nuclear arsenal would be acceptable. Stephen Van Evera, "Primed for Peace: Europe after the Cold War," *International Security* 15, no. 3 (Winter 1990/91): 54.
4. Thomas Schelling, *Strategy of Conflict* (New York: Oxford University Press, 1963), passim. Also, Bruce Bueno de Mesquita, *The War Trap* (New Haven: Yale University Press, 1981), passim. For an admirably concise summary, see Christopher Achen and Duncan Snidal, "Rational Deterrence Theory and Comparative Case Studies," *World Politics* 46, no. 2 (January 1989): 150–53.
5. Achen and Snidal, "Rational Deterrence Theory," 159.

6. Richard Ned Lebow and Janice Gross Stein are concise and unambiguous: "Proponents of 'rational deterrence theory' contend that it is conceptually sound, a good predictor of strategic behavior, and a successful strategy of conflict management. All three assumptions are unwarranted." Richard Ned Lebow and Janice Gross Stein, "Rational Deterrence Theory: I Think Therefore I Deter," *World Politics* 46, no. 2 (January 1989): 208.
7. For a good introduction, see the special issue of *World Politics* 46, no. 2, (January 1989) containing the following articles: Achen and Snidal, "Rational Deterrence Theory"; Alexander L. George and Richard Smoke, "Deterrence and Foreign Policy"; Robert Jervis, "Rational Deterrence: Theory and Evidence"; Lebow and Stein, "Rational Deterrence Theory"; and George W. Downs, "The Rational Deterrence Debate."
8. Lebow and Stein, "Rational Deterrence Theory," 208–12. Also Robert Jervis, *Perception and Misperception in International Politics* (Princeton: Princeton University Press, 1976).
9. Waltz, *The Spread of Nuclear Weapons*, 3–10, esp. 5–6. The deterrence failure school claim that their insights do in fact apply to nuclear as well as conventional deterrence. But they face the same counterfactual problem noted in the text. See, for example, Richard Ned Lebow, *Nuclear Crisis Management: A Dangerous Illusion* (Ithaca: Cornell University Press, 1987).
10. McGeorge Bundy, "Existential Deterrence and Its Consequences," in Douglas MacLean, ed., *The Security Gamble: Deterrence Dilemmas in the Nuclear Age* (Totowa, N.J.: Rowman & Allanheld, 1984), 3–13.
11. Of course, we may worry about nuclear proliferation because it threatens other goals besides war-avoidance. Even if rational deterrence theory is correct about the implications of the spread of nuclear weapons, or rather especially if it is correct, proliferation poses grave policy problems for the great powers. RDT posits that nuclear proliferation will multiply situations of mutual deterrence wherein each side can threaten the other into inaction. This condition is agreeable if the only objective is to avoid war. But war-avoidance is only one goal for great powers like the United States; in fact, at times great powers want the freedom to go to war in order to prevail over other transgressor countries. RDT suggests that nuclear proliferation would make it far easier for smaller powers to resist great powers. For instance, a nuclear-capable Iraq perhaps could have deterred the United States from launching Desert Storm. A nuclear-armed Iraq might have exercised regional hegemony secure in the knowledge that its arsenal deterred other countries from getting involved. Nuclear war is avoided, but at what cost? One reason RDT fails to convert everyone to optimism, therefore, is that nuclear proliferation vastly complicates regional power management problems. Great powers prefer that membership in the nuclear club be kept limited in order to maximize their latitude in international affairs. This is a reason to worry about nuclear proliferation, but it does not address the deeper point about whether nuclear behavior is an interesting theoretical puzzle.
12. I skip over two responses that lead away from the focus of my argument. First is the flippant answer most obvious to optimists and least persuasive to pessimists: people who worry about proliferation are benighted. A second answer is more serious, but it involves questions that go well beyond the scope of this paper: perhaps people reject proliferation optimism because they are unwilling to trust counter-intuitively rosy predictions from a social science theory. After all, as John Lewis Gaddis has been at such pains to remind us, political science (and especially the international relations subfield) is scarcely renowned for the reliability of its predictions. If intuition suggests that nuclear proliferation is bad, but a "robust and parsimonious" theory of international relations says it is good, Gaddis counsels that we disregard theory; especially when that theory has shed little or

no light on other questions of great import–say, how and why the Cold War ended. Gaddis' diagnosis is far more convincing than his confusing prescription: at points he seems to imply that we abandon forecasting entirely, but he concludes by recommending that political scientists merely augment their theories with the tools of fiction writers–"narrative, analogy, paradox, irony, intuition, imagination, and–not least in importance–style." If I understand him, I find his caution nonetheless useful, and hereby invoke all the caveats and contingencies Gaddis recommends. See John Lewis Gaddis, "International Relations Theory and the End of the Cold War," *International Security* 17, no. 3 (Winter 1992/93): 5–58.

13. Achen and Snidal, "Rational Deterrence Theory," 161.
14. The forty-five years are even less reassuring if one considers that the weapons programs of both superpowers belied any reliance on Bundy-esque existential deterrence. At the very least, neither American nor Soviet leaders acted like they thought existential deterrence sufficed. Waltz dismisses this behavior as an atavistic hangover from the 1930s. He may be right but current changes allow for a more robust test. Waltz, "Nuclear Myths," 734–35.
15. Mearsheimer, "Back to the Future," 9–10. See also the articles by John Lewis Gaddis, Robert Jervis, Carl Kaysen, John Mueller, Jack Snyder, and Steven Van Evera, compiled in Sean M. Lynn-Jones, ed., *The Cold War and After: Prospects for Peace* (Cambridge: MIT Press, 1991).
16. Achen and Snidal, "Rational Deterrence Theory," 153.
17. Waltz, *The Spread of Nuclear Weapons*, 28 and 16, respectively. While Waltz's view of the importance of this requirement is relatively consistent, his thinking on the difficulty of meeting it has undergone a curious evolution. In *Theory of International Politics*, Waltz dismisses the French nuclear deterrent as hopelessly vulnerable: "French officials continue to proclaim the invulnerability of their forces, as I would do if I were they. But I would not find my words credible." In *The Spread of Nuclear Weapons*, he is more optimistic about securing a sufficient second-strike capability: "Will such countries be able to construct and protect a deliverable force? We have found that they can readily do so." By the 1990 APSR article, he becomes positively sanguine: " ... the invulnerability of a sufficient number of warheads is easy to achieve and the delivery of fairly large numbers of warheads impossible to thwart, both now and as far into the future as anyone can see." To my knowledge, Waltz has not explained why he deems the third-ranked nuclear arsenal irredeemably vulnerable and yet any crude Third World arsenal securely invulnerable. See Waltz, *Theory of International Politics*, 181; Waltz, *The Spread of Nuclear Weapons*, 17; and Waltz, "Nuclear Myths," 732. I am grateful to Scott Sagan for suggesting this point to me.
18. Mearsheimer, "Back to the Future," 37–8.
19. "[A systemic theory like balance-of-power] ... explains why a certain similarity of behavior is expected from similarly situated states. The expected behavior is similar, not identical. To explain the expected differences in national responses, a theory would have to show how the different internal structures of states affect their external policies and actions." Kenneth Waltz, *Theory of International Politics* (Reading, Mass.: Addison-Wesley, 1979), 122.
20. Waltz notes, for instance: "Relations between the United States, the Soviet Union, and later among the United States, the Soviet Union, and China, were at their bitterest just when their nuclear forces were in early stages of development, were unbalanced, were crude and presumably hard to control." Waltz, *The Spread of Nuclear Weapons*, 16.
21. Scott D. Sagan, *The Limits of Safety: Organizations, Accidents, and Nuclear Weapons* (Princeton: Princeton University Press, 1993).
22. Even those who criticize Waltz for his conclusions about the desirability of

nuclear proliferation often invoke similarly gaunt theory, slighting the domestic factors which shape nuclear behavior. For example, Dagobert Brito and Michael Intriligator's formal model of proliferation is a useful corrective to Waltz because it considers the problem of accidental war, perhaps arising out of an unauthorized nuclear use. Consequently, it gives a more nuanced analysis of the conditions under which nuclear proliferation is desirable. But the model does not examine the operational factors which make accidental nuclear war more likely and so does not constitute a theory of nuclear operations. Brito and Intiligator, "An Assessment of the Merits." Moreover, while operational issues were raised in an exchange between Waltz and a series of critics, none advanced a comprehensive theory of how operations would vary across the field of proliferators. See exchange including Kenneth Waltz, Dagobert Brito, Michael Intriligator, George Quester, Bruce Russett, Thomas Schelling, and Dina Zinnes in *Strategies for Managing Nuclear Proliferation*, 99–162. Waltz criticizes unnamed analysts who fear proliferation on the basis of negative assessments about the domestic character of would-be proliferators. In fact, however, traditional treatments generally downplay domestic factors even while incorporating, often unconsciously, one of the few variables that structural realism deems significant, that is, the distribution of power in the system. See Bruce D. Berkowitz, "Proliferation, Deterrence, and the Likelihood of Nuclear War," *Journal of Conflict Resolution* 29, no. 1 (March 1985); Daniel S. Geller, "Nuclear Weapons, Deterrence, and Crisis Escalation," *Journal of Conflict Resolution* 34, no. 2 (June 1990); Rodney Jones, ed., *Small Nuclear Forces and US Security Policy: Threats and Potential Conflicts in the Middle East and South Asia* (Lexington, Mass.: Lexington Books, 1984); Geoffrey Kemp, *Nuclear Forces for Medium Powers: Strategic Requirements and Options*, Adelphi Papers no. 107 (London: IISS, 1974); John J. Weltman, "Managing Nuclear Multipolarity," *International Security* 6, no. 3 (Winter 1981/82); Albert Wohlstetter, et al., *Moving Toward Life in a Nuclear Armed Crowd?* (Los Angeles: Pan Hueristics, 1976). Lewis Dunn and Leonard Spector are important exceptions in that they self-consciously treat domestic factors, albeit in an inductive fashion. Lewis A. Dunn, *Containing Nuclear Proliferation*, Adelphi Paper no. 263 (London: IISS, Winter 1991); and Leonard Spector, *Nuclear Proliferation Today* (New York: Vintage, 1984); Leonard Spector, *New Nuclear Nations* (New York: Vintage, 1985); Leonard Spector, *Going Nuclear: The Spread of Nuclear Weapons 1986–1987* (Cambridge, Mass.: Ballinger, 1987); Leonard Spector, *The Undeclared Bomb* (Cambridge, Mass.: Ballinger, 1988); and Leonard Spector, *Nuclear Ambitions: The Spread of Nuclear Weapons, 1989–90* (Boulder: Westview, 1990).

23. Achen and Snidal rightly point out that lists of variables are not substitutes for theory. But they admit that the identification of important variables does constitute a significant step in theory-building. Achen and Snidal, "Rational Deterrence Theory," 155.
24. This definition of operations is modified from Stephen Biddle, "The Determinants of Offensiveness and Defensiveness in Conventional Land Warfare" (Ph.D. diss., Kennedy School of Government, 1991), 11.
25. Ashton B. Carter, John D. Steinbruner, and Charles A. Zraket, eds., *Managing Nuclear Operations* (Washington: Brookings, 1987): 1–13.
26. Elsewhere, I have sketched out a framework for estimating command and control in emerging nuclear nations. My argument here builds upon that earlier work. Feaver, "Command and Control in Emerging Nuclear Nations," *International Security* 17, no. 3 (Winter 1992/93): 160–87. Also Gregory Giles has recently assembled command and control information for certain proliferating countries, Giles, "Safeguarding Undeclared Nuclear Arsenals," 173–86.
27. Note that this discussion applies to a variety of proliferation scenarios. Motiva-

tions do not matter; operations are important regardless of why a country chooses to proliferate. The path to proliferation is also not necessarily relevant; operations are important whether a country resolutely develops a weapon, underhandly steals a weapon, or blindly stumbles from a civilian nuclear energy program into building an arsenal. Opaque and covert proliferation *does* affect operations, however, and I discuss this more fully later in the text.

28. See, for example, the analysis by Leonard Spector cited in n. 22.
29. Witness, for example, the evolution of Army battlefield nuclear doctrines. John J. Midgley, Jr., *Deadly Illusions: Army Policy for the Nuclear Battlefield* (Boulder, Colo.: Westview, 1986).
30. Thomas Schelling, *Arms and Influence* (New Haven: Yale University Press, 1966), 244–48.
31. Peter D. Feaver, "Command and Control in Emerging Nuclear Nations," 162–68. See also, Donald Cotter, "Peacetime Operations: Safety and Security," in Carter, Steinbruner, and Zraket, *Managing Nuclear Operations*, 17–74; House Committee on Armed Services, Panel on Nuclear Weapons Safety, *Nuclear Weapons Safety*, Report, 101st Cong., 2nd Sess., December 1990; and Peter Stein and Peter Feaver, *Assuring Control of Nuclear Weapons: The Evolution of Permissive Action Links* (Lanham, Md.: University Press of America, 1987).
32. For example, a country might decide to keep the warheads separate and distant from the delivery vehicles, thus making unauthorized use virtually impossible. But this severely complicates the ability to use the weapons quickly, especially if under attack.
33. For an interesting argument about how states are motivated to war by competing claims of justice see David Welch, *Justice and the Genesis of War* (New York: Cambridge University Press, 1993, forthcoming). With centuries-old rivalries and grievances reigniting throughout the world, Welch's analysis of justice claims is particularly timely.
34. Of course, a regional nuclear hegemon could be comparatively free to engage in nuclear blackmail with its hapless neighbors and so behave in an even more careless fashion. This is a possibility, but it is interesting to note that Richard Betts finds only equivocal evidence for arguing the U.S. behaved in this way during its period of nuclear "superiority." See Richard K. Betts, *Nuclear Blackmail and Nuclear Balance* (Washington, D.C.: Brookings, 1987), passim.
35. Of course, a potential attacker might fear that the intended victim would switch to a delegative command system (for example, delegate the authority to use nuclear weapons) just as a preemptive strike is underway and so be able to retaliate anyway. If this calculation discourages the potential attacker, deterrence may hold. This is one of the ways existential deterrence could work, the mere shadow of a doubt preventing an attack. As discussed before, if both sides believe in existential deterrence (and act on that belief), then the problem of nuclear behavior is moot. Several factors militate against this calculation, however: first, countries may find a last minute delegation difficult to achieve, especially if they have a rigidly assertive command structure in place; second, last minute transitions increase the chance for accidents and unauthorized activities at precisely the most dangerous time, when the nation is already in a crisis.
36. Deductively, this relationship holds if a proliferator is unable to improve the strategic environment in other ways, for example by resolving regional tensions through arms control and peace settlements or by building up its own forces to the point where it is invulnerable to an enemy first strike.
37. I use this admittedly loose understanding of strategic culture in order to capture as wide a range of factors as possible. Strategic culture here is an umbrella concept that draws from, but is not limited to, more precise formulations. See, for example, Gabriel A. Almond and Sidney Verba, eds., *The Civic Culture Revisited*

(Boston: Little, Brown, 1980); Carl H. Builder, *The Masks of War: American Military Styles in Strategy and Analysis* (Baltimore: Johns Hopkins University Press, 1989); Kenneth Booth, *Strategy and Ethnocentrism* (New York: Holmes and Meier, 1979); Colin Gray, "National Style in Strategy," *International Security* 6, no. 2 (Fall 1981); Colin Gray, *National Strategy and National Style* (Boston: Hamilton Press, 1986); Joseph Lepgold, "Cultural Arguments as Explanations of Foreign-Policy Behavior: The American Case" (Paper presented at the 1992 Annual Meeting of the American Political Science Association, Chicago, September 1992); Glenda M. Patrick, "Political Culture," in Giovanni Sartori, ed., *Social Science Concepts* (Beverly Hills, Calif.: SAGE, 1984).

38. Carl Von Clausewitz, *On War*, ed. and trans. Michael Howard and Peter Paret (Princeton: Princeton University Press, 1976); and Samuel Huntington, *Soldier and State* (Cambridge, Mass.: Belknap Press, 1957).
39 This section is adapted from Feaver "Command and Control," 170–78.
40. Not all civilians demand the same level of assertive control, but civilians in general prefer more assertive control than the military would want to give if left to its own devices. This would hold, but with lesser force, even in those cases where the civilian leader adopts a relatively delegative administrative style overall, as did Presidents Eisenhower and Reagan. This assumption holds with the United States and the Soviet Union and so has prima facie plausibility as a working assumption for proliferation studies. However, testing and if necessary modifying this hypothesis should be a priority for future research on nuclear proliferation.
41. The story of this struggle over nuclear custody is detailed at length in Feaver, *Guarding the Guardians*. It is not limited to nuclear operations, however, as the dispute over President Johnson's "micro-management" of the Vietnam War indicates.
42. Bruce Blair, *The Logic of Accidental War* (Washington: Brookings, 1993).
43. The recent dramatic changes in the Soviet Union raise interesting questions about the future of command and control in the second nuclear superpower. Most of the discussion in the West so far has focused on the problem of transition stability, i.e., whether the old and presumably reliable Soviet command system will fail during the tumultuous transition from the Soviet empire to a federation of independent republics. A secondary question, also of interest, is how evolving civil-military relations will cause the command and control system to change over time.
44. Stephen Biddle and Robert Zirkle, "Technology, Civil-Military Relations, and Warfare in the Developing World" (Paper presented at the Mershon Center Conference on Civil-Military Relations, 4–5 December 1992).
45. The link between leadership style and nuclear command and control was evident in the evolution of custody policy in the United States. Presidents with a delegative administrative style (for example, Eisenhower) tended to tolerate more delegation in nuclear command and control, whereas assertive presidents (such as Kennedy) carried this style into the control of nuclear weapons as well. See Feaver, *Guarding the Guardians*, chaps. 3, 7, and 8.
46. See Spector, *Going Nuclear*, 28–32.
47. See James Brooke, "Brazil Uncovers Plan by Military To Build Atom Bomb and Stops It," *New York Times*, 9 October 1990, pp. 1, 4.
48. The overall process is probably interactive: nuclear weapons affect civil-military relations just as civil-military relations affect the way states manage nuclear weapons. The effect of nuclear weapons on civil-military relations is discussed in Peter Feaver, "Guarding the Guardians: Civil-Military Relations and the Control of Nuclear Weapons" (Ph.D. diss., Harvard University, 1990), 46–94. See also Scott D. Sagan, "The Nuclear Sword and Scepter: Civil-Military

Relations and Nuclear War" (Paper delivered at the Mershon Center Conference on Civil-Military Relations, 4–5 December 1992).

49. See Feaver, *Guarding the Guardians*, 149–98.
50. The story of the sometimes stormy relationship is told in the official histories of the Atomic Energy Commission. Richard G. Hewlett and Oscar E. Anderson, Jr., *The New World, 1939–1946*, vol. 1 of *A History of the United States Atomic Energy Commission* (University Park, Pa.: Pennsylvania State University Press, 1962); Richard G. Hewlett and Francis Duncan, *Atomic Shield, 1947/52*, vol. 2 of *A History of the United States Atomic Energy Commission* (University Park, Pa.: Pennsylvania State University Press, 1969); Richard G. Hewlett and Jack M. Holl, *Atoms for Peace and War, 1953–61*, vol. 3 of *A History of the United States Atomic Energy Commission* (University Park, Pa.: Pennsylvania State University Press, 1989). See also Feaver, *Guarding the Guardians*, 175–91.
51. Interview with author.
52. Vice Admiral G. E. Miller (Ret.), "Who Needs PALS?" *Naval Institute Proceedings* (July 1988): 50–56.
53. Peter Katzenstein has shown how large states have more freedom to maneuver around systemic constraints on political economy issues than do small states. Peter Katzenstein, *Small States in World Markets* (Ithaca: Cornell University Press, 1985); and Katzenstein, ed., *Between Power and Plenty* (Madison: University of Wisconsin Press, 1978). Jack Snyder has argued that this phenomenon obtains in security matters as well. Snyder, *Myths of Empire*, 62.
54. See David Welch, "The Organizational Process and Bureaucratic Politics Paradigms: Retrospect and Prospect," *International Security* 17, no. 2 (Fall 1992): 120–21.
55. Joseph Nye has called the evolution of these attitudes "nuclear learning." Joseph S. Nye, Jr. "Nuclear learning and U.S.–Soviet security regimes," *International Organizations* 41, no. 3 (Summer 1987).
56. Brodie, *The Absolute Weapon*.
57. Ronald Reagan, "Address Before a Joint Session of the Congress on the State of the Union," 25 January 1984.
58. Gray, "National Style in Strategy," 37–38. Former senior policymakers aided the establishment of this consensus with their claims, many years after leaving government, that their on- and off-the-record efforts to advance war-fighting strategies were conscious efforts in futility since they knew no president would ever authorize the first strategic use of nuclear weapons. See, for example, McGeorge Bundy, *Danger and Survival: Choices About the Bomb in the First Fifty Years* (New York: Random House, 1988).
59. Catherine Kelleher and Ernest May "History of the Development and Deployment of BNW," in Stephen Biddle and Peter Feaver, *Battlefield Nuclear Weapons: Issues and Options* (Lanham, Md.: University Press of America, 1989), 13–32.
60. Betts, *Nuclear Balance and Nuclear Blackmail*. See also, Gordon H. Chang, "To the Brink: Eisenhower, Dulles and the Quemoy-Matsu Crisis," *International Security* 12, no. 4 (Spring 1988); Roger Dingman, "Atomic Diplomacy During the Korean War," *International Security* 13, no. 3 (Winter 1988/89); Michael Howard, "Nuclear Danger and Nuclear History," *International Security* 14, no.1, (Summer 1989).
61. This statement is not necessarily incompatible with Steven Kull's apparently somewhat contrary assessment. Kull traces the evolution of elite attitudes towards nuclear weapons and finds ample evidence of an embrace of a war-fighting role of nuclear weapons (what he calls the "traditional stream" of strategic thought) throughout the cold war extending into the Reagan administration. Moreover, he finds many instances of the same individual holding contradictory views on the usability of nuclear weapons. Kull's explanation,

springing from the language of psychotherapy, is that these individuals are manifesting an inner struggle between conflicting goals; national security being only one and not always the dominant ambition. Kull notes, as I do above, that procurement policies tended to reflect the view that nuclear weapons were usable, but crisis behavior was much more circumspect. Significantly, however, Kull views the inner conflict as important. Although he believes the "reality" is that nuclear deterrence makes war futile (thereby embracing one tenet of the proliferation optimists), he nevertheless worries about any inner conflict within American strategic culture (implying that strategic culture can overwhelm deterrence logic). If Kull is correct, and if his analysis is extended to proliferation studies, there is ample reason to be concerned with elite attitudes on the useability of nuclear weapons in emerging nuclear nations. Steven Kull, *Minds at War* (New York: Basic Books, 1982), passim.

62. Gray, "National Style in Strategy," 21. Gray cites as evidence Jack Snyder, *The Soviet Strategic Culture: Implications for Limited Nuclear Operations*, R-2154-AF (Santa Monica, Calif.: RAND, September 1977).
63. Although strict deterrence theorists, as noted before, claim that culture would not, in the end, matter much.
64. Stephen Peter Rosen, *Hindu Strategic Behavior*, unpublished ms.
65. An anecdote from the 1991 Gulf War makes the point sharply. A senior Indian military officer, asked to speak to the lessons of Desert Storm, remarked that he believed the most important lesson was the necessity of using nuclear weapons early in any conflict with the United States. This is clearly not conclusive evidence that the Indians would handle a large nuclear arsenal different from the way the U.S. or the Soviets did; indeed, the narrative of the cold war is replete with similarly colorful quotes from bellicose American and Soviet military officers. But it is suggestive of how strategic culture could condition the way a state would treat nuclear weapons.
66. Waltz, *The Spread of Nuclear Weapons*, 11–13.
67. The term "opaque proliferation" was coined by Benjamin Frankel in "Notes on the Nuclear Underground," *The National Interest*, no. 9 (Fall 1987). It was further developed in Avner Cohen and Benjamin Frankel, "Opaque Nuclear Proliferation," *Journal of Strategic Studies* 13, no. 3 (September 1990).
68. Yair Evron, "Opaque Proliferation: The Israeli Case," *Journal of Strategic Studies* 13, no. 3 (September 1990); and Shlomo Aronson, *The Politics and Strategy of Nuclear Weapons in the Middle East: Opacity, Theory, and Reality, 1960–1991. An Israeli Perspective* (Albany: SUNY Press, 1992).
69. Cohen and Frankel, "Opaque Nuclear Proliferation," 21–22.
70. Jonathan Schell, *The Abolition* (New York: Knopf, 1984), 114–21, esp. 118.
71. Lawrence Scheinman argues that French technocrats pushed the nuclear weapons program along with a minimum of political leadership during the Fourth Republic. Sylvia Crosbie further speculates that a small group of French military officers may have fueled the drive for an accelerated nuclear program, absent political direction, because of their disatisfaction with allied support at Dienbienphu, Suez, and the North African campaigns. Lawrence Scheinman, *Atomic Energy Policy in France Under the Fourth Republic* (Princeton: Princeton University Press, 1965), 213; Sylvia K. Crosbie, *A Tacit Alliance* (Princeton: Princeton University Press, 1974), 119–20. Political control over Brazil's nuclear program was similarly ambiguous, see Brooke, "Brazil Uncovers Plan," 1, 4.
72. Bundy, "Existential Deterrence," 9–10.
73. There are other reasons for being skeptical. For instance, the decision *not* to weaponize a bilateral opaque nuclear relationship is like a prisoner's dilemma. While it may be in both parties interests not to weaponize (not to defect), coordination problems may compel both sides to do so. Moreover, as Steven

Flank argues in his article in this volume, the process of developing nuclear proliferation is itself a political and social process. Contingent factors may push for weaponization even if the top leadership prefer the existential or Schell-esque deterrence afforded by *not* weaponizing.

74. Shlomo Aronson, *The Politics and Strategy of Nuclear Weapons*, 139–49, esp. 147; Seymour Hersh, *The Samson Option: Israel's Nuclear Arsenal and American Foreign Policy* (New York: Random House, 1991), 225–40; and "How Israel Got the Bomb," *Time*, 12 April 1976, p. 39.
75. Hersh, *The Samson Option*, 225–27.
76. Giles also suggests that this kind of opacity complicates nuclear operations, possibly making for unsafe nuclear behavior. Giles, "Safeguarding Undeclared Nuclear Arsenals," 178.
77. Hersh, *The Samson Option*, 225–29. For further speculation on Israeli nuclear operations see Peter Pry, *Israel's Nuclear Arsenal* (Boulder, Colo.: Westview, 1984), 80; and Giles, "Safeguarding Undeclared Nuclear Arsenals," 175–78.
78. Peter Gourevitch, "Second Image Reversed: The International Sources of Domestic Politics," *International Organization* 32, no. 4 (Autumn 1978): 881–911. See also, Biddle and Zirkle, "Technology, Civil-Military Relations, and Warfare in the Developing World"; Michael Desch, "Systemic Determinants of Civil-Military Relations: The Military Technical Revolution and Changing Soviet Civil-Military Relations," unpublished paper; Peter Feaver, "Civil-Military Relations and the Use of Force," a research proposal available from the author; Sagan, "The Nuclear Sword and Scepter"; Jack Snyder "International Leverage on Soviet Domestic Change," *World Politics* 42, no. 1 (October 1989); and Snyder, *Myths of Empire*, 318.

Nuclear Myths and the Causes of Nuclear Proliferation

Peter R. Lavoy

HOW MANY countries will acquire military nuclear capabilities in the new era? The experts are divided on this issue. "Proliferation pessimists" argue that the global diffusion of nuclear technology combined with the rise of new international security threats and the decline of old security structures makes rampant nuclear proliferation inevitable.[1] In contrast, "nonproliferation optimists" expect an increasingly robust regime of technology transfer controls and international arrangements for detecting, deterring, and punishing proliferation to produce a post-cold-war world of even fewer nuclear-weapons states than presently exist.[2]

At first glance, both predictions appear plausible. Each perspective points to fundamental forces that have influenced the past spread of nuclear weapons and should affect the pace and scope of nuclear proliferation long into the future.[3] The character of international politics, the availability of military technology, and the effectiveness of nonproliferation measures all matter, but so do many other factors. Nuclear weapons acquisition is a complex and historically contingent process; its causes cannot be reduced to a single feature of international life. Proliferation pessimism and nonproliferation optimism provide unreliable forecasts because they rely on overly simplified notions of what drives and inhibits national nuclear weapons activities.[4]

Understanding nuclear proliferation (and nonproliferation) is as important – and as exacting – as predicting it.[5] This article presents an original model for explaining the spread of nuclear weapons.[6] I argue that a government is likely to "go nuclear" when proficient and well-positioned individuals who want their country to build nuclear bombs, exaggerate security threats to make a "myth of nuclear security" more compelling. In contrast to familiar accounts of nuclear proliferation which emphasize the security, technology, or prestige determinants of nuclear arms acquisition, I stress the importance of nuclear myths and myth makers.[7] The strategic beliefs and political activities of highly motivated and resourceful individuals are where the sources of nuclear proliferation can be found.

This article contains three sections. The first section identifies barriers to the creation of theories about nuclear proliferation, describes the three leading arguments about the causes of nuclear arms acquisition, and reviews the empirical and logical shortcomings of each account. Next I present an explanatory model of nuclear proliferation and elaborate its conceptual components. Finally, I suggest several implications of this model for international relations theory and nuclear nonproliferation policy making.

THE CAUSES OF NUCLEAR PROLIFERATION: CONTENDING ARGUMENTS

Why do countries acquire military nuclear capabilities? This question has occupied the attention of strategic analysts and political scientists since the dawn of the nuclear era, but the conventional wisdom is far from satisfying. Scholars have failed to overcome two obstacles that impede the creation and cumulation of knowledge about the causes of nuclear proliferation.

The first barrier is empirical: only nine to twelve states are known or believed to possess military nuclear capabilities.[8] The number of relevant cases remains small even taking into account the countries that are recognized to have launched nuclear arms programs but for various reasons either stopped trying to build the bomb or have yet to test or assemble an explosive device.[9] The problem of too few cases is compounded by the paucity and unreliability of public information about many of them.[10] Although the initial motivations for and early histories of American, British, and French nuclear weapons activities are well understood,[11] studies of the Soviet, Chinese, and Swedish experiences are insightful but incomplete.[12] Coverage of the Israeli, Pakistani, and Indian cases is plentiful but far less dependable.[13] Firm details about Iraqi, Iranian, Algerian, North and South Korean, Taiwanese, Brazilian, Argentine, and South African military nuclear motives are scarce.[14]

More problematic than the lack of reliable public information about new and emerging nuclear-weapon states, however, is the dearth of carefully specified explanations of nuclear proliferation. The existing literature on the sources of proliferation is more rich than rigorous. Even the best case studies produce few enduring insights into general proliferation patterns. Descriptive knowledge about national experiences with nuclear weapons is intrinsically important. Predictive and explanatory models, however, cannot be constructed through induction alone; there are too many variables that can influence the process of nuclear arms acquisition. Innovative ideas are needed to select the relevant variables and to develop explanations of the observed and expected associations among these variables.[15]

There are three leading sets of ideas about the general sources of nuclear proliferation. The first is a technological determinist perspective: countries that *can* manufacture nuclear weapons eventually will. Second, countries that face serious military threats to their security will develop nuclear arms if they can. A variant of this *security* (or insecurity) argument is the following hypothesis: a state will attempt to acquire nuclear arms if its main military adversary obtains them. Third, countries that can build nuclear bombs will do so because of the international status and prestige that nuclear arsenals are believed to impart on proliferant countries.

Technological Determinism

> Every nation that might plausibly have started nuclear weapons programs did so: Germany, Great Britain, the United States, the Soviet Union, France, and we now know, Japan. So the case has been weakened for those who have argued that governments, or more precisely, generals, emperors, and presidents can hold back from the decision and say "No." The decision to develop nuclear weapons is not a fluke of certain governments, but a general technological imperative.[16]

Herbert York's depiction of nuclear weapons activities during the Second World War captures the essence of technological determinist accounts of military nuclear development: the universal appeal of nuclear arms and the inability of individuals and organizations to resist technological change.[17] As one author put it, "when technology beckons, men are helpless ... If a thing was technically possible, then it had to be done."[18] History and recent research, however, offer reason for skepticism about the technological imperative of nuclear proliferation and call attention to the political role of nuclear scientists and other technical specialists in shaping the beliefs that political and military officials come to hold about civil and military nuclear technologies.

The hypothesis that countries will acquire nuclear weapons if they are capable of doing so suffers from obvious empirical limitations. Many states that have the technical capacity to manufacture nuclear weapons have never attempted to do so. Examples include the Netherlands, Belgium, Germany, and Japan.[19] Some other states that are believed to have initiated efforts to develop nuclear weapons eventually terminated their programs. These countries are Sweden, Taiwan, South Korea, and probably Argentina, Brazil, and South Africa.[20] The technological imperative cannot account for notable instances of nuclear nonproliferation.[21]

A deeper problem is the neglect of individual and organizational actors and their motivations. Uranium enrichment plants, intermediate-range ballistic missiles, and atom bombs do not build themselves. The production of any large, military, technological system involves a long series of heated debates and difficult decisions about technical, economic, military, political, and moral issues.[22] Of course, proponents of the technological imperative acknowledge that humans choose to invent, engineer, and manufacture nuclear weapons; their claim is that technological momentum is so strong and the desire for nuclear arms so pervasive that decisionmakers are "pulled along."

A statement by Robert Oppenheimer in hearings before the Personnel Security Review Board in 1954 seems to support the perspective of "technological pull." When asked if his concerns about developing the hydrogen bomb increased as the feasibility of the project became more certain, Oppenheimer replied:

> I think it is the opposite of true. Let us not say about use. But my feeling about development became quite different when the practicabilities became clear. When I saw how to do it, it was clear to me that one had to at least make the thing. Then the only problem was what would one do about them when one had them. The program in 1949 was a tortured thing that you could well argue did not make a great deal of technical sense. It was therefore possible to argue also that you did not want it even if you could have it. The program in 1951 was technically so sweet that you could not argue about that.[23]

Even if we observe the power of momentum in nuclear weapons programs, however, we need not regard technology as autonomous or even as a temptation that no person conceivably can resist. Technological artifacts are invented and innovated by individuals. Similarly, a technological system "grows" and "drifts" when scientists, bureaucrats, and politicians have vested interests in the growth and durability of the system.[24] Technological change cannot be well understood by assuming the submission of individuals and the insignificance of their political interests. Because so many international powers and institutions conspire to thwart nuclear proliferation, human motivation, judgement, choice, and effort is especially crucial in the military nuclear activities of aspiring nuclear-weapons states.

National Insecurity

If nuclear proliferation is not driven by some irresistible force of technology, what conditions motivate governments to launch nuclear weapons programs? The answer most authors offer is national insecurity. Frankel argues that "a state's decision to build nuclear weapons is a result of the security equation it faces."[25] Graham contends that in nearly all cases, "a nation that has gone nuclear has faced an *acute* security threat from a nuclear-armed adversary that also had a substantial conventional military capability."[26] It is more common for authors to ascribe security motivations only to some states. For instance, George Rathjens and Marvin Miller maintain that for the United States, the Soviet Union, China, Israel, and Pakistan, "security from military threats has been the primary reason to acquire nuclear weaponry," while other states have gone nuclear for different reasons.[27] Bailey observes that "the principal motivations for nuclear proliferation vary from country to country." She then adds: "security is the principal reason a country initiates a nuclear weapons program."[28]

Some security accounts of nuclear proliferation have empirical problems. Take the hypothesis that a country will acquire nuclear arms if its chief military adversary obtains them. This is not universally correct. Israel, South Africa, Argentina and Brazil are believed to have launched nuclear bomb programs; but none of these states faces nuclear threats. More importantly, this hypothesis is superficial. Pakistan, for instance, is generally cited in support of the proposition; but Islamabad initiated its nuclear program and continues to develop its military nuclear capabilities not primarily because of fears about India's *nuclear* weapons, but because of concerns about India's *conventional* military advantages.[29]

The general point that insecure states seek nuclear weapons is more compelling, but also more elusive. On the one hand, a large number of insecure countries do not turn to nuclear force for their security. On the other hand, if we consider the concept of security broadly,[30] every country that has gone nuclear has faced some security problem. National insecurity thus seems to be necessary but insufficient to cause nuclear proliferation. But why is this so? Neorealism helps us to understand *how* military insecurity could stimulate a state's interest in acquiring military nuclear capabilities.[31]

Power balancing is the oldest concept in the literature on international relations; it is central to neorealism.[32] According to Walt and Waltz, states usually balance against the most serious foreign threats to their security; rarely do they *bandwagon*, that is, accommodate or appease the powers making these threats.[33] States balance "internally" – by relying on their own military capabilities – or "externally" – by relying on the military capabilities of allies.[34] Statesmen prefer internal balancing because it leaves less to chance and less to the will of others.[35]

While allies were crucial in the pre-nuclear era to resist foreign aggression, the advent of nuclear force has made internal balancing both more feasible and more urgent. Once the United States built the bomb, for example, the Soviet Union felt compelled to follow suit. Dean Acheson observed this in a memorandum to the president in 1945: "It is impossible that a government as powerful and power conscious as the Soviet Government could fail to react vigorously to this situation. It must and will exert every energy to restore the loss of power which this discovery has produced."[36] The logic of the cold war competition and the immense resources at the disposal of the superpowers made it important – and possible – for Washington and Moscow to keep pace with each side's nuclear force innovations.[37]

Neorealism explains not only the propensity of the superpowers to balance each other, but also the basic pattern of their balancing behavior. Washington and Moscow adopted similar strategic policies due to their fiercely competitive rivalry. Waltz makes the point more generally: "Contending states imitate the military innovations contrived by the country of greatest capability and ingenuity. And so the weapons of major contenders, and even their strategies, begin to look much the same all over the world."[38] The superpowers certainly behaved in this manner during the cold war; but how insightful is this observation in the contemporary context of nuclear proliferation?

Neorealists expect all states to become "socialized" to the global political system; thus countries that can acquire nuclear arms will do so if they wish to contend for a share of power in the international system. Not all states contend, however, and contenders can find different ways to compete for security and well-being. Geopolitical competition constrains national defense strategies; it does not determine them. "The clear perception of constraints provides many clues to the expected reactions of states," Waltz points out, "but by itself the theory cannot explain those reactions. They depend not only on international constraints but also on the characteristics of states."[39]

Waltz never suggests that neorealism explains the causes of proliferation. In fact, he recognizes that a multitude of conditions can spur states to seek nuclear weapons.[40] National programs to develop nuclear arms are a response to insecurity and a form of balancing; but the theoretical framework of neorealism is too abstract to allow precise predictions about the conditions under which states will seek nuclear weapons instead of strengthening their conventional military capabilities or allying with foreign powers.

National Prestige

Next to military security, *national prestige* is the second most frequently cited motivation for the acquisition of military nuclear capabilities. "Apart from precise applications of nuclear weapons to possible wartime scenarios," Quester writes, "the attainment of nuclear threshold status may offer a state enhanced national prestige ... because nuclear weapon capabilities are simply regarded as a trapping of national grandeur and stature, which can then translate into one or another form of security."[41] McGeorge Bundy highlights the importance of prestige in French and British nuclear proliferation:

> I am persuaded that the basic objective, historically, for both the British and French governments, has been to have a kind of power without which these two ancient sovereign powers could not truly be themselves. This requirement has been clear for each government at every moment of choice from 1945 onward, and it is not a matter of deterrent strategy as such. It is rather a matter of what Britain and France must

> have, as long as others have it, in order to meet their own standards of their own rank among nations.[42]

As in the case of security, the term prestige also suffers conceptual ambiguity. To begin with, the concept means little without reference to social or political contexts.[43] More problematic, however, is the tendency of observers to use prestige as a residual concept to account for instances of nuclear proliferation that cannot be explained by security or technological determinist motivations. If the acquisition of nuclear weapons is driven neither by military concerns nor by the imperative of technological growth, it must be motivated by a quest for political status.

The prestige argument is problematic for additional reasons. For some countries, like Israel, nuclear weapons are military weapons, not trappings of grandeur. The Israeli government refuses to acknowledge the existence of its military nuclear forces, much less boast about them. In other cases, nuclear weapons probably are desired for more than military reasons. They contribute to the international standing to which states such as Britain, France, and perhaps India aspire; but is prestige the best concept to capture this dynamic?

I am skeptical. Nuclear weapons are prestigious only if people believe so. After China exploded its first nuclear device in October 1964, many Indians perceived the political benefits of nuclear arms: "India has to have the bomb if it is to hold sway in the world. Not to make it would betray a lack of will to live – to survive – in this jungle of international strife. Not to make it would be to let the whole world treat us as some third-rate country."[44] But other Indians concluded that going nuclear would erode India's international standing: "Having asked everybody not to manufacture the bomb, can we go for it now? India's prestige would go down if it went for an atom bomb."[45] The political value – like the military security – provided by nuclear weapons is uncertain, controversial, and widely contested.[46]

Conscious of the propaganda benefits to China of its nuclear capability and concerned about mounting pressure on New Delhi to launch its own nuclear bomb program, American officials went to great lengths to help India regain and secure its position at the forefront of the developing world. Various proposals under the State Department's consideration included helping New Delhi predict the next Chinese test and attribute the prediction to Indian scientific and intelligence sources;[47] jointly sponsoring with New Delhi (and possibly the IAEA) an international nuclear science conference in India; and generally "cooperating with India in peaceful uses of atomic energy which might serve to offset the propaganda effects of the Chinese nuclear device by increasing the stature of India in the eyes of the less developed countries."[48] We now know that the United States failed to dissuade India from developing military nuclear capability; is it possible that the intense American response to the Chinese test actually *taught* New Delhi of the political and material rewards it could gain by going nuclear?

This underscores the ethnocentric premises that often inform the use of the term "prestige" in the context of proliferation. It implies that states attempting to acquire nuclear weapons to enhance their international status suffer from some weakness in national character or performance.[49] Rarely emphasized is the wider environment and the perceptions of others that make nuclear arms politically useful. In order to overcome the pitfall of ethnocentrism and the other shortcomings of prestige-centered arguments, therefore, we could accentuate the global diffusion of nuclear myths. The spread of nuclear myths is causally related to the proliferation of nuclear weapons.

Nuclear Mythology and Nuclear Proliferation

In contrast to the preceding accounts of nuclear proliferation – all of which have empirical and logical weaknesses – I offer an explanation of nuclear arms acquisition that emphasizes the strategic beliefs and political behavior of nuclear myth makers. A state is likely to go nuclear when national elites, who want the state to develop nuclear weapons, emphasize the country's insecurity or its poor international standing to popularize the myth that nuclear weapons provide military security and political power. This model also provides insight into the sources of nuclear *non*proliferation: when enterprising and well-connected individuals succeed in cultivating national consensus on the myth of insecurity through nuclear weapons, their government is less likely to initiate or continue efforts to obtain military nuclear capabilities.

This argument rests on three simple assumptions: (1) the beliefs of individuals matter for foreign policy making; (2) policymakers' beliefs about nuclear weapons are particularly important; and (3) talented and well-placed experts can help create, diffuse, and perpetuate nuclear myths. The first assertion is familiar and needs only a brief summary; the other two require more detailed elaboration.

The observation about decision making is well-known: strategic policies and choices are mediated by the policymaker's goals, judgements, and perceptions.[50] Analysis of decision making is useful to show why people in similar situations behave differently and why people react similarly to different circumstances. Analysis of decision making is not necessary for all problems of security studies, but it is impossible to explain important nuclear weapons decisions and strategies without reference to decision makers' beliefs about the political and military characteristics of these weapons. This is so because of the multiple and only partially predictable consequences of developing, deploying, and using nuclear arms.

Nuclear Myths as Responses to Uncertainty

Little is known about nuclear weapons. Although scientists understand the technical requirements for manufacturing nuclear explosives and the physical effects of their detonation, the most important questions are still unanswered. What political and military consequences can nations such as Iran, Ukraine, or Japan expect if they obtain nuclear weapons? Plausible responses come to mind, not definite answers. There are several reasons for the lack of objective answers to crucial questions about national security and power in the nuclear age; I identify three.

Shift from brute force to coercion Thomas Schelling observed long ago that military strategy in the nuclear age cannot be thought of, as it could in previous times, as the science of military victory. Nuclear strategy is the art of coercion, of intimidation and deterrence.[51] In the conventional military era, international disputes could be resolved, total wars could be won, and military objectives could be achieved through the physical employment of brute force. Conventional military battles were (and still are among non-nuclear states) determined by the cunning application of sheer strength; beliefs mattered little. When enemies can destroy one another and cannot protect themselves, however, "war no longer looks like just a contest of strength. War and the brink of war are more a contest of nerve and risk-taking, of pain and endurance."[52] The processes by which nuclear power can influence political outcomes are indirect and necessarily entail large elements of highly subjective assessments.[53]

Lack of experience with nuclear wars Nuclear strategy is a speculative enterprise. Because of the world's fortunate lack of experience with nuclear conflicts, nuclear myths are sustained by logic and faith; there is little objective information.[54] Since there are no direct precedents, indirect evidence informs beliefs about the relationship between nuclear weapons and war.[55] As Joseph Nye observes:

> Much of what passes for nuclear knowledge rests upon elaborate counterfactual argument, abstractions based on assumptions about rational actors, assumptions about the other nation's unknown intentions, and simple intuitions.[56]

Nuclear nonproliferation regime The uncertainties of nuclear deterrence, coercion, and arms control plagued the superpowers for the duration of the cold war. For prospective and emerging nuclear-weapon states, these uncertainties are compounded by the unpredictable actions of global nonproliferation actors. Countries intent on acquiring nuclear explosives must consider whether the costs supporters of the international nonproliferation regime could impose on them exceed the benefits of possessing nuclear weapons. Nuclear myths influence these and other judgements about nuclear proliferation.

The Politics of Nuclear Myth Making

The nuclear myths of a state's political and military leaders determine whether that state will launch a nuclear weapons program. When these myths change, military nuclear behavior also is likely to change. But how do policymakers acquire and revise their beliefs about nuclear weapons? When they recognize that their understanding of complex policy problems is limited, government decisionmakers frequently turn to specialists for help in understanding the problems and devising solutions to them.[57] Even when national leaders feel confident in their strategic beliefs, experts can influence the process of policy making.

A brief example can illustrate the significance of this observation for nuclear proliferation. Many Indians felt threatened by China's first nuclear explosion in October 1964. Coming only two years after China's victory over India in the 1962 Himalayan war, the nuclear test was interpreted by many Indian politicians and bureaucrats as a challenge that somehow had to be met, but the challenge initially was not seen in military terms. Because most Indian legislators and bureaucrats viewed China's bomb in political terms – as a resource for Chinese political hegemony in South and South-East Asia – only a few Indians originally advocated nuclear weapons as a solution to the problem. A leading advocate was the atomic physicist, Homi Bhabha.

The debate sparked by China's 1964 nuclear test could have gone in any direction. Through two carefully crafted interventions in the national debate, however, Dr. Bhabha succeeded in persuading Prime Minister Lal Bahadur Shastri to approve work on a nuclear weapons option. Immediately after learning from the United States of the imminence of the Chinese nuclear blast, Bhabha called a press conference to announce India's "ability to produce a nuclear bomb in eighteen months."[58] He added that China's new nuclear weapons capability demanded a commensurate Indian response. Days later, Bhabha challenged the economic argument against nuclear bombs. Citing figures produced at the Third International Conference on the Peaceful Uses of Atomic Energy, Bhabha claimed that a 10 kiloton bomb would cost $368,000 and a two megaton bomb would cost $680,000. He added that atomic explosives were some twenty times cheaper than conventional explosives.[59]

It is not important that none of these claims were completely accurate – the U.S. intelligence community estimated that India would require one to three years to produce and test its first nuclear device,[60] and Bhabha confided to an American official that "he could make and test a crude nuclear device for approximately ten million dollars."[61] The point is that Bhabha's well-timed interventions helped encourage and empower India's bomb lobby, which eventually managed to carry the day: Prime Minister Shastri authorized Bhabha and other scientists to develop a capability for producing nuclear weapons. India's efforts to launch a military nuclear program cannot be understood apart from Homi Bhabha's pivotal role as a nuclear myth maker.

Implications for Theory and Policy Making

Theory

This article presents a first-image explanation of nuclear proliferation.[62] It is a model, not a theory. Useful theories would show how second-image and third-image conditions influence the creation and the credibility of certain nuclear myths. How do different domestic political structures and traditions arouse and sway beliefs about the political and military effects of nuclear weapons? How does a state's placement in the international political system influence these beliefs? Is the plausibility of these myths at all related to domestic and international political circumstances?

More also can be learned about the cognitive and political processes of nuclear myth formation, diffusion, and perpetuation. In *Myths of Empire*, for example, Snyder treats strategic myth making as a manipulative activity: groups with clear-cut interests, monopolies of information, and other propaganda advantages concoct false arguments to mislead others about their interests and about the costs and benefits of competing policies. Snyder also describes how strategic myths that arise from domestic politics can then take on a life of their own. Mythmakers can become trapped in their own rhetoric, in the political arrangements of the myths created, and in the internationalization of myths by second-generation elites.

Policymaking

Limiting nuclear proliferation does not necessarily require understanding the motivations of the states that seek nuclear weapons. Some nonproliferation efforts succeed despite incomplete or faulty knowledge. For example, South Africa's military nuclear policies were not well understood; nevertheless, Pretoria was persuaded to join the NPT as a non-nuclear-weapon state.

Today, the nuclear aims of Iran, Iraq, Libya, and North Korea remain unclear; but intense global attention, pressure (and military activity in the case of Iraq) may keep nuclear bombs out of the control of these countries. Nonetheless, understanding the processes of nuclear proliferation should provide a lever for influencing decisions.

In the post-cold-war world the range of possible defense strategies will become more limited for most states. The end of bipolarity brings in its wake the decline of global military alliances. Unless some new and effective collective or cooperative security system emerges,[63] threatened states will increasingly have to rely on their own resources to meet their defense needs. Proliferation pessimists regard these heightened pressures for internal balancing as a recipe for rampant nuclear proliferation. These fears are premature. Although we should expect a rise in internal balancing activities, the uncontrolled spread of nuclear arms does not necessarily follow. Non-proliferation optimists are correct to stress the importance of national and international policies in influencing the scope and pace of nuclear proliferation.

Non-nuclear security assistance The argument about nuclear proliferation presented in this article has three specific policy implications. First, policy-makers must acknowledge that all states have legitimate security concerns. As long as world politics is ordered by anarchy, a primary concern of all countries will be to balance against perceived (and real) security threats. The United States and other countries supportive of nuclear nonproliferation should lead the search for non-nuclear sources of security. It will do little good to keep insisting that Pakistan, for instance, should sacrifice its military nuclear options unless other powers can advance *effective*, non-nuclear methods for Pakistan to balance Indian military might.

Nuclear-weapon-state conduct If nuclear proliferation is at least partially driven by the global diffusion of myths that nuclear weapons provide military security and political power, then the declared nuclear-weapons states ought to pay close attention to the ways in which they plan, operate, and publicize their own nuclear arms programs. Lately, it has become commonplace to claim that in the post-cold war era nuclear weapons have been devalued as a currency of international power. We should be skeptical of this conclusion. While Washington and Moscow have agreed to deep cuts in their nuclear arsenals, both sides continue to feature nuclear weapons prominently in their defense strategies. This point is not lost on other states: the myth of security through nuclear force acquisition continues to have wide appeal.

The myth of nuclear security may become even more plausible if existing nuclear powers threaten to use nuclear force against states not armed with nuclear weapons.[64] Since 1978, Washington has pledged not to use nuclear force against non-nuclear weapons states, except in the case of attack on

American territory or armed forces, by such a state allied to a nuclear-armed power.[65] In light of the Iraq experience, however, a special advisory panel has urged the director of U.S. nuclear targeting to relax this assurance.[66] Modifying Washington's negative security assurance policy in this way conceivably could discourage some states from building and brandishing mass destructive weapons, but it is just as likely that potential target states would hasten and expand the development of these forces in order to deter U.S. military nuclear action against them.[67]

The nuclear taboo If the myth of nuclear security causes nuclear proliferation, then nonproliferation officials should attempt to discredit this myth and promote the opposite myth that nuclear arms foster national insecurity (and industrial and environmental decay). Of course, the ability of any state to influence the strategic beliefs of other countries is severely limited, but three lines of action are worth examining.

First, proliferant countries should not be rewarded for their military nuclear activities. Numerous ideas for *managing* nuclear proliferation could have this effect.[68] The provision of technical assistance to increase the security, safety, and survivability of new nuclear forces may help to stabilize a regional deterrence relationship, but at the probable expense of enhancing the readiness and effectiveness of new nuclear forces and undermining the myth of nuclear insecurity.[69] If India were offered a permanent seat at the U.N. Security Council – and if this action were seen globally as a reward for Indian nuclear achievements, this too could stimulate new and existing appetites for nuclear weapons.

Second, the successful use of a nuclear weapon by *any* country probably would trigger a sharp shift in nuclear perceptions and predilections around the world. Imagine that Iraq had possessed a dozen nuclear explosive devices in 1990 and used this weaponry to deter military action by the coalition forces; would not some other states reconsider their nuclear policies?[70] Moreover, if Pakistan effectively uses – or effectively threatens to use – nuclear force as an instrument for attracting international attention and mediation for the Kashmir crisis,[71] the political value of nuclear weaponry could skyrocket. Nonproliferation officials should respond to military events in a manner that enhances the myth of insecurity through nuclear weapons.

Finally, nonproliferation policy should include international efforts to teach the costs of nuclear weapons acquisition and use. In South Asia, for example, a military conflict in which India and Pakistan detonated less than a dozen nuclear explosives would almost certainly effect unprecedented disaster for both countries.[72] Washington and Moscow could help increase popular awareness in India and Pakistan – and in other countries – of the enormous destructive power of nuclear weapons and of the disastrous consequences of their use.[73] Similarly, the declared nuclear powers could help educate presently non-nuclear-armed states about the huge economic, scientific, industrial, and environmental costs of producing, maintaining, and eventually destroying nuclear weapons.[74] This information is not mythical: it can be known, taught, and learned.

NOTES

Fellowships from the Center for International Security and Arms Control at Stanford University and the Center for Security and Technology Studies at Lawrence Livermore National Laboratory supported the writing of this article, an early draft of which was presented at the 1993 annual meeting of American Political Science Association. The views expressed are the author's and not necessarily those of the United States Government. I wish to thank Itty Abraham, Peter Feaver, Steven Flank, Ben Frankel, Neil Joeck, Scott Sagan, Amy Sands, and Jessica Stern for helpful comments on earlier versions.

1. For recent statements of "proliferation pessimism," see Benjamin Frankel, "The Brooding Shadow: Systemic Incentives and Nuclear Weapons Proliferation," *Security Studies* 2, no. 3/4 (Spring/Summer 1993): 37–78; Dagobert L. Brito and Michael Intriligator, "The Economic and Political Incentives to Acquire Nuclear Weapons," *Security Studies* 2, no. 3/4 (Spring/Summer 1993): 287–310; Lewis A. Dunn, *Containing Nuclear Proliferation*, Adelphi Paper no. 263 (London: International Institute for Strategic Studies [IISS], Winter 1991); Jed C. Snyder, "Weapons Proliferation and the New Security Agenda," in Andrew W. Marshall, J. J. Martin, and Henry S. Rowen, eds., *On Not Confusing Ourselves* (Boulder, Colo.: Westview, 1991), 261–82; Frankel, "An Anxious Decade: Nuclear Proliferation in the 1990s," in Frankel, ed., *Opaque Nuclear Proliferation: Methodological and Political Implications* (London: Frank Cass, 1991), 1–13; John J. Mearsheimer, "Back to the Future: Instability in Europe After the Cold War," *International Security* 15, no. 1 (Summer 1990): 5–56; Frankel, "International Political Changes and Nuclear Proliferation in the 1990s," in Eric H. Arnett, ed., *Science and International Security: Responding to a Changing World* (Washington, D.C.: American Association for the Advancement of Science (AAAS), 1990), 85–111.
2. "Nonproliferation optimists" include Zachary S. Davis, "The Realist Nuclear Regime," *Security Studies* 2, no. 3/4 (Spring/Summer 1993): 79–99; Thomas W. Graham, "Winning the Nonproliferation Battle," *Arms Control Today* 21, no. 7 (September 1991): 8–13; and Graham, "Winning vs. Managing the Non-Proliferation Battle," unpublished paper, no date.
3. Although today's proliferation pessimists stress the recent structural transformation of international politics, the pessimistic perspective is quite old. In 1956, for example, Harold Stassen, President Eisenhower's special assistant on disarmament, predicted that as many as twenty states soon would possess nuclear weapons. Herbert S. Parmet, *Eisenhower and the American Crusades* (New York: Macmillan, 1972), 450–51. In 1965 President Kennedy and his advisers predicted fifteen to twenty-five nuclear-armed states by the mid-1970s. *New York Times*, 23 March 1963. A decade later, the Arms Control and Disarmament Agency (ACDA) director, Fred C. Iklé, predicted a world in the mid-1980s of about thirty-five nations capable of developing several dozen nuclear weapons each. Statement before the Subcommittee on International Security and Scientific Affairs of the Committee on International Relations, U.S. House of Representatives, 5 November 1975.

4. On the problems that abstract models of the past pose for efforts to predict the future, see Robert Jervis, "The Future of World Politics: Will It Resemble the Past?" *International Security* 16, no. 3 (Winter 1991/92): 39–73; John Lewis Gaddis, "International Relations Theory and the End of the Cold War," *International Security* 17, no. 3 (Winter 1992/93): 5–58; Philip E. Tetlock, "Good Judgement in International Politics: Three Psychological Perspectives," *Political Psychology* 13, no. 3 (1992): 517–39.
5. For an attempt to forecast nuclear proliferation with a complex, explanatory model, see Stephen M. Meyer, *The Dynamics of Nuclear Proliferation* (Chicago: University of Chicago Press, 1984), 144–164. On the importance of explanation relative to prediction, see Stephen Toulmin, *Foresight and Understanding: An Inquiry into the Aims of Science* (Bloomington: Indiana University Press, 1961).
6. This model also provides a framework for understanding the sources of nuclear nonproliferation. For a detailed account of national decisions not to acquire nuclear weapons, see Mitchell Reiss, *Without the Bomb: The Politics of Non-proliferation* (New York: Columbia University Press, 1988).
7. "Nuclear myths" are unverifiable beliefs about the relationship between a state's nuclear weapons and its security. "Nuclear myth makers" are individuals who assert these myths and try to persuade others of their validity. As I use the term, myths can be believed but not "known." In the vocabulary of cognitive psychology, nuclear myths are "theory-driven" rather than "data-driven." Philip E. Tetlock and Charles McGuire, Jr., "Cognitive Perspectives on Foreign Policy," in Samuel Long, ed., *Political Behavior Annual*, vol. 1 (Boulder, Colo.: Westview, 1986), 159–61. Ken Waltz uses "nuclear myths" to represent false or "fuzzy" beliefs about the faults of nuclear deterrence. I agree that these beliefs are myths, but according to my definition so are the beliefs that favor nuclear deterrence. Kenneth N. Waltz, "Nuclear Myths and Political Realities," *American Political Science Review* 84, no. 3 (September 1990): 731–45. Like Waltz – and unlike my approach – Jack Snyder treats myths as specific sets of incorrect strategic concepts. *Myths of Empire: Domestic Politics and International Ambition* (Ithaca: Cornell University Press, 1991).
8. The United States, Russia (replacing the Soviet Union), the United Kingdom, France, and China have declared nuclear arsenals. India exploded a nuclear device in 1974 and has a nuclear weapons capability. Many experts believe that Israel, Pakistan, and South Africa have military nuclear capabilities; although by joining the Nuclear Non-Proliferation Treaty (NPT) in July 1991 and concluding a full-scope safeguards agreement with the International Atomic Energy Agency (IAEA), South Africa probably has forgone its nuclear weapons options. Not counting South Africa, the number of nuclear-armed states is eleven if one includes Ukraine, Belarus, and Kazakhstan, which inherited some Soviet nuclear weapons, but under the May 1992 Lisbon Protocol to the START treaty, agreed to remove all nuclear forces from their territory by the end of the decade and to join the NPT as non-nuclear-weapon states "in the shortest time possible."
9. Included in this category are Iraq, Sweden, Taiwan, South Korea, and probably Argentina, Brazil, Iran, and North Korea. Because Canadian scientists participated in the Manhattan project, Canada also could be added to this list. Many more states have the technical ability to produce nuclear weapons, but NPT signatories such as Japan, Germany, Netherlands, Italy, Belgium, Poland, and Hungary are now believed not to want nuclear arms. Still other countries, such as Libya and Syria, are presumed to covet nuclear bombs but presently do not possess the means to produce them.
10. Information paucity and various forms of uncertainty impede the ability of experts to know and learn about nuclear proliferation. See Peter R. Lavoy, "Learning and the Evolution of Cooperation in U.S. and Soviet Nonproliferation

Activities," in George W. Breslauer and Philip E. Tetlock, eds., Learning in U.S. and Soviet Foreign Policy (Boulder, Colo.: Westview, 1991), 735–83.

11. On the U.S. case, see Richard G. Hewlett and Oscar E. Anderson, *The New World, 1939–1946: A History of the United States Atomic Energy Commission*, vol. 1 (University Park: Pennsylvania State University Press, 1962); Richard Rhodes, *The Making of the Atomic Bomb* (New York: Simon & Schuster, 1986); and McGeorge Bundy, *Danger and Survival: Choices About the Bomb in the First Fifty Years* (New York: Random House, 1988). The best studies of the British program are Margaret Gowing, *Britain and Atomic Energy, 1939–1945* (London: Macmillan, 1964); and Gowing, *Independence and Deterrence: Britain and Atomic Energy, 1945–1952*, 2 vols. (London: Macmillan, 1974). On France, see Lawrence Scheinman, *Atomic Energy Policy in France Under the Fourth Republic* (Princeton: Princeton University Press, 1965); and Wilfred L. Kohl, *French Nuclear Diplomacy* (Princeton: Princeton University Press, 1971).
12. For the Soviet nuclear history, see David Holloway, *The Soviet Union and the Arms Race* (New Haven: Yale University Press, 1983); and Ulrich Albrecht, "The Development of the First Atomic Bomb in the USSR," in Everett Mendelsohn, Merritt Roe Smith, and Peter Weingart, eds., *Science, Technology, and the Military*, vol. 2 (Dordrecht, Netherlands: Kluwer, 1988). The best study of China's early nuclear activities and motivations is John Wilson Lewis and Xue Litai, *China Builds the Bomb* (Stanford, Calif.: Stanford University Press, 1988). On Sweden's short-lived weapons program, see Wilhem Agrell, "The Bomb That Never Was: The Rise and Fall of the Swedish Nuclear Weapons Programme," in Nils Petter Gleditsch and Olav Njølstad, eds., *Arms Races: Technological and Political Dynamics* (London: Sage, 1990), 154–74; Reiss, *Without the Bomb*, 37–77; and Christer Larsson, "Svenska K rnvapen Utvecklades Fram Till 1972!" *Ny Teknik* (Stockholm), no. 17, 25 April 1985.
13. Seymour Hersh provides a detailed account of Israel's nuclear program in *The Samson Option: Israel's Nuclear Arsenal and American Foreign Policy* (New York: Random House, 1991). See also Shlomo Aronson, *The Politics and Strategy of Nuclear Weapons in the Middle East: Opacity, Theory, and Reality, 1960–1991* (Albany: State University of New York Press, 1992); and Frank Barnaby, *The Invisible Bomb* (London: I. B. Tauris, 1989). On Pakistan, see Ziba Moshaver, *Nuclear Weapons in the Indian Subcontinent* (New York: St. Martin's, 1991); Ashok Kapur, *Pakistan's Nuclear Development* (London: Croom Helm, 1987); Neil Joeck, "Pakistani Security and Nuclear Proliferation in South Asia," *Journal of Strategic Studies* 8, no. 4 (December 1985): 80–98; Akhtar Ali, *Pakistan's Nuclear Dilemma* (Karachi: Pakistan Economist Research Unit, 1984); Steve Weissman and Herbert Krosney, *The Islamic Bomb: The Nuclear Threat to Israel and the Middle East* (New York: Time Books, 1981), 161–223; and Rodney W. Jones, *Nuclear Proliferation: Islam, the Bomb, and South Asia*, The Washington Papers no. 82 (Beverly Hills, Calif.: Sage, 1981). On India, see Ashok Kapur, *India's Nuclear Option: Atomic Diplomacy and Decision Making* (New York: Praeger, 1976); Onkar Marwah, "India's Nuclear and Space Programs: Intent and Policy," *International Security* 2, no. 2 (Fall 1977): 96–121; Shyam Bhatia, *India's Nuclear Bomb* (New Delhi: Vikas, 1979); K. Subrahmanyam, "India: Keeping the Option Open," in Robert M. Lawrence and Joel Laurus, eds., *Nuclear Proliferation: Phase II* (Lawrence: University Press of Kansas, 1974), 112–48; Shelton L. Williams, *The U.S., India, and the Bomb*, Studies in International Affairs, no. 12 (Baltimore: Johns Hopkins University Press, 1969); and G. G. Mirchandani, *India's Nuclear Dilemma* (New Delhi: Popular Book Services, 1968).
14. Leonard S. Spector provides a good overview of the military nuclear activities of these states. See Spector, with Jacqueline R. Smith, *Nuclear Ambitions: The Spread of Nuclear Weapons, 1989–1990* (Boulder, Colo.: Westview, 1990). Extensive inspections of Iraqi weapons facilities conducted by the United Nations Special

Commission and the IAEA have produced profuse information about Iraq's nuclear arms program. See David Albright and Mark Hibbs, "Iraq's Quest for the Nuclear Grail: What Can We Learn?" *Arms Control Today* 22, no. 6 (July/August 1992): 3–11; Jay C. Davis and David A. Kay, "Iraq's Secret Nuclear Weapons Program," *Physics Today* (July 1992): 21–27.

15. Laws (statements describing probable occurrences or associations between variables) can be discovered through induction. Theories need to be invented to explain *why* laws are valid and when they are likely to hold in the future. Kenneth N. Waltz, *Theory of International Politics* (New York: Random House, 1979), 2–13.
16. Herbert York's statement is cited in Deborah Shapley, "Nuclear Weapons History: Japan's Wartime Bomb Projects Revealed," *Science* 199 (13 January 1978): 155.
17. For other technological determinist perspectives, see Mark Thee, *Military Technology, Military Strategy and the Arms Race* (London: Croon Helm, 1986); Hans Bethe, "The Technological Imperative," *Bulletin of the Atomic Scientists* 41, no. 7 (August 1985): 34–36; Dietrich Schroeer, *Science, Technology and the Nuclear Arms Race* (New York: Wiley, 1984); Lord Zuckerman, "Science Advisers and Scientific Advisers," *Proceedings of the American Philosophical Society* 124 (1980): 241–55; Edward Thompson, "Notes on Exterminism, the Last Stage of Civilization," *New Left Review* no. 121 (May/June 1980); Herbert York, "Multiple-Warhead Missiles," in Bruce M. Russett and Bruce G. Blair, eds., *Progress in Arms Control? Readings from Scientific American* (San Francisco: W. H. Freeman, 1979), 122–31; and Ralph Lapp, *Arms Beyond Doubt: The Tyranny of Weapons Technology* (New York: Cowles, 1970).
18. Lapp, *Arms Beyond Doubt*, 178.
19. These states operate uranium enrichment and/or plutonium reprocessing plants and thus could produce bomb-grade fissionable material (enriched uranium or plutonium). These and many other countries also can carry out the other basic technical tasks needed to manufacture nuclear weapons, including the mastery of nuclear physics and chemistry, development of a workable weapon design, formation of fissionable material into weapon cores, testing and manufacture of high explosives, design and assembly of triggering devices, adoption of safety and control features, configuration of compatible delivery vehicles, and so on. The technical requirements and obstacles for countries developing nuclear arms are summarized by Kathleen C. Bailey, *Doomsday Weapons in the Hands of Many: The Arms Control Challenge of the '90s* (Urbana and Chicago: University of Illinois Press, 1991), 8–16.
20. Until effective international nuclear inspections are completed and their findings analyzed, it may not be prudent to presume the end of military nuclear activities in Argentina, Brazil, and South Africa. At a minimum, however, significant *policy* changes in the nuclear programs of these countries can be observed.
21. The technological determinist perspective similarly provides few insights into the sources of "opaque" nuclear proliferation. After detonating a "peaceful nuclear explosion" in 1974, for example, India has not conducted further tests, deployed nuclear weapons, or declared itself a nuclear power. To be sure, New Delhi has improved and expanded its capacity to produce nuclear arms, but probably much more tentatively than would be expected from an imperative of technological growth. On nuclear opacity, see Avner Cohen and Benjamin Frankel, "Opaque Nuclear Proliferation," in Frankel, *Opaque Nuclear Proliferation*, 14–44. On India, see David Albright and Mark Hibbs, "India's Silent Bomb," *Bulletin of the Atomic Scientists* 48, no. 7 (September 1992): 27–31; Thomas W. Graham, "Nuclear Deterrence, Arms Control, and Confidence Building in South Asia," in Eric H. Arnett, ed., *New Perspectives for a Changing World Order* (Washington, D.C.: AAAS, 1991), 123–34.

22. Langdon Winner, *Autonomous Technology: Technics-out-of-Control as a Theme in Political Thought* (Cambridge, Mass.: MIT Press, 1977), esp. 44–106; Winner, *The Whale and the Reactor: A Search for Limits in an Age of High Technology* (Chicago: University of Chicago Press, 1986); Donald MacKenzie, *Inventing Accuracy: A Historical Sociology of Nuclear Missile Guidance* (Cambridge, Mass.: MIT Press, 1990); Trevor J. Pinch and Wiebe E. Bijker, "The Social Construction of Facts and Artifacts: Or How the Sociology of Science and the Sociology of Technology Might Benefit Each Other," in Bijker, Thomas P. Hughes, and Trevor Pinch, eds., *The Social Construction of Technological Systems* (Cambridge, Mass.: MIT Press, 1989), 17–50; Hughes, "The Evolution of Large Technological Systems," in Bijker, Hughes, and Pinch, *Social Construction*, 51–82.
23. From *In the Matter of J. Robert Oppenheimer: Transcripts of Hearings before Personnel Security Review Board and Texts of Principal Documents and Letters* (Cambridge, Mass.: MIT Press, 1971), 251; Winner, *Autonomous Technology*, 72–73.
24. Hughes, "The Evolution of Large Technological Systems," 76–77. This point is explored with respect to India's nuclear and missile programs by Itty Abraham, "India's 'Strategic Enclave': Civilian Scientists and Military Technologies," *Armed Forces and Society* (Winter 1992): 231–52; and Steven Flank, "Exploding the Black Box: The Historical Sociology of Nuclear Proliferation," *Security Studies* 3, no. 2 (Winter 1993), forthcoming.
25. Frankel, "International Political Changes and Nuclear Proliferation," 90.
26. Graham, "Winning the Nonproliferation Battle," 9 (emphasis in original).
27. George W. Rathjens and Marvin M. Miller, "Nuclear Proliferation After the Cold War," *Technology Review* 94, no. 6 (August/September 1991): 26.
28. Bailey, *Doomsday Weapons*, 39.
29. This finding is discussed in the author's Ph.D. dissertation, "Learning to Live With the Bomb: The Politics of Nuclear Proliferation in South Asia," (Ph.D. diss., University of California, Berkeley, forthcoming 1993).
30. "Security in the context of the Third World does not simply refer to the military dimension, as it is often assumed in Western discussions of the concept, but to the whole range of dimensions of a state's existence which are already taken care of in the more-developed states, especially those of the West." Caroline Thomas, *In Search of Security: The Third World in International Relations* (Boulder, Colo.: Lynne Reinner, 1987), 1. See also Barry Buzan, *People, States and Fear*, 2nd ed. (Boulder, Colo.: Lynne Reinner, 1991); Mohammed Ayoob, "The Security Problematic of the Third World," *World Politics* 43, no. 2 (January 1991): 257–83; and the essays in Edward E. Azar and Chung-In Moon, *National Security in the Third World* (Aldershot, England: Edward Elgar, 1988).
31. Neorealism's main expectations are (1) the recurrence of balances of power in the international political system; (2) the tendency of states to balance, that is, to strengthen themselves in the face of external military threats; and (3) the inclination of states to imitate one another and to become socialized to the world political system. Waltz, *Theory of International Politics*, 128. The first prediction concerns an *outcome* of international interaction and thus is less directly relevant to the causes of nuclear proliferation than to the strategic consequences of this process. On this issue, see Waltz, *The Spread of Nuclear Weapons: More May Be Better*, Adelphi Paper no. 171 (London: IISS, Autumn 1981). The two predictions about state *behavior*, however, have direct implications for studies of the sources of nuclear proliferation.
32. On balancing, see Waltz, *Theory of International Politics*; and Stephen M. Walt's applications of neorealism to the problem of alignment: *The Origins of Alliances* (Ithaca: Cornell University Press, 1987); "Alliance Formation and the Balance of Power," *International Security* 9, no. 4 (Spring 1985): 3–43; "Testing Theories of Alliance Formation: The Case of Southwest Asia," *International Organization* 43, no. 2 (Spring 1988): 275–316; and "Alliance, Threats, and U.S. Grand Strategy: A

Reply to Kaufman and Labs," *Security Studies* 1, no. 3 (Spring 1992): 448–82.

33. Contrary to Waltz's structural balance of power theory, Walt argues that states balance not because of an unfavorable redistribution of international political, economic, or military capabilities, but mainly in response to threats, which are treated as a function of the identity, aggregate power, geographic proximity, offensive military capabilities, and perceived intentions of adversaries. *The Origins of Alliances*, 21–26, 263–66; "Testing Theories of Alliance Formation," 279–82, 311–13. See also Inis L. Claude, Jr., *Power and International Relations* (New York: Random House, 1962), 64–65. For an argument that states more often bandwagon, see Robert G. Kaufman, "To Balance or to Bandwagon? Alignment Decisions in 1930s Europe," *Security Studies* 1, no. 3 (Spring 1992): 417–47. Walt responds in "Alliance, Threats, and U.S. Grand Strategy."
34. Waltz, *Theory of International Politics*, 168. On the tradeoffs states make in choosing between arms and alliances in their security policies, see James D. Morrow, "Arms versus allies: trade-offs in the search for security," *International Organization* 47, no. 2 (Spring 1993): 207–33; Benjamin Most and Randolph Siverson, "Substituting Arms and Alliances, 1870–1914: An Exploration in Comparative Foreign Policy," in Charles F. Hermann et al., eds., *New Directions in the Study of Foreign Policy* (Boston: Allen & Unwin, 1987), 131–57; Michael N. Barnett and Jack S. Levy, "Domestic Sources of Alliances and Alignments: The Case of Egypt, 1962–73," *International Organization* 45, no. 3 (Summer 1991): 369–95; and James D. Morrow, "On the Theoretical Basis of a Measure of National Risk Attitudes," *International Studies Quarterly* 31, no. 4 (December 1987): 423–38.
35. Waltz, *Spread of Nuclear Weapons*, 2. For inventories of the many balancing options available to states, see Edward Vose Gulick, *Europe's Classical Balance of Power* (Ithaca: Cornell University Press, 1955), ch. 3; and Glenn H. Snyder, "Alliances, Balance, and Stability," *International Organization* 45, no. 1 (Winter 1991): 128.
36. U.S. Department of State, *Foreign Relations of the United States, 1945*, vol. 2, *General: Political and Economic Matters* (Washington, D.C.: Government Printing Office, 1967), 49; Robert Jervis, *The Meaning of the Nuclear Revolution* (Ithaca: Cornell University Press, 1989), 231.
37. Some observers see the recurrence of this pattern in new nuclear regions. India's retired Chief of Army Staff, General K. Sundarji, remarks: "It should have been seen as axiomatic that if either India or Pakistan produced nuclear weapons the other side had to as well. It does not matter who started the process." "Nuclear Realpolitik," *India Today*, 31 August 1991, p. 81. As mentioned above, I do not agree that strategic imitation caused Pakistan's decision to develop nuclear arms.
38. Waltz, *Theory of International Politics*, 127.
39. Ibid., 122.
40. Waltz notes that nations may want nuclear weapons for one or more of seven reasons: (1) for great powers to counter the weapons of other great powers; (2) for weaker states that doubt great-power security guarantees; (3) for nonaligned states to match the nuclear weapons of adversaries; (4) for nations that face conventionally superior adversaries; (5) for countries that wish to avoid costly conventional arms races; (6) for nations that want nuclear arms for offensive purposes; and (7) for states that hope to improve their international standing. *The Spread of Nuclear Weapons*, 7–8.
41. George Quester, "Conceptions of Nuclear Threshold Status," in Regina Cowen Karp, ed., *Security With Nuclear Weapons?* (Oxford: Oxford University Press, 1991), 217.
42. Bundy, *Danger and Survival*, 501.
43. Meyer attempts to remedy this problem by distinguishing four political incentives for proliferation: (1) regional power status and pretensions; (2) global power

status and pretensions; (3) pariah status: the acquisition of nuclear arms may demonstrate national viability; and (4) military alliance with nuclear power: the nation may wish to enhance its bargaining position in an alliance and it may wish to assert political and military independence from its strategic patron. *The Dynamics of Nuclear Proliferation*, 51–56.

44. Anonymous editorial, *Organizer*, 28 December 1964.
45. Former Defense Minister Krishna Menon, *Indian Express*, 11 January 1965.
46. Nuclear myths are also changeable. For example, the recent moves by Argentina and Brazil to cease military nuclear activities reflect new consensual understandings that greater political autonomy and well-being would accrue from enhanced economic performance and more amicable relations with the world's leading economic powers. Argentina and Brazil terminated their bomb programs not because of changes in objective conditions – political and economic threats to their sovereignty remain. Rather, their perceptions of nuclear arms as the solution to these threats changed – as did the individual decision makers perceiving, believing, and promoting nuclear myths.
47. Chester Bowles, cable to United States Department of State, 10 February 1965, National Security Archives (NSA), Nuclear Nonproliferation Collection, record number 39945.
48. United States Atomic Energy Commission, "Discussion Paper on Prospects for Intensifying Peaceful Atomic Energy Cooperation with India," 23 November 1964, NSA.
49. "The prestige incentive for nuclear weapons can ... be seen as *compensatory*: the 1974 [Indian nuclear] explosion shows that despite the modesty of industrial and economic progress, in the nuclear field India is first class." Richard K. Betts, "Incentives for Nuclear Weapons: India, Pakistan, Iran," *Asian Survey* 19, no. 11 (November 1979): 1054 (emphasis added).
50. From the vast literature on foreign policy decision making, see especially Robert Jervis, *Perception and Misperception in International Politics* (Princeton: Princeton University Press, 1976); and Harold and Margaret Sprout, *The Ecological Perspective on Human Affairs* (Princeton: Princeton University Press, 1965).
51. Thomas C. Schelling, *Arms and Influence* (New Haven: Yale University Press, 1966), esp. 1–34.
52. Ibid., 33.
53. Jervis, *Meaning of the Nuclear Revolution*, 182. Waltz argues that nuclear weapons actually reduce military uncertainties. See *Spread of Nuclear Weapons* and "Nuclear Myths and Political Realities," 731–45.
54. Jervis, *Meaning of the Nuclear Revolution*, 182–3; Emanuel Adler, "The Emergence of Cooperation: National Epistemic Communities and the International Evolution of the Idea of Nuclear Arms Control," *International Organization* 46, no. 1 (Winter 1992): 107–9.
55. See Phil E. Tetlock, Charles B. McGuire, and Gregory Mitchell, "Psychological Perspectives on Nuclear Deterrence," *American Review of Psychology* 42 (1991): 239–76.
56. Joseph S. Nye, Jr., "Nuclear Learning and U.S.-Soviet Security Regimes," *International Organization* 41, no. 3 (Summer 1987): 382.
57. See Ernst B. Haas, *When Knowledge is Power* (Berkeley and Los Angeles: University of California Press, 1990); and the special issue of *International Organization* 46, no. 1 (Winter 1992) on "Knowledge, Power, and International Policy Coordination." See especially Peter M. Haas, "Introduction: Epistemic Communities and International Policy Coordination," pp. 1–35.
58. *New York Times*, 5 October 1964.
59. *Hindu*, 27 October 1964; *New York Times*, 27 October 1964.
60. Dean Rusk, cable to United States Consulate General, Bombay, 21 January 1965, NSA, Nuclear Nonproliferation Collection, record number 22993.

61. Jerome Weisner, cable to United States Secretary of State, 21 January 1965, NSA, Nuclear Nonproliferation Collection, record number 22456.
62. Beliefs about the nature of variables that influence world politics lead scholars to different levels of analysis. Waltz distinguishes three levels, or "images": the first image is human psychology and the decision-making process, the second is the nature of states and societies, and the third is the structure of the international political system. *Man, the State, and War* (New York: Columbia University Press, 1959).
63. For innovative ideas on this subject, see Ashton B. Carter, William J. Perry, and John D. Steinbruner, *A New Concept of Cooperative Security*, Brookings Occasional Paper (Washington, D.C.: Brookings, 1992).
64 I develop this point in "Peaceful Proliferation: The Spread of Nuclear Arms in the New Era" (Paper presented at the International Institute for Strategic Studies and the Arms Control Association, New Faces Conference, Bellagio, Italy, 6–10 July 1992). See also Thomas W. Graham and A. F. Mullins, "Arms Control, Military Strategy, and Nuclear Proliferation," in David Goldfisher and Graham, eds., *Nuclear Deterrence and Global Security in Transition* (Boulder, Colo.: Westview, 1992).
65. Washington made this commitment earlier for Latin America in the 1968 Treaty of Tlatelolco protocol, and American officials have restated it in recent years.
66. "In cases of clear aggression which threaten fundamental U.S. interests, especially where weapons of mass destruction or other highly lethal weapons are involved, the United States should retain an option to leave ambiguous whether it would employ nuclear weapons in retaliation to gross acts on the part of the aggressor." Thomas C. Reed and Michael O. Wheeler, "The Role of Nuclear Weapons in the New World Order," report to Air Force General Lee Butler, 13 January 1992, p. 19. See also Eric Schmitt, "Head of Nuclear Forces Plans for a New World," *New York Times*, 25 February 1993, p. B–7.
67. An Indian Defense Planning Staff official told an American colleague that Iraq's mistake was not having nuclear weapons to face the international coalition in 1990–91: a poor country that cannot develop advanced conventional munitions must have nuclear weapons.
68. See Mearsheimer, "Back to the Future;" Dunn, *Containing Nuclear Proliferation*, 47–51; Colin Gray, "Arms Control in a Nuclear Armed World?" *Annals of the American Academy of Political Science Studies*, no. 430 (March 1977): 110–21; Shai Feldman, "Managing Nuclear Proliferation," in Jed C. Snyder and Samuel F. Wells, Jr, eds., *Limiting Nuclear Proliferation* (Cambridge, Mass.: Ballinger, 1985), 301–18; Richard Haass, *Conflicts Unending: The United States and Regional Disputes* (New Haven: Yale University Press, 1990), 91.
69. For these reasons, Dunn suggests helping new nuclear powers solve doctrinal rather than technological problems. *Containing Nuclear Proliferation*, 49–53.
70. Even though Iraq did not acquire nuclear weapons, the Persian Gulf war taught some officials the myth of security through nuclear proliferation.
71. According to some reports Pakistan attempted this in 1990. See Seymour M. Hersh, "On the Nuclear Edge," *New Yorker*, 29 March 1993.
72. See S. Rashid Naim, "Aadhi Raat Ke Baad (After Midnight)," in Stephen Philip Cohen, ed., *Nuclear Proliferation in South Asia: The Prospects for Arms Control* (Boulder, Colo.: Westview, 1991), 23–61.
73. Stephen Philip Cohen, "Policy Implications," in Cohen, *Nuclear Proliferation in South Asia*, 353–4; Dunn, *Containing Nuclear Proliferation*, 53–54.
74. For a detailed discussion of these costs, see Kathleen C. Bailey, "Proliferation: Why Not?" in Bailey, *Improving the Nuclear Nonproliferation Regime* (Boulder, Colo: Westview Press, forthcoming, 1993).

Understanding Nuclear Proliferation:
Theoretical Explanation and China's National Experience

Avery Goldstein

TO EXPLAIN nuclear proliferation one must answer a central question: Why do states choose to develop nuclear weapons? Even without knowing its precise content, one suspects a theory answering this question must include system-level (that is, structural) and unit-level (that is, national) variables. On the one hand, if systemic constraints alone explained the decision to "go nuclear" then proliferation should already have been more rapid than it has been since 1945. Despite facing the universal problem of providing for their own security in an anarchic realm, many states with the requisite capabilities have chosen not to deploy nuclear weapons. On the other hand, if national attributes alone explained proliferation, then there should have been greater uniformity in the sorts of states who have pursued a nuclear weapons capability. Instead one finds in this group states that are liberal-democratic as well as authoritarian, capitalist as well as socialist, secular as well as religious (Islamic and Jewish), advanced-industrial as well as poor-developing.

Unfortunately, these observations may suggest dim prospects for anything like a grand theory of proliferation. Structural causes lead to greater similarity in the behavior of actors (i.e., the pursuit of nuclear weapons) than would be expected on the basis of their individual attributes. Yet these attributes are important enough to result in significant variations not easily explained by the constant constraint of structure. Can one do more than sound the familiar refrain that both internal and external influences must be considered? Does this lead to the methodologically unsatisfying conclusion that one must develop a separate theory of proliferation for each case?

If it is difficult to conceive of a general theory to cover all cases, it may nevertheless be possible to aspire to greater theoretical coherence than much of the work on proliferation displays. It may be possible to push beyond case-specific historical description and to identify the most important incentives affecting any state's decision about pursuing a nuclear capability. In this essay I attempt to do this by considering the interdependence of structural and unit-level influences on state behavior relevant to the proliferation question. First, I present the general argument, indicating the way in which the anarchic structure of the international system together with a state's definition of national interest, its available resources and the current military technology affect the incentives to become a nuclear power. Next, I examine the extent to which the causes identified help account for the Chinese decision to develop, deploy, and maintain nuclear weapons. In the conclusion, I briefly suggest lessons from this theoretically informed case study for thinking about nuclear proliferation in the post-cold-war world.

Structural Causes

Neorealist theory highlights the constraining consequences of two aspects of international structure that shape the behavior of states – ordering principle and polarity.[1] The first, though formally a variable, has in practice remained constant throughout the history of the modern international system. Anarchy, not hierarchy, is the label used by neorealists to identify the way the society of nation-states is arranged. International politics is a self-help realm in which there is no *reliable* superordinate authority to make and enforce rules or resolve disputes among states. Ultimately states must be prepared for the possibility that they will have to fend for themselves. Polarity is a variable that further describes the arrangement or structure of the system by identifying the distribution of capabilities across its principal parts. International relations scholars count the number of great powers populating a given anarchic realm to decide whether the system is best defined as bipolar or multipolar (three or more). Neorealists argue that variations in polarity have important consequences for the ways in which states, both the great powers and others, behave in an anarchic, self-help system. Specifically, how do these two structural causes affect choices relevant to the study of nuclear proliferation?

Anarchy

The condition of anarchy encourages states interested in ensuring their survival to prepare to counter perceived threats to their security. The preparations may include appeals to other states (to form alliances) or international organizations (to establish collective security regimes) as well as attempts to develop the means independently to cope with the recognized threat. Though either external or internal methods of amassing the power to balance against prospective adversaries may be pursued, the uncertainty of life in an anarchic realm leads states to discount the value of unenforceable security guarantees from others, especially international organizations, but even allies. As will be seen in the Chinese case, doubts about the reliability of allies in a self-help system encourage states to provide for themselves.

Competition among states coexisting in a condition of insecurity further encourages each to exploit the most strategically effective forms of military power it is able to deploy. In the contemporary era, this provides an incentive for states to develop nuclear weapons as a hedge against the dangers posed by prospective adversaries and the inevitable risks of abandonment by diffident allies or indecisive international organizations.

Polarity

Variations in polarity modify, but do not fundamentally alter the security imperatives of anarchy. Scholars have paid particular attention to differences in the dynamics of alliance politics in multipolar as opposed to bipolar systems. Although alliances are never coalitions of equally capable actors, in a multipolar world each ally can make a significant contribution to dealing with the common concerns that brought them together. Because of the condition of anarchy, doubts about the usefulness and reliability of allies exist under multipolarity, but self-interest constrains partners to look out for each other's needs.[2] In a bipolar system, however, the gap in capabilities between the strongest actor and all others in an alliance against a common superpower adversary weakens this constraint and fundamentally alters the alliance dynamics. Lacking the military value of great powers in a multipolar era, lesser powers in a bipolar era must more seriously worry that a self-interested superpower ally might not fulfill peacetime pledges of strategic support, especially if this entails sharing grave risks in confronting the other superpower over matters other than its own national interests. A self-regarding superpower would have little incentive to take dangerous steps simply to ensure the continued loyalty of a militarily marginal ally. In short, the "fear of abandonment," as Glenn Snyder dubbed it, though a problem for alliances in all anarchic realms because they lack a reliable mechanism for enforcing international commitments, is especially acute for lesser powers in a system of such severely skewed capabilities.[3]

Bipolarity, then, reinforces the consequences of competition under anarchy that encourage states to provide for their own security. Yet for lesser powers involved in conflicts with the system's duopolists, bipolarity also makes it harder to act on this self-help imperative. Their interest in doing so provides a strong incentive for developing nuclear weapons because the revolutionary strategic implications of nuclear weapons, described below, increase the feasibility of fulfilling the requirements of an independent security policy.

Though perhaps most obvious under bipolarity, with the advent of nuclear weapons, states able to develop them are likely to see their possession as a prudent response to the fear of abandonment in an anarchic international system, regardless of the distribution of capabilities. Vulnerability to thermonuclear destruction compounds unavoidable doubts about the wisdom of depending on self-interested allies as a means for countering the gravest threats a nation might face. In an anarchic, nuclear world such dependence is neither sensible, as fulfilling international commitments entails unprecedented risks, nor necessary, as self-help is made easier.

Unit-Level Causes

Although all states face the constraints of international structure, their response to these constraints varies according to their national attributes. Three unit-level variables are relevant to this discussion of security policy – definitions of national interest, available resources, and military technology.

National Interest

Neorealist theory relies on a simplifying assumption that states, at a minimum, will attempt to ensure their survival. Neorealists, of course, understand that the states' definitions of national interest are in reality more varied and often more expansive. Nevertheless, it is a useful starting point to recognize that each begins with an interest in providing for its national security in this relatively narrow sense. After considering the implications of this core concern for the study of proliferation, I briefly discuss the relevance of additional concerns that sometimes color definitions of national interest.

All states have a national interest in preparing to deal with threats to their territorial integrity and political autonomy. Because national interest is defined partly in terms of political autonomy, self-reliance is to be preferred over dependence in formulating policies to ensure security. Reliance on others (most importantly allies, conceivably international organizations) is less attractive not only because of the structurally generated reliability concerns just raised, but also because dependence on allies entails a compromise of autonomy.

States strive to provide for their security partly by acquiring the military means to dissuade a prospective aggressor or to protect themselves against attacks from one who cannot be dissuaded. Combining forces with others who profess similar security concerns is an alternative for mustering the capabilities needed to deal with such threats. Inasmuch as it requires contingency planning and strategic coordination by allies, however, this approach places constraints on a state's foreign policy. Absent the ability to provide for its own security, a state cannot easily ignore these constraints because its destiny partly depends on maintaining the loyalty of its allies. It is possible that the benefits (in terms of reduced economic burdens and increased military effectiveness) of accepting such constraints on national autonomy may be judged worth the costs (that may include limits on domestic as well as foreign policy in order to maintain the integrity of the alliance). To the extent a nation judges the necessary compromises unacceptable, however, nuclear weapons may well be viewed as a solution to the problem of discovering the military means for a more self-reliant security policy. They enable states independently to deal with prospective threats against which the conventional forces within their reach are deemed inadequate. As will be further explained below, nuclear weapons enhance national autonomy by facilitating independent security strategies of dissuasion by deterrence against a wide range of prospective threats.

If a national interest in self-reliantly assuring security in the face of a threatening adversary provides an incentive to develop nuclear weapons, does a more expansive definition of national interest reflecting an ambition to extend political and territorial hegemony provide an even stronger incentive? It seems unlikely. Nuclear weapons are an effective means to dissuade an adversary by threatening punitive retaliation (deterrence) and possibly by threatening to annihilate invading armies (defense). They seem a much less effective means to persuade others to comply with demands by threatening to initiate and persist in inflicting punishment (compellence) or threatening actually to use force to seize one's objectives (offense).

Compellence, as Schelling explained, generally tends to be more difficult than deterrence, mainly because one seeks to do more than just maintain the status quo.[4] It is often necessary to take action indicating the credibility of threats to inflict unacceptable punishment before compliance can be obtained. Unless he can be certain that the targeted state has no nuclear weapons *and* that there is no chance a nuclear patron will respond on its behalf, an aggressor's nuclear weapons are not easily married to compellent strategies. Absent a nuclear monopoly, the risks of initiating even limited military action are enormous. Most importantly, the burden of demonstrating credibility by engaging in the first step of a potentially dangerous escalatory process falls to the party practicing compellence. The aggressor must worry about the possibility, however small, that he is initiating a chain of events that could prove catastrophic. And although the aggressor can attempt to manage the risk by choosing only clearly safe and very limited steps, this is not likely to make *nuclear* pressure more persuasive. Indeed, such action could easily be interpreted as reflecting a lack of resolve, weakening rather than strengthening the credibility of compellent threats to resort to nuclear force if compliance is not forthcoming.

Nuclear weapons are also difficult to integrate with meaningful offensive strategies, even against a nonnuclear adversary. Serious difficulties arise whether one is attempting to rely on the indirect use of force to convince others to yield without a fight because the outcome of battle is clear in advance, or one is attempting to plan for the direct use of force to fight against an unyielding adversary until one's objectives are achieved. Because the use of nuclear weapons would degrade the value of the prize sought (both in terms of natural and human resources), it is difficult to figure out how to use nuclear weapons to subdue others.[5] Knowing this, it is hard to persuade others to comply with demands because they believe you will be able to achieve your objectives through the direct use of nuclear force.

In short, to the extent the definition of national interest provides an incentive to develop nuclear weapons, the motivation is more convincingly tied to the limited, universal core concerns about national security, dissuading threats to territorial integrity and political autonomy, than to the pursuit of ambitiously expansive goals.

There is, however, also a nonstrategic aspect of national interest that requires attention. In addition to worrying about coping with external threats, states also concern themselves with perceptions of their role on the international stage. To put it bluntly, beyond the minimal interest in security, states' leaders (even if only for domestic political reasons) are attentive to matters of national prestige. Leaders may believe that their status requires demonstrating an ability to do what others with international status can do. Not surprisingly, this has been cited as an incentive to develop nuclear weapons by those who view such a capability as a way for countries to retain (Britain and France) or earn (China) recognition as a great power, elevate their standing as a regional power (India, Israel, Pakistan, Iraq, Iran), or achieve prominence within a transnational movement (the Arab, Islamic, or developing world). Indeed, the British and French openly asserted the value of their national nuclear forces as a credential necessary for playing their proper international role.[6]

Yet, on closer consideration, the prestige motive lacks the explanatory power of some of the other incentives for becoming a nuclear state. Status concerns may be helpful for understanding the determination to demonstrate the capability to develop and deploy nuclear weapons. It is not as helpful, however, for explaining the determination to maintain and modernize the arsenal once the technical virtuosity has been displayed. Perhaps more important, even if leaders in the 1950s and 1960s sincerely believed that a nuclear capability was essential to a nation's status, since then the connection has been tenuous at best. In fact, during the cold war the status of nuclear Britain and France declined while that of nonnuclear Germany and Japan rose.[7] Fluctuations in China's international prestige are also hard to link with the maintenance and modernization of its nuclear forces after 1964. More plausibly, China's international prestige was linked in the 1950s–70s with perceptions of its geopolitical significance during the cold war, and in the 1980s with its economic successes.

Moreover, as nonproliferation norms have taken hold since the late 1960s, it seems likely that states' leaders worry as much about the possible negative, as the positive, consequences for a country's status that follow from deploying nuclear weapons.[8] The anticipated adverse reaction of international opinion, in addition to fears of possible preventive military reactions, may explain the reluctance of more recent nuclear-capable states to admit their achievement. As others have noted, "opaque proliferation" may be the route taken especially by developing states who have motives for deploying nuclear weapons that outweigh the potentially damaging consequences to their status if discovered. The widely embraced norms against proliferation also serve to weaken, if not eliminate, status as a likely incentive for nonnuclear advanced industrial states (for example, Japan, Germany) who could deploy such weapons from fully exploiting their potential. On balance, then, even if it once was a strong motive for deploying and maintaining nuclear weapons, a national interest in international status or prestige no longer provides clear incentives for doing so. On the contrary, it can plausibly be suggested that a decision to pursue a nuclear weapons capability must now be taken *despite* the price that will be paid for violating a widely-shared international norm (by refusing to sign, failing to adhere to, or withdrawing from the regime rooted in the Non-Proliferation Treaty).

Available Resources

States differ not only in their national interests, but also in the resources they can muster to pursue their interests, however defined. Strategies and the forces that fulfill their requirements are not formulated in an economic vacuum. Available resources are partly determined by immutable facts of geography, but are also affected by human ingenuity at exploiting those natural endowments. National wealth varies with the efficiency of economic management. Even wealthy states, however, face opportunity costs. National resources devoted to dealing with military security concerns are resources not devoted to raising the standard of living or maintaining a state's international economic competitiveness.[9] To the extent states confront such tradeoffs, they have an incentive to be sensitive to the return on their security investment. One way to bolster security on the cheap is to free-ride on the military efforts of an ally. For reasons already highlighted, however, such free-riding will be limited by the need to purchase private insurance (a national military capability) against the possibility of abandonment, especially in the nuclear age.[10] The amount of insurance needed will, of course, be determined by the risks confronted. For some, a modest and inexpensive conventional military capability may suffice. For those facing more capable adversaries, however, greater expenses must be incurred. As perceived threats dictate selecting from among expensive alternatives, the incentives to spend scarce national wealth on nuclear, rather than conventional, forces become stronger.[11]

Nuclear weapons are a relatively affordable means for providing national security in large measure because even their primitive versions are capable of inflicting horrendous damage on an adversary or his military forces. This sharply distinguishes them from conventional forces in terms of their military usefulness for fulfilling the requirements of dissuasive strategies of either deterrence or defense. Dissuasion by defense requires a state to deploy forces so formidable that a prospective aggressor is convinced an attack cannot accomplish its military objectives. Dissuasion by deterrence requires deploying forces that threaten to inflict such severe punishment on the prospective aggressor that he will not attack even if he has the ability to accomplish his immediate military objectives. Relying on conventional forces for either of these dissuasive strategies involves gearing one's deployment to that of the adversary. The value of an investment in expensive conventional forces, then, depends on ratios of military power (adjusted for qualitative differences) and the skillfulness with which its use is planned. Not only is it difficult to be confident about the outcome of engagements between known forces, but technological change constantly threatens the military value of one's investment.[12] In addition, quantitative and qualitative improvements in the adversary's capabilities will have to be countered. Precision-guided munitions, electronic countermeasures, stealth technology, and sophisticated armoring are only some of the elements in the clash of modern conventional forces that make them an expensive gamble and volatile investment. By contrast, where nuclear weapons are involved the calculus is bluntly simple, not agonizingly complex, and the value of the investment is relatively stable. The enduring ability of the most basic nuclear weapons quickly to inflict catastrophic damage on military forces or society, even in the teeth of massive deployments of technologically sophisticated defenses, remains unchallenged.[13] Until this distinctive characteristic is neutralized, nuclear weapons will be an economical hedge against obsolescence for states seeking an affordable way to fulfill the requirements of strategies for dissuading highly capable adversaries.

Nuclear forces also represent a prudent security investment because they offer insurance against dramatic shifts in threat perceptions. Investment in conventional forces deployed specifically to dissuade aggression by a land-based military power, for example, may be of little value in the event political developments shift one's attention to the threat posed by a naval adversary. By contrast, changes in the adversary's identity are much less likely to diminish the security value of a state's investment in nuclear weapons. Given the requisite delivery capability, states relying on nuclear forces can deal with dramatic shifts in threat perceptions simply by retargeting.

Finally, beyond these sorts of financial considerations, two other links between national resources and the incentive to develop nuclear weapons can be identified. One is geography. The feasibility of strategies relying on the deployment and possible use of conventional forces is significantly affected by considerations such as room for maneuver, available lines of communication, choke points, climate, and location of national assets. States without the territory necessary to trade space for time, with few natural obstacles to impede or channel an aggressor's attack, or whose key economic centers are few in number or closely concentrated face daunting challenges in devising strategies relying on conventional forces. Although these challenges may be surmountable, the effort will require a diversion of national resources that returns one to the sorts of financial concerns already addressed. By contrast, the geographic constraints on the usefulness of nuclear weapons are looser. At least with regard to maintaining a punitive retaliatory capability as part of a deterrent strategy, ingenuity with regard to deployment and doctrine can offset problems relating to size and terrain. As will be seen quite clearly in the Chinese case, even without resorting to the relatively expensive option of deploying weapons at sea, states can confront prospective adversaries with the fear of devastating retaliation by land-based nuclear forces through combinations of mobility and deception that weaken a prospective aggressor's confidence in the effectiveness of preemption. And even without resorting to an explicit launch-under-warning or launch-under-attack doctrine, indeed even if one espouses a no-first-use policy (as do the Chinese), national control of nuclear weapons requires a prospective adversary to confront the concern that the timing for the use of nuclear weapons is ultimately the victim's prerogative in an anarchic world, previously espoused doctrine notwithstanding.

The other national resource to be considered is manpower. Its quantity and quality both affect the feasibility of alternative strategies. To fulfill the requirements of strategies relying on increasingly sophisticated conventional forces, the available pool of recruits must be large enough and skilled enough to support the deployment of units adequate to dealing with the prospective adversary. A small or ageing population relative to the adversary, a poorly educated population, or political constraints on conscripting those who are qualified but would otherwise opt for more attractive civilian employment are some of the manpower problems states may face. Managing a nuclear deterrent force, while certainly requiring skilled manpower, is less likely to bump up against limits set by the size and age distribution of the available pool of recruits.

In sum, then, for states with the security motive and the economic wherewithal for the necessary research and development program, the tighter the resource constraints (available wealth, geography, or manpower) they confront, the greater the incentive to deploy nuclear weapons.

Military Technology

The military technology available to states in a particular historical era shapes decisions about how national resources can best be employed to pursue national interests. This assertion should not be mistaken for a technological determinist argument: National resources constrain a state's ability to exploit the frontiers of scientific knowledge for military purposes, and perceptions about the need independently to provide for national security affect its interest in doing so. Nevertheless it may still be fruitful to discern the distinct consequences of variations in technology. For those states with adequate resources and sufficient reason to want to provide for their own military security, what are the implications of the advent of nuclear weapons? The most important implication is the strategic revolution they have produced. Recognition of the way they transform military strategy provides another incentive for states to deploy nuclear weapons.

As authors from Brodie through Jervis have argued, nuclear weapons revolutionized strategy by displacing the old sort of offense/defense calculus as the main military concern of great powers.[14] Deterrence took center stage. Once the great powers possessed survivable nuclear forces, the principal concern would no longer be determining how to use military force to win, or avoid losing, war should it occur, but rather how to forestall a conflict in which either party could at any time destroy the other.[15] Because even small numbers of crude nuclear weapons could be used to inflict catastrophic damage on an adversary, in a world without leakproof defenses close consideration of the likely military results of a conflict between nations with survivable nuclear forces was less important than considering how the fear of a catastrophic outcome could prevent the initiation of hostilities. Consequently, among nuclear states scenarios for using force or the threat of force to serve offensive, or even defensive, strategies were subordinate to the dominant strategic logic of dissuasion by deterrence. A prospective aggressor might hope that a nuclear adversary could be disarmed in a first strike or somehow convinced to submit despite retaining a retaliatory capability. But the nuclear revolution meant that lingering doubts about the possibilities for miscalculation, accident, or irrationality (discussed in the Chinese case below) sharply reduced the expected utility of initiating a conflict that could escalate to the use of weapons of mass destruction.[16]

It should be noted that the consequences of the nuclear revolution reinforce the self-help incentive to deploy nuclear weapons that derive from coexistence in a condition of anarchy. First, as mentioned in the discussion of structural constraints, vulnerability to thermonuclear destruction subverts confidence in alliances because it exacerbates the fear that self-interested states will abandon one another rather than risk national survival over their partner's vital interests. Second, the relative ease with which horrifying punishment can be inflicted using nuclear weapons supports strategies of deterrence that do not depend on joint action. In a world where offense/defense balances are key, adding together the capabilities of allies may matter a great deal. In a world where the fear of retaliation turns less on information about the order of battle and more on the choices of national decision-makers, it is neither politically sensible nor militarily necessary to share ultimate control over one's destiny with others.

The advent of nuclear weapons technology, then, led to a strategic revolution that put considerations of deterrence first. Though perhaps forgotten by the superpowers who felt they could do more, this simple strategic consequence of the nuclear revolution was clearly grasped by the others who sought and deployed nuclear weapons – Britain, France and China.[17] Because nuclear weapons introduce the risk of catastrophic escalation, they provide a robust solution to what has historically been a central task of foreign policy, preventing aggression against the states' vital interests.[18] Moreover, against foes of comparable or even much greater capability, they do so without placing an undue strain on resources and without requiring a dependence on allies that is both potentially unreliable and politically unseemly.

Is Nuclear Deterrence Robust?

Before examining the Chinese experience more closely, it may be helpful to clarify the sense in which nuclear weapons married to strategies of dissuasion by deterrence provide for a robust security guarantee. Nuclear deterrence is robust in the sense that it covers a wide variety of possible threats to vital interests. It is sometimes asserted, however, that nuclear weapons do not provide a very robust security guarantee because they are only useful for discouraging others from launching an unrestrained nuclear assault on one's national territory by threatening retaliation in kind. If this were the only security benefit, nuclear weapons would still be valuable, but only for dealing with one scenario, and arguably one that a modestly sensible adversary could avoid. Nuclear weapons, then, would simply be a costly addition to the conventional military-security establishments states have always needed to ensure their vital interests. The incentive to pursue a nuclear capability reflects the belief that such weapons are not simply a supplement, but rather a complement and indeed a partial substitute for forces states might otherwise feel they need. Nuclear weapons enable states to dissuade others from threatening their vital interests, even in scenarios where the prospective aggressor has no plans to resort to nuclear weapons at all. Their dissuasive effect is so robust because the nuclear revolution has transformed the strategic significance of uncertainty about the future.

In a world of conventional weapons, dissuasion may be difficult because an aggressor, even an uncertain and cautious one, can plan carefully to monitor the developing costs and benefits of his military action. Hoping for victory he may actually be prepared to settle for much less. In the worst of circumstances he faces only humiliating defeat. At a minimum, the costs of failure are unclear, may be reversible, and are gradually accumulated as a confrontation unfolds. By contrast, in a confrontation between nuclear states, the risks of military action are not so readily managed. The foreseeable costs of miscalculating the worst case response are clear, irreversible, and not necessarily distant. Thus, uncertainty in a nuclear world means that any military threat to a nuclear state's vital interests carries an unknowable risk that confrontation might escalate with disastrous results.[19]

Some correctly point out that this fear is felt by both parties in a nuclear dyad but surprisingly conclude that this means nuclear weapons cancel one another out. Security, such nuclear skeptics would argue, requires the ability to compete with a prospective adversary at lower rungs on the ladder of escalation. Although it is true that the possibility of nuclear disaster constrains both parties, the key point about the robustness of the nuclear deterrent strategy is that it succeeds insofar as it makes it difficult for the adversary to "go first." Between nuclear states, the knowledge that going first means initiating a sequence of events that contains the possibility of disaster exerts a powerfully dissuasive effect. The use of force, nuclear or conventional, in ways that jeopardize a nuclear state's vital interests entails risks that simply did not obtain before the revolution in military technology. This is not an absolute guarantee of security for nuclear states. Others may yet act in the belief or hope they can control escalation. But the difficulty of entertaining such beliefs or relying on hope is fundamentally transformed by the victim's possession of nuclear weapons.[20]

Having suggested how the structure of the international system, national interest, resources, and military technology provide incentives for states to develop nuclear weapons, I now consider the relevance of these influences in the Chinese case.

The Case of China

China's determination to develop nuclear weapons followed from two beliefs: first, that their alliance with the Soviet Union did not provide adequate security; and, second that a self-reliant strategy of dissuasion by nuclear deterrence would better serve China's national security interest than the alternatives of dissuasion by conventional deterrence or dissuasion by conventional defense.[21] Explaining Beijing's decision, then, entails answering two questions: Why did the Chinese lack confidence in the practical value of the Sino-Soviet alliance? Having decided on the importance of self-help, why did the Chinese choose nuclear deterrence over the conventional deterrent and defensive alternatives? The answers to these questions reveal the importance of the incentives for proliferation identified above.

Alliance Politics and the Self-Help Imperative

The Sino-Soviet treaty of alliance signed in 1950 is easily understood as a response to the structural constraints of the bipolar postwar international system. As balance-of-power theory would lead one to expect, given the scope of the threat the United States posed to China, and given the limited resources a war-ravaged, poverty-stricken nation could muster in response, Beijing pursued the only viable external means for countering one superpower, the armed might of the other.[22] Although the impressive Anti-Japanese resistance mounted by the Chinese Communist Party (CCP) during the Second World War made Beijing reasonably confident that the United States, despite its vast military superiority, would recognize the impracticality of any attempt to invade and occupy the mainland, the newly founded regime faced other plausible and quite serious security risks to which alone it could not respond effectively. American power could be employed directly or through allied proxies in ways that threatened the People's Republic of China's (PRC) vital interests (for example, renewed hostilities on the Korean peninsula could threaten the industrial base in Manchuria; the U.S.-backed KMT on Taiwan could threaten the urban centers concentrated on the east China coast). Most troubling, however, was the Eisenhower administration's coercive diplomacy during the closing months of the Korean conflict, as well as during the 1954 and 1958 crises in the Taiwan Straits that compelled Beijing to face the possibility that the United States might actually resort to its declared option of employing nuclear weapons against China to realize American foreign policy goals.

Lacking the ability independently to deal with the conventional power projection capabilities of the United States or counter its nuclear threats, China depended on its Soviet superpower patron. Events early in the 1950s, at least, seemed to confirm the sensibility of this approach. But if bipolarity dictated dependence on a powerful ally, the enduring international condition of anarchy left the Chinese leaders nervous about this arrangement. Specifically, although the Chinese recognized that the partnership was warranted because they faced a common enemy in the United States, they also recognized that the allies' national interests threatened by the adversary differed in important ways. The principal Chinese concern was the threat to the East Asian seaboard posed by U.S. forces in the Pacific. In a bipolar world, however, the principal Soviet concern was the strategic threat to their homeland posed by American power, especially airpower, based mainly in Europe and the United States. Short of an unlikely American occupation of China, Washington's actions threatening the PRC, then, would not fundamentally affect the most important Soviet interests. Beijing, therefore, could not be confident that Moscow would choose to face down the United States over Chinese interests if this meant risking a war unnecessarily jeopardizing Soviet national survival. Given the self-enforcing nature of agreements in an anarchic realm and especially with the emerging danger of nuclear catastrophe, it was instead reasonable to conclude that during a Sino-American crisis Moscow would have strong incentives to renege on commitments whose fulfillment might prove disastrous, treaties and peacetime rhetoric notwithstanding.

As long as the Chinese lacked a viable alternative to depending on the alliance, however, they prudently tried to cope with this fear of abandonment in two ways. Each aimed at focusing Soviet attention on possible links between the credibility of alliance commitments and the credibility of their principal deterrent threats against the United States. One was to maintain a posture of deference to their superpower patron, including distasteful compromises of Chinese autonomy, despite the fact that ending infringements on national sovereignty was touted as one of the major achievements of the communist-led revolution.[24] By establishing themselves as one of the Soviet's most loyal allies, the Chinese increased the reputational price the Soviets would pay for abandonment even if China's intrinsic military value might be small.[25] The other way in which the Chinese tried to ensure alliance solidarity was to emphasize the severity of the American threat and to encourage concern in Moscow with the implications of tolerating even small losses to the common enemy.

Despite Beijing's deferential behavior towards Moscow and its promotion of an unyielding socialist bloc stand against American imperialism, China soon began to have serious doubts about the viability of this approach. In a bipolar world, attempts to gain leverage over a superpower ally by artfully managing perceptions could not overcome the basic problem of China's small practical importance for Soviet security. Soviet foreign policy would be made with an eye first to dealing with its superpower rival, even if pursuit of this principal national interest jeopardized the alliance, since its practical contribution to Soviet security was marginal at best.[26] By the mid-1950s, Khrushchev was promoting a foreign policy emphasizing peaceful coexistence and competition with the capitalist world, reducing his willingness to confront the United States over anything less than vital Soviet interests. Chinese concern that this signalled a flagging of Moscow's alliance commitment was decisively reinforced by Khrushchev's muted and belated support for China when it faced U.S. nuclear coercion during the 1958 Taiwan straits crisis.[27] Believing their fears of abandonment were now fully justified, the leaders in Beijing resolved as quickly as possible to reduce their dependence on a potentially irresolute ally. For reasons addressed below, they decided that the key to a viable independent security policy, was development of a national nuclear arsenal.

China had initiated its nuclear weapons research program in the wake of what Beijing saw as U.S. nuclear blackmail during the Taiwan straits crisis in 1954–55. At the outset, the program was envisioned as consistent with a strong Sino-Soviet alliance. China expected and received aid in this effort from its superpower patron. After the 1958 crisis, however, Chinese and Soviet views dramatically changed. Khrushchev worried about his Chinese ally precipitating conflicts over issues of little importance to the Soviets with an American adversary that evinced a disconcerting willingness to escalate to the level of general nuclear war. Mao worried about abandonment by an irresolute, if capable, Soviet ally precisely at those moments when support was most needed. Soviet footdragging, stonewalling, and ultimately reneging on its pledges of assistance to China's atomic bomb project in this period further frustrated and finally infuriated Beijing.[28] Moscow's attempts to condition continued aid on arrangements that would grant them some control over the PRC's military activities especially aggravated Chinese suspicions by touching a raw nationalist nerve made hypersensitive by the country's century-long history of international humiliation.[29] As the Chinese perception of the American threat diminished somewhat after 1958, the political price of Soviet support seemed unacceptably high and the strategic benefits unacceptably low. In 1959, as Khrushchev announced the termination of Soviet aid, China accelerated its nuclear weapons program which it now saw as a totally self-reliant crash effort to provide the nation with a security guarantee that could substitute for dependence on what Beijing had decided was a diffident, meddling ally.[30]

Selecting a Self-reliant Alternative

The international structural constraints of bipolarity and especially anarchy, as well as a strong national interest in autonomy combined to highlight the practical importance and political preferability of a self-reliant security policy. But why did the Chinese focus on nuclear weapons as the key to such a policy? China embraced a strategy of dissuasion by deterrence rather than defense because of the resource constraints it faced. China emphasized nuclear rather than conventional forces as the means for fulfilling the requirements of its deterrent strategy because the revolutionary implications of this new military technology suggested that such forces would provide for a more robust yet still affordable means for dissuading powerful foes.

Dissuasion by Deterrence Rather than Defense China's available resources in the two decades after the Second World War precluded an effective defensive capability against its principal adversary. The costs of deploying forces in a quantity and of a quality sufficient to dissuade a superpower from threatening Chinese vital interests because its assault would be militarily stymied were simply too high. Indeed, establishing credible defenses against even the most plausible limited conventional attacks by the United States would have required Beijing substantially to divert national resources from the pressing need to nurture an economy that could provide for the basic needs of the Chinese people. To do so would have had serious implications for the legitimacy of the young communist regime. The CCP's ability to rule without relying solely on coercion depended on fulfilling the twin nationalist promises of their revolution – at last to provide security not just against the sort of foreign interference that had burdened China during the preceding century but also against what had become a domestic condition of grinding poverty.[31] Beijing, then, could not incur the steep opportunity costs of employing much of the nation's scarce resources to pursue the vast air and naval capabilities that would be necessary for defense against a superpower.[32]

Thus, in the 1950s there was simply no prospect of the People's Liberation Army's (PLA) conventional capabilities matching superior U.S. forces, even with the modest modernization facilitated by Soviet aid. China could not afford to buy, and the Soviets were unwilling to donate, quantities of conventional arms sufficient to offset the qualitative American advantage.[33] Following the Sino-Soviet split, China's disadvantage in conventional forces became more acute. Cut off from its only source of significant foreign military aid, China's leaders knew that for the foreseeable future they would lack not just the funds, but also the industrial and scientific base essential to an independent program to modernize its conventional defensive capabilities.[34]

Two additional resource problems mentioned above that are related to but distinct from financial concerns, manpower and especially geography, compounded the difficulties China would face if it tried to pursue the strategic alternative of an independent conventional defense. The PLA recruited its soldiers mainly from China's vast supply of poor peasants. Although this ensured the availability of large numbers at low cost, this pool of recruits represented the least educated segment of society. In an era of increasingly technical requirements for successful performance in combat, the shortfall in quality, if not quantity, of personnel posed a daunting problem for the regime if it would seek to deploy forces able to fend off the assault of a modernized superpower's army. Wise government policy could have addressed the problem of personnel quality through investment in education. But cost aside, in China during the late 1950s and for much of the remaining two decades of Mao Zedong's life, the leaders in Beijing were not prepared to make the political and financial commitment to an educational system that would significantly elevate the quality of its graduates. Instead, the educational system became a hotbed of political controversy in the ongoing debate about the relative importance of redness and expertise.[35]

A regime determined to develop a modern conventional capability might have altered its educational policy. But there was not much China could do about certain fundamental geopolitical realities that would complicate a strategy of dissuasion by defense. Beijing faced two key problems as it contemplated a self-reliant security policy. First, Manchuria, the nation's heavy industrial base during the era of Stalinist economics, was an inviting salient vulnerable to being severed from the rest of Chinese territory through a quick strike for this limited purpose launched by an adversary with conventional superiority. Second, outside Manchuria, China's most developed regions were concentrated along the east coast. Though less easily seized, these cities, and Manchuria too, were highly vulnerable to punishing air strikes against which the Chinese lacked effective defenses.[36] In short, China could not deal with at least two quite plausible superpower threats to these areas of vital national interest by convincing the prospective adversary of the PLA's ability to deny him his military objectives. Indeed, beginning in the 1950s China seemed to accept the indefensibility of its most valuable urban assets and devoted scarce resources to a futile development plan that aimed at reducing their vulnerability by dispersing the industrial base in the country's vast hinterland.[37]

In short, although the PLA's capabilities might enable China to cope with assorted regional adversaries, such as India, it did not hold much promise as a defensive force that could reliably dissuade the more important superpower threats China faced during the cold war. Instead, once China decided it needed to hedge against the failure of alliance security guarantees, the realistic choice for dealing with this principal national security concern was between two forms of dissuasion by deterrence, that relying on conventional and that relying on nuclear forces. Deterrence, of whichever sort, was more attractive because it would not require China to develop and deploy the massive capabilities necessary to frustrate the military assault by a superior adversary. Deterrence required only developing a capability to inflict punishment on the adversary sufficient to convince him that success on the battlefield would not enable him to realize his political objectives at an acceptable price. Having decided upon self-reliance, Beijing's relatively limited resources encouraged an emphasis on deterrence; having decided upon deterrence, Beijing's threat perceptions would encourage the development of nuclear weapons as an affordable means for making the strategy more robust.

Conventional Deterrence as an Alternative to Nuclear Weapons In the 1950s, the Chinese communist leaders already had an ability to threaten with unacceptable punishment any adversary who might contemplate invading and occupying the Chinese mainland. Mao's writings on People's War provided the doctrinal underpinning for this strategy of dissuasion by deterrence employing conventional forces.[38] During the war against Japan in the 1930s and 1940s, the CCP had impressively established the credibility of its threat to confront even the most powerful invader's occupation with a sustained national resistance relying on the most meager material assets. If Beijing had been satisfied with the effectiveness of this dissuasive strategy, it would not have needed to develop nuclear weapons to provide for national security. For a state like China, one incapable of fully exploiting modern military technology, this sort of conventional deterrent seems an eminently sensible option because it allows the regime to translate nationalist political credentials into military power. Moreover, in principle, the prospect of dealing with protracted popular resistance could provide a threat sufficient to dissuade even a superpower adversary from attacking, and at reasonable cost.

But despite publicly touting its effectiveness for three decades, the CCP in practice began to respond to what it recognized as the inadequacy of People's War for ensuring Chinese security. Four shortcomings of this strategy of conventional deterrence account for China's doubts about its usefulness and help explain its pursuit of the deterrent alternative based on threats of nuclear retaliation.

First, it may be difficult to convince a strongly motivated aggressor that he can secure only a pyrrhic victory. As long as he knows he need only engage conventional forces, a relatively powerful adversary is likely to believe that the risks of testing the balance of capabilities and resolve are manageable. Thus, a determined and moderately risk-acceptant decisionmaker may see this as a chance worth taking. However devastating they may eventually prove to be, conventional, as opposed to nuclear, deterrent threats do not force a prospective aggressor to confront the risk of almost instantaneously catastrophic punishment.[39]

History in fact would seem to justify the concern that great powers may not easily be dissuaded from acting against much weaker states by the risk of confronting popular resistance.[40] One *can*, of course, point to impressive examples of a great power eventually retreating in the face of punishment inflicted by a minor power's protracted "people's war" (for example, France and the United States in Indochina, the Soviets in Afghanistan). But the fact remains that these great powers were not deterred from first attempting to impose their will. Rather, these examples suggest that protracted popular resistance has sometimes effectively served a strategy of *compellence* (relying on the threat of continuing to inflict pain in order to persuade an enemy to withdraw when forcible eviction by military offensive is not feasible), not that it has been a powerful deterrent.[41] Indeed, however impressive the resistance organized by the CCP during the Second World War, Beijing knew that it had not succeeded in compelling a Japanese withdrawal from the mainland before the United States obtained Tokyo's unconditional surrender.

Second, conventional deterrence of this sort typically involves trading space for time rather than forward defense to repulse the adversary's attack. But once the Chinese Communist Party was in charge of a national government, as opposed to a "revolutionary guerrilla base area," it was more reluctant to embrace a security policy one of whose central features might require accepting the loss (even if temporary) of territory under its administration. While Beijing might be prepared to bend to the need for strategic retreat once a battle had been joined, peacetime planning to do so was politically unattractive.[42] By contrast, nuclear forces would provide China with a deterrent that made large-scale strategic retreat unnecessary.[43]

Third, China's economic geography reduced the practical, as well as political, appeal of a conventional deterrent approach that entailed trading space for time. Though far from modernized, China did possess an array of fixed industrial assets which, as noted above, were concentrated in a small number of urban settings. Strategic retreat from these vulnerable cities in preparation for national resistance would involve exercising one of two options. One would be to abandon these assets to the aggressor. But doing so would make the initiation and maintenance of an armed national resistance more difficult. It might even enhance the invader's capacity to prosecute a war of suppression. The other alternative would be to deny the aggressor these valuable urban assets by sabotage in conjunction with the strategic retreat. This might increase the logistical problems confronted by the invader. It would not be likely, however, to have a decisive military impact since the vast capabilities of China's superpower adversary would enable it to supply his forces adequately by relying on materiel transported from home.

The fourth, and arguably most important, problem with China's conventional deterrent reflected what Luttwak terms the paradoxical logic of strategy. A strategy's effectiveness may well lead the adversary to revise his plans in ways that render it irrelevant.[44] Whatever the potential weaknesses of conventional deterrence, the prospect of a protracted Chinese resistance was almost certainly credible enough that the United States would never have seriously contemplated the idea of a total war of invasion and occupation (that is, a strategic offensive) as a means to achieve its political objectives. Instead Beijing had to worry about two quite different, but more plausible, threats against which its conventional deterrent would not be very helpful.

First, relying on its ability to project massive military force, China's superpower adversary could attempt to coerce or blackmail Beijing (that is, adopt a compellent rather than an offensive strategy). Whether or not such efforts achieved their intended purposes, China would be hard pressed to dissuade the United States from making the attempt and even inflicting the threatened punishment. The Chinese had no effective counter to such threats as long as the People's Liberation Army lacked the ability defend against or to reach out and punish the adversary. Washington's nuclear saber rattling in the last months of the Korean War and in the subsequent Taiwan Straits Crises revealed China's vulnerability to this sort of coercion. The unattractive options for China were to yield to American pressure, to refuse and then suffer the threatened punishment, or to turn to their Soviet patron and hope it would receive support sufficient to dissuade the common enemy. Most prominently during the 1958 Taiwan Straits Crisis, the Chinese, disappointed with the lukewarm backing provided by Moscow and unwilling to engage U.S. military forces, chose to back down.[45] Although they covered their retreat with Maoist rhetorical flourishes about the indecisiveness nuclear weapons, on the heels of this experience China sought to develop an effective independent counter to superpower coercion by accelerating its own efforts to acquire some of these "indecisive" weapons.[46]

The second important threat against which China's vaunted conventional deterrent of People's War would be of little use, was the possibility of a rapid military strike by the adversary seeking *limited* territorial or political objectives. The awesome U.S. military might take advantage of surprise, speed, and a concentration of overwhelming force to present China with a fait accompli – seizing territory and then sealing off the area by exploiting its small size or isolation. Under this scenario, China could be deprived of the time and space essential to organizing and then sustaining a durable and effective resistance. More troubling still, mounting a counteroffensive to retake the area would be an especially daunting task for China because the overall material advantages of the superpower would be reinforced by the tactical advantages associated with fighting on the defensive.[47] Given the geopolitical realities described above, China's most attractive, valuable assets were in fact vulnerable to precisely such limited military operations. However unlikely such scenarios seem, offensive operations to seal off Manchuria or gain a foothold on the eastern seaboard (and when the Soviets become a focus of concern, to detach western portions of Xinjiang province) were always more plausible than attempts to occupy China's heartland.

Conventional deterrence, then, although compensating for material inferiority by converting the CCP's nationalist credentials into a military asset, was a less than satisfactory solution to China's actual security problems. For addressing the real dangers China might face, People's War was inadequate. China's vaunted conventional deterrent was a strategy geared to a largely nonexistent peril.[48] China required an economically sound deterrent strategy that more effectively enabled the state to cope with the spectrum of plausible threats it confronted from a militarily superior adversary. Because it would provide for a more robust, yet affordable, security guarantee, nuclear deterrence was the preferred solution to China's problem of dissuasion.[49]

Nonstrategic National Interests and China's Nuclear Weapons

Before examining the sense in which the strategic revolution of the nuclear era provided an incentive for China to develop and maintain a nuclear arsenal, a few brief comments on some of the "nonstrategic" national interest motives is necessary. In discussing the PRC's response to the fear of abandonment, I have already suggested the way in which China's preference for autonomy was driven not solely by the self-help imperatives of life in an anarchically ordered realm, but also by a strong, historically conditioned national preference for political independence. Even if they had been confident of the Soviet military guarantee against the U.S. threat in the 1950s, Beijing's uneasiness with its junior status and the compromises of its sovereignty the alliance entailed would have encouraged the pursuit of an independent national means to provide for Chinese security. In this sense national pride, as opposed to narrowly military security interests, was an incentive for self-reliance that ultimately required a Chinese nuclear capability.

An interest in bolstering China's international prestige also contributed to Beijing's decision to pursue nuclear weapons in the 1950s and retain and modernize them thereafter.[50] Serving this interest was not simple, however, since it required Beijing to limit the harm to their reputation that might follow from the deployment of weapons they themselves had belittled and that global public opinion was coming to disdain. In part they dealt with this difficulty through rhetoric that placed the moral burden for China's program on the shoulders of the superpowers whose nuclear threats required a self-defense capability.[51] The fear of negative reactions was also addressed by portraying the weapons program as another element in China's international effort to champion the rights of the world's developing states threatened by superpower hegemony. In the interest of world peace, not just national security, China would shatter the superpowers' nuclear monopoly.[52] On these terms, Beijing believed it could square its early anti-nuclear rhetoric with pursuit of its own capability as well as its initial disdain for the nonproliferation regime.[53] Whatever the persuasiveness of China's arguments that aimed to ensure that becoming a nuclear weapons state would only help, rather than hurt, China's international status, these prestige motives were not as important as the security motives. Certainly this must be the judgement about Beijing's decision to move beyond early testing to the arduous task of deploying and maintaining a viable, if small, delivery capability. The link between a demonstrated nuclear capability and China's international prestige would hardly have warranted this costly effort in and of itself since the nuclear weapons tests in the mid-1960s did not provide any obvious boost in the country's status.

Nuclear weapons status did not enable China to achieve political leadership of the developing world. Instead, contention with noncommunist nationalists and with Soviet-loyal communist rivals continued. Nuclear weapons status did not enable China to supplant Taiwan in the United Nations and on the Security Council. Instead, entry would await the decision by the United States to refrain from blocking China's admission because improved relations with Beijing would give Washington diplomatic leverage in its relations with Moscow. Nuclear weapons status did not enable China to wield influence over the arms control talks between the "hegemonic superpowers." Instead, ironically, the threat of a future Chinese nuclear capability provided a pretext for the United States to plan deployment of a small force of anti-ballistic missiles despite their ineffectiveness against the Soviet threat and the envisioned limitations of an Anti-Ballistic Missile treaty. Nuclear weapons status did not even enable China to exercise greater regional influence. Instead, as China suffered through the turmoil of the Cultural Revolution and the stagnation of its failed Maoist development policy, Japan and the East Asian newly industrialized countries became the regional economic luminaries, India broke China's regional nuclear monopoly, and Vietnam acted as something of a Southeast Asian regional hegemon.

Nevertheless, China's determination to invest its scarce defense modernization resources on establishing the viability of its nuclear deterrent would not wane. Whatever the payoff in terms of international prestige, the security benefits were deemed to be of overriding importance. The expected payoff from deploying nuclear weapons reflected the Chinese leaders' grasp of their value in fulfilling the requirements of a robust yet affordable strategy of dissuasion by deterrence.

China and the Nuclear Revolution

Given their comparatively limited resources, China's leaders clearly understood they would not be able to compete with their superpower adversary in terms of quantity or quality of nuclear weaponry. Despite this anticipated disadvantage, they believed a heavy investment in developing, deploying, and maintaining nuclear weapons was wise. They concluded that when the balance in question involved nuclear weapons, traditional notions of inferiority and superiority were simply not relevant.[54] Small nuclear forces would suffice to dissuade an adversary possessing a much larger arsenal. In confrontations between nuclear adversaries, they were confident that deterrence dominated offensive and defensive considerations and that the threat of nuclear retaliation would provide a relatively robust means for dissuading even a vastly mightier adversary.[55] The key to success would be to deploy nuclear weapons capable of inflicting terrible pain and suffering and then to maintain uncertainty in the adversary's mind – uncertainty about his prospects for a completely successful preemptive strike, and uncertainty about the circumstances under which China's clearly horrifying destructive force might be employed.

China's early nuclear weapons program led to the deployment of powerful, if highly inaccurate, bombs well suited to inflicting terrifying levels of pain and suffering on an adversary.[56] Though unable to provide for the levels of weapons survivability or warhead accuracy that superpower arsenals were achieving, China was confident that an adversary would have to worry that some of these multimegaton citybusters might be missed in any preemptive strikes.[57] From the start, China's military leaders dedicated themselves to discovering economical means for complicating the strategic planning of prospective aggressors and increasing the likelihood that retaliatory forces would survive even a determined attack. Relatively inexpensive methods of concealment and mobility for land-based systems and planning for a limited sea-based system were part of the solution.[58] China also deployed decoys that cheaply enabled them to confound any adversary's attempt at first-strike targeting.[59]

Aside from such measures that explicitly aimed at cultivating doubt in the adversary's mind about the sorts of punishment China might be *able* to inflict in response to aggression, the dissuasive effect of China's nuclear weapons derived also from uncertainties about behavior, in particular the likelihood Beijing would actually use the capabilities at its disposal. Even if China was presumed to have a partially survivable capability, an adversary would not be dissuaded if it believed the decision to retaliate would not be made. Indeed, those analysts of the superpowers' nuclear relations who focused on the importance of offensive advantages for the logic of preemption, would almost certainly conclude a fortiori that China's small and crude nuclear arsenal could not provide a credible deterrent.[60] In a nuclear exchange, after all, China's primitive command and control seemed highly vulnerable to decapitation that would render any surviving weapons capability useless. And even if a national command survived, what purpose could rationally be served by launching nuclear weapons when China would be outgunned at every rung on the ladder of escalation? A small nuclear force would not enable China to engage in a tit-for-tat limited competition in inflicting and absorbing pain, and a massive strike would be suicidal by inviting retaliation in kind from an ever capable superpower adversary. Even if not disarmed, China, to use Paul Nitze's term, would be "self-deterred."[61] By this reasoning, China's nuclear investment was far from adequate and, therefore, could not provide a very robust security guarantee. Adequate investment would require military budgets well beyond China's reach.

Why then, were the Chinese convinced that the nuclear deterrent they could afford was in fact worth the investment? The answer is that they seemed to appreciate that states necessarily operate with incomplete information about the future. Chinese nuclear forces would be integrated with a strategy that saw "the essence of warfare," in Sun Zi's terms, as "the art of ambiguity."[62] If its adversaries could be certain that Beijing would always behave as a rational actor, then it would be true that China's mere *ability* to threaten horrifying retaliation would not enable it to dissuade any attack short of an unrestrained first strike. Under any other scenario, the aggressor could be confident that a rational leadership would recognize that escalation could only worsen China's position. But it is precisely because there could be no certainty that China would behave as if it were a rational actor that Beijing was confident nuclear deterrence and the concomitant risk of escalation to a catastrophic exchange provided so robust a security guarantee, even for a state like China with a relatively small collection of inaccurate, though powerful, weapons.

More specifically, an aggressor would have to worry that, in response to serious provocation, China would turn out to be an "unsafe" actor, an actor whose behavior was inconsistent with the rationality assumption for one of two reasons.[63] First, the adversary would have to worry about the possibility of an accidental launch, as explained below. In this case China's behavior might not reflect rational choice because decisionmakers lacked full control over events or lacked necessary information. Second, an adversary who would bank on cost-benefit analysis inhibiting a nuclear China from choosing escalation would have to worry about the small, but everpresent, possibility that despite having the necessary information and control over their forces, the leaders in Beijing might not choose rationally. The extreme time pressures and the enormous stakes involved in managing a nuclear crisis, particularly given China's small margin for error and its historically rooted sensitivity to international humiliation, could well undermine rational decision-making in Beijing.[64]

China would benefit from the dissuasive effects of such uncertainty about its future behavior simply by acquiring a nuclear retaliatory capability. There would be no need to *ensure* that China would be an unsafe actor in a crisis. On the contrary, China like all nuclear states, had strong incentives to establish peacetime controls that minimized the obvious risks that accident and irrationality present. Nevertheless, any adversary would have to consider the ways in which their provocative actions might elevate the normally low probability that China would cease to behave as if it were a safe, rational actor for the reasons cited. The effect of increased military pressure on the Chinese leaders' cognitive processes may be hard to anticipate.[65] Its likely effect on the risk of accidental launch seems more obvious.

Especially for a country like China, with less sophisticated and durable forces and c^3I, an adversary would have to worry about Beijing's ability to engage in attack assessment and to maintain negative control over its forces during hostilities, in particular if the aggressor's actions create a less forgiving nuclearized environment.[66] An accident that confounded rationality assumptions might take the form of either an authorized or unauthorized nuclear launch. An *authorized* strike could result from Beijing's mistaken belief, based on human or technical intelligence failure, that the adversary had initiated an unlimited nuclear offensive, the only circumstance under which even a rational Chinese leadership would be indifferent to the suicidal consequences of massive retaliation. In addition, an adversary would have to worry that its actions could provoke an accidental *unauthorized* launch as control over forces placed on crisis alert eroded due to failures in China's command system or organizational confusion after nuclear release authority had been delegated to lower levels.[67] The sorts of durability concerns analysts raised about the superpowers' command and control organization and hardware would, of course, be magnified in the case of the technologically less advanced Chinese systems.[68] Although an adversary might hope that the result of disruption and destruction would be paralysis, he could not ignore the possibility that retaliatory strikes might be launched by those to whom authority had been delegated or predelegated. In fact, prudence would require an adversary to assume that some sort of early delegation in a crisis would be China's sensible response to the durability problems deriving from the marked technological inferiority of their c^3I.[69]

Beijing's commitment to develop, deploy, and maintain a nuclear deterrent, then, reflected a belief that the strategic consequences of the nuclear revolution enabled China to provide itself with a credible deterrent by taking advantage of an adversary's uncertainty about the retaliatory forces the PRC would have at its disposal as well as the circumstances under which they might be employed. Could China be sure to retain its punitive capability and would the leaders choose to use it? Even if the adversary's answer to this question was "probably not," Beijing planned to exploit the fact that this was equivalent to answering "possibly yes." Where nuclear weapons would be involved, a slender possibility powerfully inhibits the initiation of hostilities.

Given this strategic logic emphasizing the art of ambiguity, it is ironic that China's most prominent declaratory nuclear policy seemed to prevent it from fully capitalizing on the combination of terror and uncertainty about future behavior that provides for a robust deterrent. At the time of its first nuclear weapons test, and with each subsequent test, China publicly espoused a nuclear no-first-use pledge.[70] Though consistent with Beijing's argument that it deployed nuclear weapons only out of a need to counter superpower hegemony, this doctrine made little strategic sense for a country whose nuclear weapons were fewer in number and more vulnerable than those of its adversary. Instead a doctrine threatening to launch on warning would better dissuade a nuclear attack and a doctrine reserving the right of nuclear first use would better dissuade even serious conventional challenges to Chinese security.[71] A shift away from a no-first-use policy would help address the problem of weapons vulnerability and further complicate an adversary's strategic planning. But Beijing apparently did not see these practical benefits as outweighing the symbolic damage to its status as a reluctant nuclear power, especially since it knew that China's no-first-use (NFU) stance probably would have little actual effect on an adversary's planning or behavior. Precisely because of the incongruity between a no-first-use doctrine and China's relatively vulnerable retaliatory capability, an aggressor could not safely count on Beijing adhering to its peacetime pledge in the heat of conflict over vital interests. Absent an international authority capable of enforcing contracts, an adversary would have to discount the no-first-use pledge as "cheap talk."[72]

Conclusion

This essay began with a question: Why do states choose to develop nuclear weapons? Relying neither on a purely deductive nor a purely inductive methodology, I have attempted to answer this question by identifying a cluster of important variables that affect this decision and considered their usefulness by examining the Chinese case. The explanation I have offered points to the interdependence of structural and unit-level influences that combine to encourage the pursuit of a nuclear capability. Structural constraints, especially the condition of anarchy, encourage self-regarding behavior that undermines confidence in international guarantees, even ostensibly firm military alliances, as a means for ensuring security. For both practical strategic and normative political reasons, states prefer autonomy to dependence and, therefore, search for ways to provide for their own security.

The search, however, is influenced by resource constraints (economic, geographic, and human) that may limit the possibilities for security independence and shape the strategy and forces that can realistically be selected. The constrained search for a practical means to provide for a self-reliant policy is also influenced by beliefs about how current military technology shapes the strategic environment. I argued that states seek to develop nuclear weapons because they provide a hedge against important threats to national security that is more reliable than dependence on external assistance, and yet both more strategically robust and economically affordable than the available alternatives given the current state of military technology.

The Chinese experience was then introduced to provide an example of the way these general causes operated in a particular case. Despite tight resource constraints and the challenges of dealing with superpower-sized threats in a bipolar world, the Chinese, disappointed with the benefits from an alliance that necessitated compromises of the regime's autonomy, sought a self-reliant alternative for ensuring their national security. Given the awesome capabilities of China's adversary and its own economic, geographic and human resource constraints, the leaders in Beijing could not muster the means necessary for fulfilling the requirements of a defensive strategy based on conventional forces. The self-reliant nonnuclear alternative that China could afford, deterrence based on the threat to organize and sustain a national war of resistance against any invader who would occupy China, was unsuitable for dealing with other serious threats the country faced. Nuclear weapons, though by no means cheap or easy for China to develop, were expected to provide Beijing with a deterrent that was affordable. Equally important, as a consequence of the strategic revolution this new military technology had produced, nuclear weapons promised a deterrent that was more robust than People's War.

In both the general discussion and specific application of the argument, I focused mainly on explaining why states pursue nuclear weapons. What about explaining why states may *not* pursue them? Instead of attempting to provide an exhaustive list of possibilities, one can answer that the same set of causes should be relevant to explaining nonproliferation as well as proliferation. With regard to the principal causes I have cited, a state would not pursue a nuclear capability if it is a state (1) satisfied with its security, either because it perceives no serious threats or is comfortable with existing international security guarantees (from allies or international organizations) and any compromises of autonomy that may be necessary to ensure sufficient levels of such external support; (2) whose economic, geographic, and human resources make it feasible to deploy nonnuclear forces to cope with the most important threats to its national security; or, (3) whose industrial and scientific capacity is inadequate for exploiting nuclear technology. Brief consideration of these reasons why a state may not pursue nuclear weapons will illuminate the implications of this essay's argument for the possibilities of predicting and controlling their spread.

The third in the list of reasons a state may not develop nuclear weapons, reflects common sense as much as theoretical argument. Despite the revolutionary strategic implications of nuclear weapons, some states may be unable to produce them. As the absolute costs of exploiting the technology drop and knowledge disseminates, however, one would predict that the number of states that simply cannot produce nuclear weapons will shrink. If so, controlling the diffusion of nuclear capabilities requires denying states essential materials, radioactive fuels as well as engineering capabilities. Elements of the current nonproliferation regime are part of just such an effort to deal with the options available to states rather than their choice among options. So, too, are preventive or preemptive military strikes against incipient nuclear powers. Given the difficulty both of enforcing international agreements and ensuring the effectiveness of military action in an anarchic realm, these approaches seem destined to slow, rather than stop, the spread of a capability some states may have good reason to pursue.[73] Nevertheless, attempts to ensure that membership in the nuclear club expands at a measured pace may be desirable if one believes that keeping the barriers to entry high increases the care with which this most valuable national security asset will be managed.[74]

The other reasons listed as to why states may not pursue nuclear weapons, bear on the choices of those who *could* make the effort. My central argument in this essay is that insecurity is the principal motive for developing such a capability. Obviously, then, a state unconcerned about external threats would lack this key reason to pursue nuclear weapons. Few states, however, behave as though they believe they can afford the luxury of indifference, even during periods of low international tension. The vast majority instead hedge their bets against the risks of an uncertain future. More interesting, then, is the possibility that security-conscious states might decide not to exploit their nuclear potential because they are satisfied with external security guarantees or alternative nonnuclear, but self-reliant, strategies.

If a state believes alliances or international organizations ensure its national interests against foreseeable threats, it may not seek much of an independent capability, nuclear or nonnuclear. The theoretical arguments outlined in this essay, however, explain why one would predict that states will be wary of relying on external guarantees. The undependability of international commitments leads states to worry about what might happen if their interests and those of their security patrons diverge. Since this concern ultimately results from the constraint of anarchy, short of a transformation of the international system's structure, there is little that can be done decisively to control the structural incentive for self-reliance that contributes to nuclear proliferation.[75]

It is also possible, however, that a state may believe its conventional defensive or deterrent capabilities will be sufficient to dissuade the threats prospective adversaries pose. If so, the state could forego nuclear weapons. In this case, the leaders may conclude that challenges to its security do not merit diverting additional national resources from more productive peacetime uses, facing the risks of international military, economic, or diplomatic sanctions, or confronting domestic political opposition. But the argument of this essay suggests that as the costs of modern conventional forces escalate, the expense of countering an adversary's capabilities will encourage states to consider the economic virtues of the strategically robust nuclear alternative. If so, controlling the spread of nuclear weapons requires bolstering the affordability and reliability of dissuasion by relying on conventional forces. In principle, at least, limitations on advanced weapons technology transfers and conventional arms control that yielded defense-dominant balances could reduce this incentive to go nuclear. Unfortunately, the difficulties of maintaining an international technology control regime and the complexities involved in stabilizing a conventional arms competition between mutually suspicious adversaries constrained to coexist in an anarchic realm (that is, fostering cooperation under the security dilemma) must temper any hopes for exercising such control.[76]

In sum, then, as long as the causes identified in this essay are at play, one would predict that the number of nuclear states will continue to grow and that the prospects for controlling this growth are not bright. Only a change in the anarchic structure of the state system that rendered international commitments enforceable, a technological breakthrough that negated the strategic consequences of the nuclear revolution, or a transformation of interests that led all states to rank adherence to global norms against possessing nuclear weapons above preserving national autonomy would decisively weaken the incentives to deploy nuclear weapons. Short of such unlikely changes, policies to limit the spread of nuclear weapons will address the symptom rather than the causes of proliferation. As already noted, such symptomatic treatment, even if only partially successful, may be advisable. But given the underlying causes, few major states with important international interests should be expected to eschew nuclear weapons indefinitely. One instead should expect that among nonnuclear states able to produce nuclear weapons, abstinence will endure only until serious threats to vital interests are perceived.

What specifically does this analysis suggest about the spread of nuclear weapons in the post-cold war world? If, as seems likely, the structure of the international system becomes multipolar, this will have several relevant implications. First, competition among the great powers should be neither as intensive nor extensive under multipolarity as it was under bipolarity. In a bipolar world, there is a clear incentive for each great power to respond to any change that might work to the advantage of its adversary.[77] Consequently, lesser powers in the cold war system could often exploit Soviet-American rivalry to garner military aid. In a multipolar post-cold war world, such states may find they must figure out how most effectively to draw on their own resources. According to the logic of the argument presented here, for some the self-help imperative is likely to foster an interest in nuclear weapons as an affordable security guarantee. At the same time, however, in the absence of the close competition between superpowers that transformed lesser powers into clients and proxies, preventive military action against nuclear aspirants will pose fewer risks of escalation. Although this might suggest a sturdy coercive fallback for the diplomacy of the nonproliferation regime, it must be added that multipolarity results in a diffusion of responsibility as well as competition. Thus, in line with the logic of collective action, unless the threat posed by an aspiring nuclear state is clearly focused, potential enforcers may rationally delay undertaking costly preventive military measures, hoping that others will do the job for them. Finally, a shift to a multipolar system may mean that strategic alignments will be less clearcut than under bipolarity, for the system's great powers as well as regional powers. Such indeterminacy may increase the importance of the robustness of nuclear deterrence. As noted in this essay, easily retargetable nuclear weapons provide a valuable hedge against changes in the source of threats to national security and the obsolescence of a force structure tailored to a specific adversary's current capabilities.

Aside from a change in the international system's polarity, the post-cold war world will likely be distinguished from its predecessor by a greater emphasis on economic issues. States seem to have grasped the obsolescence of conquest as a means of enrichment.[78] Instead, the route to prosperity is understood to lie with domestic economic development in order to be able to cope with ever more intense international competition. In such a world, states will have strong incentives to pay close attention to the opportunity costs of high levels of military spending. To the extent nuclear weapons satisfy basic security requirements at lower cost than conventional defenses, they will remain attractive to states determined to focus their attention and resources on nonmilitary concerns.[79]

Finally, although the end of the cold war changes much, it does not seem likely to change two aspects of international politics that, I have argued, have important implications for the spread of nuclear weapons. First, the system's structure, though no longer bipolar, will remain anarchic. Consequently, states will still have incentives to provide for their own security as best they can. Second, the strategic consequences of the nuclear revolution endure.

Consequently, this essay's arguments about the robustness and affordability of nuclear deterrence that make it attractive to self-regarding states remain relevant.

These comments about the spread of nuclear weapons in the post-cold war world are stated in quite general terms. Two specific cases, however, may seem to defy the logic of the essay's argument and, given their significance, to raise doubts about its implications for the future – Japan and Germany. During most of the cold war, the combination of restrictions imposed by the victors after 1945, strong alliance commitments from a motivated American patron, as well as easily anticipated domestic and international reaction to a national nuclear capability, discouraged Bonn and Tokyo from pursuing this option. In the immediate post-cold war period, the absence of obvious major threats to vital interests means that they still have little incentive to question existing policy, however obsolete its basis. But now that the common threat perception that motivated U.S. support for anti-Soviet alliances has disappeared, and now that Germany and Japan are reemerging as fully independent international actors with a strong interest in maintaining their economic competitiveness, the argument of this essay leads to the expectation that as yet unforseen challenges to security will prompt serious consideration of the nuclear alternative. At that point, the principal barrier against Germany and Japan deploying nuclear weapons as an economical and robust national security guarantee (short of improbable preventive military action) will be domestic and international opinion. Both my general argument and the Chinese experience, however, raise doubts about the effectiveness of this constraint. If the security concerns are genuine, the force of international opinion will not be decisive. And if the security concerns are genuine, domestic opinion against a nuclear capability that other states have deemed necessary for themselves may well crumble, especially as the original justification for Japan's and Germany's nuclear abstinence becomes ever more remote.

Notes

1. Neorealism's classic formulation is the systemic balance-of-power theory set forth by Kenneth N. Waltz, *Theory of International Politics* (Menlo Park, Calif.: Addison-Wesley, 1979). Steve Walt presents a refinement of the theory properly emphasizing that it is threats, rather than power per se, against which states balance. Though power in the hands of others may be inherently threatening in an anarchic realm, mediating influences at both the unit- and system-level (ideological beliefs, reputations, geopolitics) preclude a simple direct link between relative power and balancing behavior. Stephen M. Walt, *The Origins of Alliances* (Ithaca: Cornell University Press, 1988).

2. Glenn H. Snyder, "The Security Dilemma in Alliance Politics," *World Politics* 36, no. 4 (July 1984): 461–95; See also, Thomas J. Christensen and Jack Snyder, "Chain Gangs and Passed Bucks: Predicting Alliance Patterns in Multipolarity," *International Organization* 44, no. 2 (Spring 1990): 137–68. These authors note variations in the likely solidarity of alliances in multipolar realms, but see an even more fundamental distinction between all multipolar realms and their bipolar cousins.
3. Such concerns on the part of NATO European allies during the cold war led them to seek visible signs of U.S. commitment (weapons and troops) because these were seen as methods of more firmly linking American interests with their own.
4. On the difficulties of compellence compared to deterrence, see Thomas C. Schelling, *Arms and Influence* (New Haven: Yale University Press, 1966), 69–91.
5. Offensive strikes with nuclear weapons simply to destroy the adversary, that is, preventive war, may be viable against a nonnuclear foe. Under such circumstances a ruthless leadership unmoved by the fate of the human and natural resources of an adversary could act. Two minimal conditions would have to be met, however. The environmental consequences of attack would have to be acceptable to the attacker (in terms of radioactive fallout, effects on the water supply, refugee flows) and the attacker must be certain that the victim will not be able to call on a nuclear ally to carry out retaliatory strikes on its behalf. Although the allies of nuclear states, for reasons stated above, naturally worry most about abandonment, aggressors worry most about the possibility, however small, that a nuclear patron will in fact respond. One could add other obvious considerations, such as the political fallout, both domestic and international that would also affect a leadership's plan to execute a preemptive offensive war plan against a nonnuclear adversary.
6. For British and French statements to this effect, see Lawrence Freedman, *The Evolution of Nuclear Strategy* (London: Macmillan, 1981), 311, 313, 321.
7. One could of course deal in counterfactuals and assert that Britain and France would more quickly have declined had they foregone nuclear weapons. Though debatable, one might simply note that their most obvious badge of international prestige, permanent seats on the UN Security Council, was not a consequence of nuclear status.
8. This may be the most important consequence of the Non-Proliferation Treaty (NPT) and associated documents. It is the nonproliferation regime, only partly defined by the letter of international law, that influences behavior. The NPT itself cannot prevent a determined state, even a signatory, from developing and deploying nuclear weapons. Even a "law-abiding" signatory can withdraw from the treaty on short notice. For the text of the treaty, see Leonard S. Spector with Jacqueline R. Smith, *Nuclear Ambitions* (Boulder, Colo.: Westview, 1990), 424–29.
9. Going beyond the old guns vs. butter tradeoff, the importance of such opportunity costs is highlighted in the general arguments offered by Kennedy on the rise and fall of great powers and addressed in the subsequent debate about the alleged decline of the United States in the late twentieth century Paul Kennedy, *The Rise and Fall of the Great Powers* (New York: Vintage, 1987); cf. Samuel P. Huntington, "The U.S.– Decline or Renewal?" *Foreign Affairs* 67, no. 2 (Winter 1988/1989); Joseph S. Nye, *Bound to Lead* (New York: Basic Books, 1990).
10. The risks of free-riding limit the usefulness of the pubic goods theory of alliances. See Avery Goldstein, "Discounting the Free Ride: Alliance-based Security in the Postwar World," unpublished MS. Of course states without the resources to do much for themselves will have little alternative to depending on others or accommodating prospective adversaries.
11. This assumes military budgets exceeding the minimum needed to develop and

deploy a modest nuclear arsenal. Through the late 1980s, this probably limited the candidate states to those with annual military spending in excess of two or three billion dollars. Of the known nuclear powers in 1988, Israel's budget of $5.71 billion was the smallest, threshold state Pakistan was spending $2.58 billion. See Avery Goldstein, "Robust and Affordable Security: Some Lessons from the Second-Ranking Powers During the Cold War," *Journal of Strategic Studies* 15, no. 4 (December 1992), Appendix A (p. 519). On the costs of developing a viable retaliatory force, see Gene I. Rochlin, "The Economic Burden of a Nuclear Force: No Data in Search of a Theory," in Dagobert L. Brito, Michael D. Intriligator, and Adele E. Wick, eds., *Strategies for Managing Nuclear Proliferation* (Lexington, Mass: Lexington Books, 1983).

12. The difficulty of the calculation was reflected in the heated debate about the actual balance of NATO and Warsaw Pact forces in Central Europe during the cold war. See John J. Mearsheimer, "Numbers, Strategy, and the European Balance," *International Security* 12, no. 4 (Spring 1988): 174–85; Barry Posen, "Is NATO Decisively Outnumbered?" *International Security* 12, no. 4 (Spring 1988): 186–202; Joshua M. Epstein, "Dynamic Analysis and the Conventional Balance in Europe," *International Security* 12, no. 4 (Spring 1988): 154–65 ; Eliot A. Cohen, "Toward Better Net Assessment: Rethinking the European Conventional Balance," *International Security* 13, no. 1 (Summer 1988): 128–79; Joshua M. Epstein, "The 3:1 Rule, the Adaptive Dynamic Model, and the Future of Security Studies," *International Security* 13, no. 4 (Spring 1989): 90–127; and John J. Mearsheimer, "Assessing the Conventional Balance: The 3:1 Rule and Its Critics," *International Security* 13, no. 4 (Spring 1989): 54–89.

13. This argument is made obsolete if a leakproof defense against nuclear weapons is devised. Such a strategically revolutionary system would have to provide confidence in its ability to cope both with the variety of traditional delivery methods (missiles and planes) as well as creative delivery methods (smuggling, hidden assembly and deployment). Current technologies offer little hope for addressing either. For a recent calculation of the destructiveness of even a limited number of very small nuclear detonations that might result despite imperfect defenses, see Steve Fetter, "Ballistic Missiles and Weapons of Mass Destruction: What Is the Threat? What Should Be Done?" *International Security* 16, no. 1 (Summer 1991): 5–42. For a useful summary of the controversies in the debate over ballistic missile defenses, see Steven E. Miller and Stephen Van Evera, eds., *The Star Wars Controversy* (Princeton: Princeton University Press, 1986).

14. Military historians have long discussed the shifting advantage between offense and defense as weaponry has evolved through the ages. Such analysis entails considering both the objective characteristics of the hardware as well as the subjective beliefs of those who formulate strategies that would employ it. The two may sharply conflict as is frequently noted in discussions of mistaken beliefs about the advantage of the offense many asserted on the eve of the First World War. The assertion that the nuclear revolution resulted in deterrence supplanting offense and defense is most relevant after the brief U.S. nuclear monopoly ended and the age of mutual vulnerability dawned. But even before a balance of terror was established, plans for an offensive strike as part of a preventive war against the Soviet Union had to consider the implications for deterrence in the future if the initial salvo failed to produce a surrender. The worry was not simply technical, given the limited number and power of the nuclear warheads available at the time, but also psychological as the indecisive use of nuclear weapons might undermine the terror associated with them since Hiroshima. On the cautiousness of U.S. crisis behavior despite great nuclear superiority during the mid-1950s through mid-1960s see, for example, Richard K. Betts, "A Nuclear Golden Age? The Balance Before Parity," *International Security* 11, no. 3 (Winter 1986/87): 3–

32 ; Betts, *Nuclear Blackmail and Nuclear Balance* (Washington: Brookings, 1987).

15. Thus, Brodie's now classic 1945 statement that the chief purpose of military establishments in the nuclear age was to avert rather than win wars. See Bernard Brodie, *War and Politics* (New York: Macmillan, 1973), 377. For a recent summary of the debate and historical experience that followed in the decades after Brodie's initial comments, see Robert Jervis, *The Meaning of the Nuclear Revolution* (Ithaca, N.Y.: Cornell University Press, 1989).
16. This inhibiting effect does not depend on close comparison of the qualitative and quantitative balance of forces, important considerations when offensive and defensive strategies are weighed. Nevertheless, during the cold war, debate among some scholars and policymakers continued to emphasize the importance of offensive and defensive considerations. But despite contrary appearances, even in the most prominent of these arguments the new core consideration of deterrence remained in a position of logical priority. Colin Gray, for example, argued that the Soviet leaders would only refrain from initiating a war that advanced their interests if they calculated that the United States was able to defeat their communist regime, not just devastate the country's population and industry. Colin Gray, "Nuclear Strategy: A Case for a Theory of Victory," *International Security* 4, no. 1 (Summer 1979): 54–87. His entire argument, however, depended on *first* establishing that the Soviet leaders were indifferent to the horrifying damage that the United States could at any time inflict on the "socialist homeland" by relying on a small fraction of its nuclear arsenal. Only *after* asserting this premise of indifference (a heroic assumption even about brutal authoritarian rulers who presumably wished a society over which to wield power) could one logically debate the need to move beyond a strategy of deterrence to more traditional considerations of offense and defense. The plausibility of this premise is open to debate, but its central importance for Gray's argument is clear.

    Paul Nitze, too, offered a nightmare scenario that seemed to suggest the continuing importance of a focus on offense and defense, rather than deterrence, in the nuclear age. Paul Nitze, "Deterring our Deterrent," *Foreign Policy*, no. 25 (Winter 1976/77). Nitze's concern was the possibility that the Soviets might initiate war by undertaking a well conceived, limited, counterforce first strike leaving the United States with the choice of triggering a suicidal countervalue exchange or submitting to a Soviet aggressor whose action had assured itself of military superiority. This scenario, however, required accepting *first* that the Soviets would be so confident of a moderate, rational American reaction that they would opt to carry out the plan. This assumption was particularly heroic given the fragility of C^3I in a nuclearized environment. See Desmond Ball, *Can Nuclear War Be Controlled?*, Adelphi Papers no. 169 (London: International Institute for Strategic Studies, 1981); Paul Bracken, *The Command and Control of Nuclear Forces* (New Haven: Yale University Press, 1983); Bruce Blair, *Strategic Command and Control: Redefining the Nuclear Threat* (Washington: Brookings, 1985). See also Robert Jervis, *The Illogic of American Nuclear Strategy* (Ithaca: Cornell University Press, 1986); Fred Kaplan, *The Wizards of Armageddon* (New York: Simon & Schuster, 1983), 377–79.
17. It is not surprising that the more elaborate nuclear scenarios emphasizing offensive and defensive strategies and the force structures for carrying them out were the special privilege of the cold war superpowers. Having easily satisfied the basic requirements of deterrence, they could (or believed they could) afford the luxury of supplementing this with forces to cover a wide variety of expensive contingencies: low-intensity conflict, limited and full-scale conventional war, as well as limited nuclear war. The collapse of the Soviet Union and the weakening

of the United States economy relative to its competitors may well suggest that this was not an affordable luxury.

18. See Robert J. Art, "To What Ends Military Power?" *International Security* 4, no.4 (Spring 1980): 3–35.
19. See Kenneth N. Waltz, "Toward Nuclear Peace," in Robert J. Art and Kenneth N. Waltz, eds., *The Use of Force*, 3rd ed., (Lanham, Md.: University Press of America, 1988), 688. See also, Avery Goldstein, "Robust and Affordable Security: Some Lessons from the Second-Ranking Powers During the Cold War."
20. The way in which this affects the decision making of prospective aggressors is given a formal treatment in Goldstein, "Robust and Affordable Security," Appendix D (524–27). The inhibiting fear of escalation that makes nuclear deterrence robust as well as relatively affordable, is a result of the strategic consequences of the nuclear revolution, not the particular policies of states. States who have decided to deploy nuclear weapons can, however, act to reinforce its effects. This can be done in part by exploiting the complementarity of nuclear and conventional forces. By denying an adversary the certainty of unresisted aggression, conventional weapons compel an adversary who would challenge vital interests to confront grave risks. Not because they are a tripwire triggering an automatic response, but as Schelling noted because they increase the likelihood of *some* response that might eventually result in unforseen escalation, conventional deployments by nuclear states serve to dissuade serious, even if initially limited, threats to vital interests. Another step by which states can cause a prospective aggressor to recognize the potentially catastrophic consequences of initiating hostile military action, is to respond to crises by placing nuclear forces on a heightened alert status. While of course reducing the vulnerability of weapons, more importantly, this step also raises the inhibiting specter of catastrophe by weakening without eliminating, the tighter peacetime negative controls over nuclear weapons. Regardless of whether control is seriously compromised, the move fosters the sorts of doubts in the adversary's mind that enhance deterrence.
21. This analysis omits treatment of China's strategic interests that may go beyond dissuasion. During the period under discussion, China's principal, though not sole, foreign policy goal was to ensure the political and territorial integrity of the mainland's newly founded revolutionary regime. A viable national security policy had to deal with this problem before addressing others. To the extent Beijing had an interest in expanding its de facto borders, especially by resolving territorial disputes with its regional neighbors and an interest in assisting communist revolutionaries in the developing world, investment in nuclear forces actually diverted funds that might have been employed to enhance China's conventional power-projection capabilities. Not surprisingly, by the 1980s, *after* China was confident it had satisfied the requirements of its dissuasive strategy, the regime more aggressively began to pursue conventional forces that would support Chinese claims in various maritime disputes with its neighbors.
22. The centrality of this security logic, as opposed to one emphasizing ideological affinity, was underscored by China's decision to seek the strategic leverage of improved relations with the politically repugnant United States in the early 1970s as a means for addressing the threat from the Soviet superpower. Because China lacked the resources to defend against the variety of military threats the Soviets posed, and did not yet have confidence in the reliability of its nuclear retaliatory forces, Beijing once again felt constrained to cultivate one superpower to counter the other. With regard to the nonstrategic determinants of China's foreign policy, it should be remembered that despite their common marxist heritage, ideology had been a divisive element in relations between the Soviet and Chinese Communist Parties since the early 1930s. It might be added that personal

relations between Mao and the Soviet communist leaders were far from close. On the more personal tensions, see especially Harrison E. Salisbury, *The New Emperors* (Boston: Little, Brown, 1992).

23. American rhetoric about rolling back the postwar gains of communism, its signing a security treaty with and arming a regime on Taiwan dedicated to reversing the verdict of the Chinese Civil War, and its proclaimed willingness to resort to nuclear weapons if it seemed militarily advisable all fed China's threat perceptions. Nuclear threats were reflected in both declaratory policy during the late Truman and Eisenhower administrations and the deployment of nuclear capable forces to the East Asian theater: John W. Lewis and Litai Xue, *China Builds the Bomb* (Stanford: Stanford University Press, 1988), 229; Chong-pin Lin, *China's Nuclear Weapons Strategy* (Lexington, Mass: Lexington Books, 1988), 77; Betts, *Nuclear Blackmail and Nuclear Balance*, chaps. 1, 2; Marc Trachtenberg, "A 'Wasting Asset': American Strategy and the Shifting Nuclear Balance, 1949–1954," *International Security* 13, no. 3 (Winter 1988/89), 5–49; Roger Dingman, "Atomic Diplomacy During the Korean War," *International Security* 13, no. 3 (Winter 1988/89), 50–91; Rosemary J. Foot, "Nuclear Coercion and the Ending of the Korean Conflict," *International Security* 13, no. 3 (Winter 1988/89), 92–112. See also Yuanchao Li, "The Politics of Artillery Shelling: A Study of the Taiwan Strait Crises," *Beijing Review* 35, 7–13 September 1992, pp. 32–38. Beijing's brief attempts to deal with the U.S. threat through conciliation, represented by the spirit of Bandung, bore no fruit. See Harvey W. Nelsen, *Power and Insecurity: Beijing, Moscow, and Washington, 1949–1988* (Boulder, Colo.: Lynne Rienner, 1989), 38–40.

24. China's overwhelming interest in ensuring that the Soviets would continue to provide security for socialist bloc members induced the CCP to agree to distasteful compromises of Chinese independence. During the early 1950s Beijing agreed to military and economic concessions for the Soviets in China (Nelsen, *Power and Insecurity*, 7, 11). And despite the fact that the CCP had long insisted on independence for itself in the international communist movement and apparently supported it for others, throughout most of the 1950s Beijing maintained its public endorsement of the Soviet party's absolute leadership of the socialist bloc. See Donald Zagoria, *The Sino-Soviet Conflict, 1956–1961* (Princeton: Princeton University Press, 1962), chaps. 1, 2, 4; also Nelsen, *Power and Insecurity*, 11. As long as the PRC believed it required Moscow's backing in recurrent crises with a hostile and powerful United States, disagreements with the Soviets were muted and principled independence, however reluctantly, was compromised.

25. On the distinction between intrinsic military value and reputation value, see Glenn H. Snyder, "Deterrence and Defense," in Art and Waltz, *The Use of Force*, 36–43. On the interdependence of commitments, see Thomas C. Schelling, *Arms and Influence* (New Haven: Yale University Press, 1966), 55–59, 116–25. On the complexity of manipulating perceptions of a state's reputation, see Nalebuff's discussion of Jervis's reputational paradox, Barry Nalebuff, "Rational Deterrence in an Imperfect World," *World Politics* 43, no. 3 (April 1991), 313–35. The reputational stakes are raised not only by the strategic effects of the nuclear revolution, but also by bipolarity that enhances the apparent zero-sumness of international political competition between the duopolists (Waltz, *Theory of International Politics*, 170–72).

26. See Waltz, *Theory of International Politics*, 169–70.

27. On Khrushchev's troubling shift to peaceful coexistence, see Gordon H. Chang, *Friends and Enemies* (Stanford: Stanford University Press, 1990), chaps. 4, 5. Comments by analysts I interviewed in Beijing in 1991 suggest there is room for debate over exactly what the Chinese expected the Soviets to do during the crisis

and the extent to which they felt abandoned. Clearly, however, the crisis raised doubts in Beijing about Moscow's willingness to stick with an ally determined to retain its foreign policy independence. See Ibid., 190–94; Nelsen, *Power and Insecurity*, 41–45.

28. Lewis and Xue, *China Builds the Bomb*, 64, 160–63.
29. During a January 1960 meeting addressing the problems encountered, Nie Rongzhen impugned Soviet motives by stating that their "technical aid has become untrustworthy" and that they want "to maintain a considerable gap between China and the Soviet Union in scientific research on the development of new types of weapons and military equipment" (Ibid., 72). Moscow's demands included telecommunication facilities within China to enhance Soviet C^3I and perhaps (the Chinese claimed, the Soviets denied) a jointly manned Sino-Soviet Pacific Fleet. In a review of China's foreign policy independence, Li Dai chronicles Beijing's resistance to Soviet attempts to interfere in China's domestic affairs and the rejection of Khrushchev's proposal for a "'joint flotilla' along the Chinese coast and a 'long-wave radio' communications network to command the Soviet flotilla in an effort to exercise military control over China." Li Dai, "Independence and China's External Relations" *Shijie Zhishi* no. 19 (10/1/81) in FBIS, 19 November 1981, p. A2; see also Lewis and Xue, *China Builds the Bomb*, 63–64.
30. Following China's refusal to accede to Moscow's demands for closer military integration under the Kremlin's leadership, "[t]he Soviet Union then perfidiously tore up agreements and contracts, recalled its experts and carried out large-scale subversive activities in our Xinjiang region." (Li, "Independence and China's External Relations," A2). China's nuclear scientists would name their first atomic bomb "596" in reference to the date of the letter that announced there would be no Soviet prototype weapon delivered to China (Lewis and Xue, *China Builds the Bomb*, 72, 150).
31. For further consideration of the linkages between domestic and international security, see Barry Buzan, *People, States, and Fear*, 2nd ed. (Boulder, Colo.: Lynne Rienner, 1991).
32. Chinese analysts have asserted that nuclear weapons offer a more economical alternative to huge conventional forces, freeing up resources for the essential task of domestic development. See Sun Mingming and Cai Xiaohong, *Dongdangzhong de Guojia Anquan* (Beijing: Jiefangjun Chubanshe, 1988), 189; Peng Guangqian, Wang Guangxu, et al., eds., *Junshi Zhanlüe Jianlun* (Beijing: Jiefangjun Chubanshe, 1989), 168; and Chen Chongbei, Shou Xiaosong, and Liang Xiaoqiu, *Weishe Zhanlüe* (Beijing: Junshi Kexue Chubanshe, 1989), 29.
33. Indeed, Soviet aid was limited not only in quantity but also in quality. Lewis and Hua noted that it was Soviet policy not to "transfer state-of-the-art weapons to allies before it had deployed at least two types of more advanced systems." John Wilson Lewis and Hua Di, "China's Ballistic Missile Programs: Technologies, Strategies, Goals," *International Security* 17, no. 2 (Fall 1992): 13.
34. Similar problems confronted China in the late 1960s when it faced a serious threat from the Soviet Union. Beijing could have little confidence that Moscow would view a defensive capability based on modified Korean-war vintage arms as much of a match for arms with which they had been replaced in the Soviet Red Army. One estimate put the cost of U.S. aid that would be needed to prepare China to withstand a Soviet conventional attack at $41–$63 billion. This sum almost certainly exceeded the entire annual defense budget of the PRC and may represent two to three times actual defense spending. Moreover, the figure would likely have been even higher if the Chinese instead sought to undertake the bulk of the conventional modernization independently. See Banning N. Garrett and Bonnie S. Glaser, *War and Peace: The Views from Moscow and Beijing* (Berkeley:

Institute of International Studies, 1984) 23. See also Leo Yueh-yun Liu, "The Modernization of the Chinese Military" *Current History* 79 (September 1980): 11.

35. This was the upshot of the collapse of the short-lived cultivation of intellectuals during the Hundred Flowers movement in 1956–57. Thereafter, despite brief pauses in the ideological assault on educational standards, the regime failed to sustain support for scholarship until the 1980s. Indeed, even in the era of Deng's reforms, many have criticized the lack of a financial, as opposed to a rhetorical, commitment to education.
36. The geopolitical difficulties that diminished the appeal of relying on conventional defense were compounded when the Soviets became China's principal security concern. At that point, Beijing not only had to worry about the vulnerability of Manchuria and the eastern cities, but two new problems. First, vast resource-rich territories in the Chinese west and northwest bordering the Soviet Union were sparsely populated by minorities of dubious loyalty. Second, Beijing itself was relatively close to heavily armored Soviet divisions that could strike from the north and northwest. In an era when even a fully restored great wall would have little obstructive value, topography provided the capital city with few defensive advantages along the likely axes of Soviet advance.
37. This was the now notorious Third Front. Ironically, even without committing themselves to costly conventional defense, this dispersal that aimed at enhancing the viability of China's conventional deterrent, though perhaps not as costly as full-scale modernization of the PLA, contributed to the stagnation of the Chinese economy that ultimately undermined the legitimacy of the Maoist development program. Nelsen, *Power and Insecurity*, 89–90. See also Salisbury, *The New Emperors*, esp. ch. 14.
38. The classic essays explaining how to rely on the "masses" as part of a strategy to prevail over a militarily superior adversary were written by Mao Zedong in the 1930s. Mao Zedong, *Selected Works of Mao Tse-tung*, vol. 2 (Peking: Foreign Languages Press, 1965), 79–194. For a discussion in China's recent strategic studies literature, see Yang Xuhua and Cai Renzhao, *Weishe Lun* (Beijing: Guofang Daxue Chubanshe, 1990), 384–86, 388, 389, 390; Chen, Shou, and Liangü, *Weishe Zhanlüe*, 69.
39. This also suggests the disadvantage of conventional deterrent strategies based on threats other than protracted popular resistance. Where punishment can be inflicted with only conventional forces, the special dissuasive effects associated with uncertainty about the outcome of a confrontation with a nuclear-armed state would not obtain.
40. Here one again confronts the notorious problem with assessing the effectiveness of deterrence. Its success is reflected in nonevents. How many great power interventions never occurred, or were never even contemplated because of the fear of dealing with protracted popular resistance? Answering such a question is difficult and requires close consideration of the foreign policy preferences of particular states. This important matter falls outside the scope of this essay. It can be noted, however, that there are at least examples of conventional deterrence failures, something fortunately lacking with respect to nuclear deterrence. For discussion of the vexing problem of demonstrating the effectiveness of deterrence by citing counterfactuals, see "The Rational Deterrence Debate: A Symposium," *World Politics* 41, no. 2 (January 1989); Patrick M. Morgan, *Deterrence: A Conceptual Analysis* (Beverly Hills: Sage Publishers, 1977).
41. On compellent strategies see Schelling, *Arms and Influence*, 69–91. On the feasibility of small powers successfully compelling the withdrawal of militarily superior adversaries see Andrew Mack, "Why Big Nations Lose Small Wars: The Politics of Asymmetric Conflict," *World Politics* 27, no. 2 (January 1975): 175–200.

42. Threats to take actions unpleasant for oneself as well as the adversary, of course, also characterize nuclear deterrence. The problem with the nonnuclear alternative is the greater likelihood deterrence will fail and that the unpalatable threat will have to be executed.
43. In the late 1980s, China would view a national war of resistance as a last resort in the event their nuclear deterrent and improving defensive capabilities proved insufficiently dissuasive. Once Mao died in 1976, Chinese military leaders more openly articulated their objections to a conventional deterrent strategy based on luring the enemy in deep and waging a protracted armed struggle. At a time of heightened Sino-Soviet tensions in 1979, one of China's Marshals, Nie Rongzhen, "... directed the urgent deployment of all available strategic weapon systems, saying that 'though a bit backward ... [they] would still be better than millet plus rifles in fighting a war'." Lewis and Hua, "China's Ballistic Missile Programs," 19. In an attempt to preserve as much of the PLA's glorious revolutionary legacy as possible, the emerging alternative doctrine after 1978 was referred to as People's War Under Modern Conditions. Symbols notwithstanding, China's program of military modernization was "designed specifically to avoid a protracted war." Paul H. B. Godwin, "The Chinese Defense Establishment in Transition: The Passing of a Revolutionary Army?" in A. Doak Barnett and Ralph N. Clough, eds., *Modernizing China* (Boulder, Colo.: Westview, 1986), 73. See also Gerald Segal, "China's Strategic Posture and the Great-Power Triangle," *Pacific Affairs* 53, no. 4 (Winter 1980/81): 685–87, 690, 691; Harlan Jencks, "People's War Under Modern Conditions: Wishful Thinking, National Suicide, or Effective Deterrent?" *China Quarterly* 98 (June 1984): 313. Such planned changes have been widely analyzed in the literature on military modernization and reform in post-Mao China. See Paul H. B. Godwin, *The Chinese Defense Establishment: Continuity and Change in the 1980s* (Boulder, Colo.: Westview, 1983); Charles D. Lovejoy and Bruce W. Watson, eds., *China's Military Reforms* (Boulder, Colo.: Westview, 1986); Ellis Joffe, *The Chinese Army after Mao* (Cambridge: Harvard University Press, 1987); Larry M. Wortzell, ed., *China's Military Modernization* (New York: Greenwood, 1988).
44. Edward N. Luttwak, *Strategy: The Logic of War and Peace* (Cambridge: Harvard University Press, 1987), 18–21.
45. On Mao's cautiousness during the crisis, see Li "The Politics of Artillery Shelling: A Study of the Taiwan Strait Crises," 32–38.
46. Lewis and Xue, *China Builds the Bomb*, 40. In 1955, Mao disparaged U.S. nuclear threats by asserting, "the Chinese people are not to be cowed by U.S. atomic blackmail ... .The United States cannot annihilate the Chinese nation with its small stack of atom bombs." Cited by John W. Lewis, "China's Military Doctrines and Force Posture," in Thomas Fingar, ed., *China's Quest for Independence: Policy Evolution in the 1970s* (Boulder, Colo.: Westview, 1980), 163. In 1969, when China still had not deployed an effective nuclear retaliatory capability, Beijing again felt compelled by superpower saber-rattling to back down, taking the initiative in defusing the Sino-Soviet border conflict before it got out of hand. See Richard Wich, *Sino-Soviet Crisis Politics* (Cambridge: Harvard University Press, 1980), chaps. 9, 10.
47. See John J. Mearsheimer, *Conventional Deterrence* (Ithaca: Cornell University Press, 1983), 53–56. The advantages of fighting on the defensive were prominently stressed by Mearsheimer, "Numbers, Strategy, and the European Balance"; Posen, "Is NATO Decisively Outnumbered?" Epstein, "Dynamic Analysis and the Conventional Balance in Europe," and other "optimists" in the debate over NATO prospects against a conventional Soviet attack during the 1970s and 1980s. For the "pessimists'" response, see Cohen, "Toward Better Net Assessment." For debate over the theoretical and empirical warrant for offense-

defense ratios that determine outcomes of battles see also Epstein, "The 3:1 Rule," and Mearsheimer, "Assessing the Conventional Balance."

48. As Paul Godwin put it in analyzing China's readiness in the 1970s to deal with the Soviet threat, "[W]hat happens if the Chinese hold a people's war and nobody comes?" ("The Chinese Defense Establishment in Transition," 73).
49. Chinese analysts describe the country's "self-defensive finite deterrent" as being comprised of two components– finite nuclear deterrence, and People's War. The nuclear component is listed first, and the People's War component is described as a deterrent against a full-scale invasion of the homeland (Chen, Shou, and Liang, *Weishe Zhanlüe*, 214–15).
50. See Peng and Wang, *Junshi Zhanlüe Jianlun*, 166–167.
51. Beginning in the mid-1950s Mao asserted that China would need atomic bombs in order to avoid being bullied by others. See Ibid., 166; Chen, Shou, and Liang, *Weishe Zhanlüe*, 209. This justification was clearly set forth in Chinese statements following its first nuclear test in October 1964, Peng and Wang, *Junshi Zhanlüe Jianlun*, 167. In the 1980s, Deng Xiaoping would reassert the self-defensive justification (Yang and Cai, *Weishe Lun*, 409; Chen, Shou, and Liang, *Weishe Zhanlüe*, 209). And in response to growing international anti-nuclear opinion in the 1980s, Chinese analysts sympathized with the sentiments of Greenpeace and others but noted " ... as long as the danger of war still exists, in circumstances where there is no essential guarantee of world peace, our country's nuclear weapons are still an important force for deterrence during peacetime." Yang and Cai, *Weishe Lun*, 408; see also 407, 410.
52. Peng and Wang, *Junshi Zhanlüe Jianlun*, 167.
53. China has always proclaimed its willingness to go along with abolition of all nuclear weapons, but until this could be achieved asserted that "China will steadfastly ... strengthen its national defenses, safeguard the motherland, and safeguard world peace."(Ibid., 168).
54. For Chinese analysis reflecting on the significance of the nuclear revolution, see Ibid., 160; Chen, Shou, and Liang, *Weishe Zhanlüe*, 47, 71.
55. See Yang and Cai, *Weishe Lun*, 410.
56. Several characteristics of the Chinese nuclear triad stand out. First, during the cold war, other than an ageing bomber force of dubious penetrability, China had relatively few long range ballistic missile systems (six to thirteen ICBMs). China's first generation SSBNs each carried 12 missiles with limited range (1700km). See Lewis and Hua, "China's Ballistic Missile Programs," 26–27. Second, the yields of the warheads on the longer range systems (carrying 1, 3, and 5 megaton warheads) were relatively large. While such enormous yields are not suitable for limited nuclear options, they pose a terrifying threat of massive retaliation. Third, the obsolescence of its bomber force, the limited effectiveness of its costly SSBN fleet, and geographic considerations constrained the PRC to rely heavily on land-based ballistic missiles.
57. See Chen, Shou, and Liang, *Weishe Zhanlüe*, 3; Peng and Wang, *Junshi Zhanlüe Jianlun*, 169. Yang and Cai, *Weishe Lun*, 411. Chinese analysts assert that the aggressor would have to worry about some surviving nuclear retaliatory capability even if he launched a "large scale surprise nuclear attack." Ibid., 410. They also insist that the fear of devastating retaliation can deter an aggressor despite his deployment of very effective, even 90 per cent effective, ballistic missile defenses. See Sun and Cai, *Dongdangzhong de Guojia Anquan*, 193.
58. To ensure the survivability of its land-based missiles, the Chinese emphasized the importance of "dispersal, concealment, and mobility." Lin, *China's Nuclear Weapons Strategy*, 52. John Lewis and Hua Di cite a Central Military Commission and State Council directive (20 May 1977) that followed up on a request from Mao Zedong (25 May 1975) to adopt a policy of "in-cave storage/preparation and

out-cave erection/filling/firing" for the DF-4. This cave-basing system for the DF-4 (dubbed "shooting a firecracker outside the front door" by Zhang Aiping) became operational 2 August 1980. "China's Ballistic Missile Programs," 24.

59. Hiding weapons and exposing decoys frustrated satellite reconnaissance by the superpowers who could not be sure they had succeeded in locating and targeting all of the PRC deterrent arsenal. Lewis and Hua recently confirmed that in deploying their only genuine long range ICBM, the DF-5, the Chinese sought to enhance survivability by building "a large number of bogus silos," "shallow holes disguised to look like the real thing." This was the alternative selected after consideration had been given to ship- or rail-mobile basing and camouflage of fixed silos. Lewis and Hua assert that shortcomings in China's early warning system made the option of launch on warning an impractical solution to the vulnerability problem. "China's Ballistic Missile Programs," 24–25. See also Lin, *China's Nuclear Weapons Strategy*, 52, 69; On the role of deception and uncertainty in targeting, see also Waltz, "Toward Nuclear Peace," 695–99.
60. Given their unsophisticated force posture, some doubted the usefulness of China's small nuclear deterrent. As one analyst put it, the PRC had only a "tentative second-strike capability," and an "assured second-strike capability" would not be attained until it deployed more than 100 ICBMs, preferably MIRVed, as well as a fleet of at least six SSBNs. See Robert S. Wang, China's Evolving Strategic Doctrine," *Asian Survey* 24, no. 10 (October 1984): 1050.
61. On the logic of "self-deterrence" by rational actors see Paul H. Nitze, "Deterring our Deterrent," *Foreign Policy*, no. 25 (Winter 1976/77).
62. Lin, *China's Nuclear Weapons Strategy*, 69.
63. The role of a potentially "unsafe actor" in deterrence of the strong by the weak is explored at length in Goldstein, "Robust and Affordable Security."
64. On problems with assuming durable rationality in a nuclear crisis, see Kurt Gottfried and Bruce G. Blair, *Crisis Stability and Nuclear War* (New York: Oxford University Press, 1988), 265–68; also Edward Rhodes, *Power and MADness: The Logic of Nuclear Coercion* (New York: Columbia University Press, 1989), 135–40.
65. This is especially true since it is not sufficient that the challenger worry about the victim's rationality in general or even harbor some vague belief about unpredictability associated with "the fog of war." He must believe there is an unacceptable risk that the victim will engage in this specific irrational act. See Robert Powell, "The Theoretical Foundations of Strategic Nuclear Deterrence," *Political Science Quarterly* 100, no. 1 (Spring 1985): 85; Rhodes, *Power and MADness*, 72–77, 164–66, 191–92. Thus a reputation for irrationality may not be helpful, and in any case is difficult for government leaders to cultivate. See Schelling, *Arms and Influence*, 41.
66. See Ball, *Can Nuclear War Be Controlled?*; Bracken, *The Command and Control of Nuclear Forces*; Blair, *Strategic Command and Control*; and Rhodes, *Power and MADness*, 139–40.
67. Robert Powell, *Nuclear Deterrence Theory* (New York: Cambridge University Press, 1990), ch. 2; Thomas C. Schelling, *The Strategy of Conflict* (New York: Oxford University Press, 1960), 188–90, 201–3; Rhodes, *Power and MADness*, 78–81.
68. Ball, *Can Nuclear War Be Controlled?*; Bracken, *The Command and Control of Nuclear Forces*; Blair, *Strategic Command and Control*.
69. Not surprisingly, little is known about China's nuclear command and control. But one of the PRC strategists may have intentionally provided a clue as to what Beijing wanted others to believe about the difficulties of successful preemption: "[The adversary] who cannot preempt all of China's nuclear missiles ... would have to continue to worry about Chinese retaliation 'perhaps hours, days, weeks,

months, or even years later.' Even if China's leadership is destroyed in a decapitating nuclear attack, 'the Chinese people ... will be able to wait ... until a new leadership is formed ... capable of ordering the retaliation ... .. China does not need an invulnerable C^3 system' to ensure the viability of its nuclear deterrent." see Garret and Glaser, *War and Peace*, 129.

70. China's no-first-use pledge is: "We will not attack unless we are attacked: If we are attacked, we will certainly counterattack. China will counterattack only when the enemy uses nuclear weapons first." See Zhang Jianzhi, "View on Medium-Sized Nuclear Powers' Nuclear Strategy," from *Jiefangjun Bao* (3/20/87) in FBIS (April 1987): K33. See also Peng and Wang, *Junshi Zhanlüe Jianlun*, 167.
71. The reasoning during the cold war that rejected a no-first-use pledge by NATO and instead encouraged the United States to retain its prerogative of launch on warning or under attack, would seem even more persuasive for a comparatively small nuclear state like China. In an analysis of the nuclear deterrents of the medium powers, Chinese analysts note that "In order to strengthen the credibility of the deterrent, some countries reserve the right to use nuclear weapons first." (Chen, Shou, and Liang, *Weishe Zhanlüe*, 97). Their specific examples are Britain and France, even though the broader discussion lumps China with them as a medium power.
72. Indeed, China's leading military newspaper during the 1980s provided clues about the range of actual conditions that might precipitate China's use of nuclear weapons. After restating China's no-first-use pledge the author asserted that China's nuclear forces enabled her to "check an enemy from rashly launching *any war* of aggression *and* from using nuclear weapons against our country ... ." (Zhang, "View on Medium-sized Nuclear Powers," 33, emphasis added). He also listed four functions of nuclear weapons, including their usefulness for countering superior conventional forces and discouraging superpower attempts at nuclear blackmail. See also Lewis, "China's Military Doctrines and Force Posture," 163. Chinese strategic analysts I interviewed in spring 1991 all insisted that China's no-first-use pledge was sincere. In elaborating on their no-first-use views, however, they also acknowledged that a prospective aggressor would have to worry about Chinese sincerity and that such worries were strategically valuable. See also comments by Peng and Wang that China's nuclear weapons provide a peacetime deterrent of possible superpower invasion without reference to the aggressor's military means (*Junshi Zhanlüe Jianlun*, 167).
73. Israel's air strike against Iraq in 1981, and more massive U.S. military and international diplomatic efforts in the early 1990s, not to mention Iraq's accession to the NPT, have slowed, but may yet not prevent Baghdad from realizing its nuclear ambitions.
74. The relative optimism of two of the more prominent visions of the consequences of the continued spread of nuclear weapons derives in part from such a belief. See Kenneth N. Waltz, "Toward Nuclear Peace," in Art and Waltz, *The Use of Force* ; John J. Mearsheimer, "Back to the Future: Instability in Europe After the Cold War," *International Security* 15, no. 1 (Summer 1990): 5–56.
75. Indeed, during the bipolar cold war, allies' threat perceptions, though not identical, were at least largely similar. Nevertheless, China, as well as Britain and France, were concerned enough about possibly conflicting interests to want a hedge against the potential unreliability of alliance guarantees. Such worries will likely be even stronger for states that might depend on diffuse collective security guarantees against unspecified adversaries in the post-cold war era. Those who believe that international commitments to Ukrainian security, for example, can substitute for their possession of a nuclear capability must explain why Ukrainians will comfortably conclude that their international patrons would run grave risks, perhaps facing off against the Russians, to ensure Ukrainian

interests. Second thoughts about foregoing nuclear weapons have already emerged in Kazakhstan as well as Ukraine. See Edith M. Lederer and Sergei Shargorodsky, "New Reluctance on Giving Up Their Nuclear Arms," *Philadelphia Inquirer*, 14 June 1992, p. A3; John F. Cushman, Jr., "Senate Endorse Pact to Reduce Strategic Arms," *New York Times*, 2 October 1992, pp. A1, A6.

76. See Robert Jervis, "Cooperation under the Security Dilemma," *World Politics* 30, no. 2 (January 1978): 167–214.

77. See Karl W. Deutsch and J. David Singer, "Multipolar Power Systems and International Stability," in James N. Rosenau, ed., *International Politics and Foreign Policy*, 2d ed. (New York: Free Press, 1969), 315–24; Richard N. Rosecrance, "Bipolarity, Multipolarity, and the Future," in Rosenau, ed., *International Politics and Foreign Policy*, 325–35; Kenneth N. Waltz, "International Structure, National Force, and the Balance of World Power," in Rosenau, *International Politics and Foreign Policy*, 304–14.

78. In the developing world, the sorry experiences of the great powers trying militarily to impose their control after the spread of nationalism has clarified the costs of this approach. In the developed world, the nature of post-industrial economics has reduced the possibilities for and efficiency of tapping the productivity a conquered people. See Steve Van Evera, "Why Europe Matters, Why the Third World Doesn't: American Grand Strategy after the Cold War," *Journal of Strategic Studies* 13, no. 2 (June 1990): 1–51. John Mueller argued that the growing recognition "war doesn't pay" rendered war itself obsolete by the mid-twentieth century. Although one may take issue with Mueller's bold conclusion, and instead point to the influence of the nuclear revolution or bipolarity as contributing factors to the "long peace," his assertion that historical experience has led modern states to focus on the costs, rather than benefits, of war-fighting is rarely disputed by his critics. See John Mueller, "The Essential Irrelevance of Nuclear Weapons: Stability in the Postwar World," in Sean M. Lynn-Jones, ed., *The Cold War and After* (Cambridge: MIT Press, 1991), 45–69; Robert Jervis, "The Political Effects of Nuclear Weapons: A Comment," in Lynn-Jones, *The Cold War and After*, 70–80; Carl Kaysen, "Is War Obsolete? A Review Essay," in Lynn-Jones, *The Cold War and After*, 81–103; and John Lewis Gaddis, "The Long Peace: Elements of Stability in the Postwar International System," in Lynn-Jones, *The Cold War and After*, 1–44.

79. These economic considerations also reduce the chance that current nuclear states will move beyond prudent reductions and give up their weapons, disarmament rhetoric notwithstanding. Recent Russian views explicitly argue the economic advantages of a nuclear, rather than conventional emphasis. Fred Hiatt, "Russians Are Leaning Toward Nuclear Reliance for Security," *Philadelphia Inquirer*, 26 November 1992, p. A22. This mirrors the economic logic of the strategic approach pushed by Khrushchev in the late 1950s.

The Power of Suggestion: Opaque Proliferation, Existential Deterrence, and the South Asian Nuclear Arms Competition

Devin T. Hagerty

SOME OBSERVERS believe South Asia is "the likeliest place in the world for a nuclear war."[1] Relations between India and Pakistan are as strained today as they were at independence forty-six years ago. The two countries fought three wars between 1947 and 1971, the last of which dismembered Pakistan and carved from it the independent nation of Bangladesh. The Kashmir dispute, which played a central role in the other two Indo-Pakistani wars, has since 1989 erupted into a full-scale insurgency that threatens to undermine the very legitimating ideologies on which the two states were founded in 1947 – secularism in India and Islam in Pakistan. Although neither government is eager to fight another war over Kashmir, powerful and vocal constituencies in both countries agitate against compromise, making it unlikely that the Kashmir problem will be resolved soon.[2]

The specter of nuclear weapons hovers ominously over these tense Indo-Pakistani relations. India and Pakistan have been developing nuclear weapon capabilities since at least the early 1970s.[3] Both countries are now de facto nuclear weapon states: each is "thought able to deploy one or more nuclear weapons within months or to have already deployed them."[4] India exploded a nuclear device in 1974; Pakistan has not done likewise, but its foreign secretary admitted in early 1992 that Islamabad has clandestinely acquired the capability to build a least one nuclear weapon.[5] In addition, both countries deploy nuclear-capable aircraft and are developing or trying to acquire nuclear-capable ballistic missiles.[6] If a fourth Indo-Pakistani war breaks out, New Delhi and Islamabad will assume that the other can deliver at least one nuclear weapon on short notice.[7] This "would be the first major military conflict in history between two states with ready access to nuclear arms."[8]

Two well-established positions dominate the literature on the consequences of nuclear proliferation in the developing world. The majority position or conventional wisdom holds that the spread of nuclear weapons to "conflict-prone regions" like South Asia is inimical to international peace. From this perspective, the acquisition of nuclear weapons by more Third World countries increases the likelihood that they eventually will be used in war.[9] A complex web of international regimes and national policies has institutionalized this proposition since the 1960s. It is a rare analyst who will challenge this orthodoxy to suggest that nuclear proliferation may have less negative, or even positive, implications for international security.

Contrary to this mainstream "logic of nonproliferation," one variant of nuclear deterrence theory, arguably the best-developed area of international relations scholarship, suggests that the spread of nuclear weapons may *contribute* to international security rather than threaten it.[10] According to this logic, the most hostile of adversaries will be dissuaded from initiating military conflict if each can threaten the other with unacceptable damage, even after absorbing a nuclear first strike. The model underpinning nuclear deterrence theory was the nuclear balance between the United States and the Soviet Union during the cold war. Many international relations scholars argue that the "long peace" in Europe between 1945 and 1991 was due partly to the superpowers' deployment of nuclear weapons: the prospect of nuclear punishment helped to deter the United States and the Soviet Union from war with each other.[11] Proponents of this minority position on the consequences of nuclear proliferation believe this mutual deterrent dynamic can be replicated in regional nuclear balances like the one in South Asia.

The "logic of nonproliferation" and the "logic of nuclear deterrence" cannot comfortably co-exist. Viewed as theories about the likely consequences of nuclear proliferation, the former predicts that the spread of nuclear weapons will be destabilizing and the latter predicts that it will be stabilizing (defined here, respectively, as promoting vs. dampening tendencies toward war). In this article, I use this theoretical tension as a framework to examine the consequences of regional nuclear proliferation, via a case study of the strategic relationship that seems to have the most potential for escalation to the use of nuclear weapons – that between India and Pakistan.[12] In particular, I address one central question: how has the development of nuclear weapon and delivery capabilities by India and Pakistan affected the likelihood of war in South Asia during the last few years?

Building on the work of Benjamin Frankel and others, I apply the recently formulated "opaque proliferation" concept to the South Asian nuclear arms competition.[13] Prior to this advance in nuclear proliferation scholarship, analysts conceptualized the proliferation process along the lines of the "declared" nuclear weaponization of the five original nuclear powers, despite the fact that the new nuclear weapon states of the past twenty years – among them India and Pakistan – have adopted a very different approach. Unlike their predecessors, opaque proliferants follow a policy of calculated nuclear ambiguity, mixing a public posture of restraint with the covert development of nuclear weapons, or at least the option to build them quickly. The nuclear weapon capabilities of opaque proliferants are "suggested," rather than openly acknowledged.

Before the development of the opaque proliferation concept, the declared proliferation universe of the five original nuclear powers drove the analysis of regional nuclear proliferation. Scholars debated whether nuclear proliferation *will be* stabilizing or destabilizing based on historically reasonable but necessarily speculative assumptions about the new proliferants' expected nuclear force postures.[14] This approach yields indeterminate expectations about the effects of proliferation: on one hand, the logic of nonproliferation portrays nuclear proliferation as extremely dangerous, and on the other, the logic of nuclear deterrence suggests the potential for regional nuclear deterrent balances. The outcome of this debate is inconclusive; in each individual case the actual consequences of nuclear proliferation will lie somewhere between the two positions. The crux of the question is the efficacy of nuclear deterrence in regional security environments, which can be expected to vary.[15]

Most analyses of the consequences of nuclear proliferation regard the transition to nuclear weapon capability as the most dangerous phase of the proliferation process; in reality, however, the transition to nuclear weapons is a more or less permanent "phase" in that process. India and Pakistan provide a good example: they have been in a state of perpetual transition to nuclear weapon status since at least 1974. Neither is willing to forgo its nuclear option, but neither wants to be the first over the threshold of acknowledged nuclear weapon deployment. They are, in a sense, racing to be last. Rather than imputing an artificial set of nuclear capabilities to new proliferants like India and Pakistan, the opaque proliferation approach recognizes this transition period not as an outcome, but as a continuing *process* of regional nuclear proliferation. This generates better explanations of regional nuclear dynamics.

Following Neil Joeck, I conceptualize the nuclear competition between India and Pakistan not as an unrestrained arms race, but as a mixed-motive interaction between adversaries.[16] The two countries continue to feel their usual assortment of imperatives toward conflict, like the insurgency in Kashmir, but nuclear weapon capabilities introduce a new set of incentives to cooperate. Among these are the desire to avoid mutual devastation and to forestall an expensive and possibly destabilizing nuclear arms race. These new incentives for cooperation may have sharply modified the behavior of the Indian and Pakistani governments as they try to cope with the Kashmir problem.

In this article I address an important question that has been briefly raised but not resolved in the opaque proliferation literature: can opaque proliferants derive deterrence value from nuclear capabilities they publicly deny having? If so, how?[17] My main theoretical argument is that the logical form of nuclear deterrence under conditions of opaque proliferation is "existential deterrence," a term coined by McGeorge Bundy[18] and defined by Marc Trachtenberg as a situation in which:

> the mere existence of nuclear forces means that, whatever we say or do, there is a certain irreducible risk that an armed conflict might escalate into a nuclear war. The fear of escalation is thus factored into political calculations: faced with this risk, states are more cautious and more prudent than they otherwise would be.[19]

A simple deterrence hierarchy clarifies the concept of existential deterrence. The table below depicts the various relationships between proliferation type (declared vs. opaque) and deterrence type. In the superpower balance, declared nuclear capabilities contributed to a deterrence relationship framed by the debate among U.S. analysts between deterrence through "mutual assured destruction" and a "warfighting" posture. In this debate, nuclear deterrence was perceived to rest at least in part on relative capabilities. At the next level, the British, Chinese, and French nuclear postures have been based on the logic of minimum deterrence. Nuclear deterrence at this level is based less on calculations of a relative balance, which is always lopsided, than on the concept of "first strike uncertainty," in which the country with much greater nuclear capabilities could still not be confident of its ability to launch a perfectly successful first strike against a lesser nuclear power. At the third level, in the absence of meaningful nuclear competition, at least for now, Israeli nuclear deterrence is based on the perception of overwhelming nuclear superiority. Finally, at the fourth level, I argue that nuclear deterrence in South Asia today may rest not on calculations of a relative balance, which are all but impossible for either side to make, but on the notion that nuclear capabilities *exist* and might be used, deliberately or not, in a future conflict. In short, my hypothesis is that existential deterrence now characterizes the nuclear competition between two opaque proliferants, India and Pakistan.[20]

Table 1

TYPES OF PROLIFERATION AND DETERRENCE

Proliferant	Prolif. Type	Main Adversary	Deterr. Type
Superpowers	Declared	Each other	MAD/warfighting
Medium Powers	Declared	Superpower	Minimum
Israel	Opaque	Hostile neighbors	Nuclear hegemony
India/Pakistan	Opaque	Each other	Existential

The chief question raised by the nexus between opaque proliferation and existential deterrence concerns the credibility of nuclear weapon postures that are shrouded in secrecy. How can each side threaten the other with nuclear punishment in the absence of convincing evidence that it can carry out such a threat? I argue that in its efforts to make the South Asian nuclear balance transparent, and thus to pressure India and Pakistan into nuclear restraint, the international non-proliferation community inadvertently lends credibility to the very nuclear postures it is trying to discourage. Existential deterrence operates in South Asia due to a great irony: by increasing the transparency of nuclear weapon programs in India and Pakistan, the United States and other nonproliferation advocates actually enhance the credibility of capabilities they are trying to suppress.

My argument proceeds as follows: first, I provide a brief historical overview of the South Asian nuclear arms competition. I then survey the literature on the consequences of nuclear proliferation, and show that both the logic of nonproliferation and the logic of nuclear deterrence are inadequate because they arrive at indeterminate conclusions about the consequences of proliferation. I argue that the opaque proliferation concept is more productive because it better reflects the prevailing pattern of nuclear proliferation, and that the deterrent effect derived from opaque nuclear postures is logically of the existential variety described above. Finally, I discuss the dynamics of the South Asian nuclear competition during two recent crises over Kashmir. I conclude that an existential deterrent balance may indeed be operating between India and Pakistan, and that in the South Asian case nuclear weapon capabilities seem to have had a dampening effect on conflict.

South Asian Nuclear Arms Competition

The South Asian nuclear arms competition has developed as a two-phase historical process. In the first phase, the Indian and Pakistani nuclear weapon programs began as links in a global proliferation chain extending back to the Second World War. The first link in this chain was the Manhattan Project, the American effort to build atomic bombs, whose urgency stemmed from a fear that Hitler's scientists would master nuclear weapon technology first. After the U.S. atomic bombings of Hiroshima and Nagasaki in 1945, the Soviet Union vigorously pursued its own nuclear weapon capability, which Moscow achieved in 1949.

China's 1964 nuclear explosion and subsequent weapon program stemmed from Beijing's concern over first the United States and later the Soviet Union as threatening superpower adversaries. In turn, India's 1974 "peaceful nuclear explosion" was the result of a sustained national debate over Indian nuclear weaponization in the wake of the Chinese test. Finally, Pakistan's 1972 decision to pursue its own nuclear weapon program was precipitated by its devastating defeat in the 1971 Bangladesh war, and confirmed by India's 1974 nuclear test.[21]

The Indian and Pakistani decisions to pursue nuclear weapon capabilities stimulated the second phase of the South Asian nuclear weaponization process.[22] Although China remains significant for Indian nuclear planners today, the main dynamic in the second phase has been an intensified *bilateral* nuclear competition between New Delhi and Islamabad. India's 1974 nuclear test strengthened Pakistan's resolve to pursue its own nuclear weapon capability. Pakistan's clandestine but widely reported efforts to build nuclear weapons in the late 1970s and early 1980s provoked a renewed Indian debate over nuclear weaponization.[23] More recently, the Indo-Pakistani nuclear competition has encompassed efforts by both countries to develop nuclear-capable ballistic missiles. The rest of this section will briefly summarize the development of each country's nuclear weapon program.

Indian nuclear research began in 1944, three years prior to independence. An Indian Atomic Energy Commission was created in 1948, but it was not until 1954 that "steady but modest funding" flowed into nuclear research and development under the aegis of a Department of Atomic Energy.[24] The Indian nuclear program's primary objective until the 1962 Sino-Indian war and the 1964 Chinese nuclear explosion was "development for long-term civilian needs."[25] Even so, there were hints of India's subsequent "nuclear option strategy" from the start. As early as 1946, in response to a question about the possibility of a future Indian deployment of nuclear weapons, Indian nationalist leader and soon-to-be prime minister Jawaharlal Nehru reportedly "stated his hope that India would develop atomic power for peaceful uses but warned that, so long as the world was constituted as it was, every country would have to develop and use the latest scientific devices for its protection."[26] During the 1950s, a confidential consensus emerged among a small core of Indian leaders that New Delhi should reserve its right to develop nuclear weapons if threatened by a future Chinese nuclear weapon capability.

The 1964 Chinese test intensified and made public the debate over India's nuclear option. Four considerations drove this debate: the nature of the threat posed by Chinese nuclear weapons, the availability and implications of external nuclear guarantees vis-à-vis China, the cost of an Indian nuclear weapon program, and the morality of nuclear weapons.[27] Indian analysts' main strategic concern was not an immediate military threat from the Chinese, but the prospect of future political intimidation by Beijing. In 1964 Indian prime minister Lal Bahadur Shastri reportedly "launched a program to reduce the time needed to build nuclear arms to six months," thereby giving "official sanction to the development of an Indian nuclear weapons option."[28] The result in the late 1960s was an Indian "nuclear-option but nonweapon stance."[29]

Prime Minister Indira Gandhi decided in 1972 to proceed with an Indian nuclear explosion as soon as possible, and gave her final approval to the test in 1974. Her decision likely was based primarily on India's changed security environment after 1971, and secondarily on domestic politics. India's security planning vis-à-vis China was jarred by President Richard M. Nixon's 1971 "opening" to Beijing. Pakistan's role as a conduit in the China initiative and the U.S. "tilt" toward Pakistan during the 1971 Bangladesh war may have added to New Delhi's perception that India was the target of a U.S.-Pakistan-China security axis. Nixon's dispatch of an aircraft carrier task force toward the Bay of Bengal during the Indian liberation of Bangladesh was especially galling to Indian officials, who regarded the U.S. move as a loathsome modern variant of gunboat diplomacy. Others have suggested that by 1974 Indira Gandhi's domestic political problems (which led eventually to the "Emergency" of 1975–77) were so acute that the Indian nuclear test was partly designed to bolster her own standing at home.[30] Whatever its exact causes, India's "peaceful nuclear explosion" has allowed New Delhi to demonstrate its nuclear weapon capability to potential adversaries without suffering the international opprobrium associated with the actual deployment of nuclear weapons.

Pakistan's nuclear research program was stimulated by President Dwight D. Eisenhower's 1953 "Atoms for Peace" proposal and a subsequent exhibition which toured Pakistan touting the advantages of atomic energy in socio-economic development. Pakistan established its own Atomic Energy Commission in 1956, and its nuclear program during the 1950s and early 1960s was marked by an absence of military considerations. This changed in the mid-1960s, when then-Foreign Minister Zulfikar Ali Bhutto precipitated Pakistan's own "nuclear option" debate by arguing for a Pakistani response to India's gains in the nuclear field. Bhutto's precise motives, like Indira Gandhi's, remain obscure, but appear to have arisen from a similar combination of international and domestic political imperatives. The 1965 Indo-Pakistani War, though inconclusive, revived questions about Pakistan's security, especially in view of the country's lack of strategic depth. These concerns were exacerbated by the knowledge that India had begun reprocessing plutonium in 1964. Also, Bhutto began to mount his own challenge to President Ayub Khan in the wake of the 1965 war, and perceived that "nuclear nationalism" would be an effective way to shore up his domestic support at a time when Pakistan's security problems seemed especially acute. Indications are, however, that Bhutto's position on nuclear weapons remained a minority viewpoint until his rise to power after the Bangladesh war.

The Bangladesh debacle re-oriented Pakistan's nuclear program, and India's 1974 nuclear test intensified this revised posture. In 1972, given the gaping asymmetry that had emerged in Indian and Pakistani military capabilities with the loss of East Pakistan, President Bhutto reportedly ordered Pakistan's scientific community to begin developing nuclear arms. Bhutto believed a Pakistani nuclear weapon capability would increase not only Pakistan's international security and his own domestic standing, but also Pakistan's stature among the newly wealthy nations of the Islamic, oil-producing Middle East. India's 1974 test did not cause, but rather increased, Pakistan's nuclear resolve.[31]

The 1970s marked the shift from the first, global, stage of the South Asian nuclear proliferation process to the second, regional, stage. Some analysts believe China remains Indian nuclear planners' primary concern,[32] while others think that since India's 1974 nuclear test and Pakistan's successful efforts to acquire nuclear weapon technology, the India-Pakistan dynamic has become the central element in South Asian nuclear proliferation. As Richard P. Cronin writes, although India's nuclear ambitions originally derived from other factors, "the crux of the proliferation threat lies in the India-Pakistan rivalry."[33] By the mid-1980s, "concerns over nuclear developments in Pakistan had become by far the dominant factor in Indian nuclear decision making."[34]

Steady improvements in nuclear weapon and delivery capabilities have marked the second phase in South Asian nuclear weapon development. Stephen P. Cohen calls this the "first time in the short history of the nuclear world that two competing states, unrestrained by alliances, moved simultaneously to military nuclear status."[35] Leonard S. Spector estimates that India can produce 15 nuclear devices annually, and may have enough plutonium for 40–60 nuclear weapons. These are potentially deliverable on some 200 Indian Air Force Jaguars, MIG-23s, MIG-25s, MIG-27s, and Mirage 2000s. In 1989, India successfully tested a nuclear-capable intermediate-range ballistic missile (IRBM), the Agni. When deployed, Indian IRBMS will improve New Delhi's nuclear delivery capabilities in two important ways: (1) Indian forces will for the first time be able to strike strategic targets in eastern China; and (2) India's ballistic missiles will lie beyond the range of Pakistan's delivery systems, thus providing a more reliable second-strike capability.[36]

Pakistan was judged to have produced weapons-grade nuclear material by a 1986 U.S. Special National Intelligence Estimate. In 1987 Pakistani President Mohammed Zia ul-Haq declared that "Pakistan has the capability of building the Bomb. You can write today that Pakistan can build a bomb whenever it wishes. Once you have acquired the technology, which Pakistan has, you can do whatever you like."[37] In February 1992, Pakistani Foreign Secretary Shahryar Khan said that Islamabad has "elements which, if put together, could become a (nuclear) device."[38] Spector estimates that Pakistan can deploy five to ten nuclear bombs on some 40–50 F-16s, or Mirage IIIs and Mirage Vs. Pakistan also tested two potentially nuclear-capable ballistic missiles in 1989, one with a range of nearly 200 miles. These missiles could reach targets in northwestern India.[39]

The Logic of Nonproliferation

Most writing on nuclear weapons until the late 1950s concerned the U.S.-Soviet nuclear balance. Sustained thinking about the wider spread of nuclear weapons – or "nuclear proliferation" – was stimulated by the British, French, and Chinese nuclear weapon programs. The spread of nuclear weapons was originally referred to in the scholarly literature as the "Nth country problem." Since each of the early nuclear proliferants was perceived to be allied with one or the other superpower, the overriding concern about "Nth countries" during the late 1950s and early 1960s was the threat nuclear weapons posed to the stability of the central, or U.S.-Soviet, balance. The diffusion of nuclear weapons to a "non-bloc area of world politics" was alarming insofar as it raised the probability of "a central nuclear exchange between major blocs."[40] Albert Wohlstetter viewed nuclear proliferation as the "principal menace lending a sense of urgency to our negotiations on arms control and as a trend to be fought," and argued that "wide nuclear diffusion would be gravely disruptive. It would increase the likelihood of the use of nuclear weapons both by accident and by deliberation."[41]

The focus of the nuclear proliferation debate shifted in the 1970s from the central balance to conflict-prone regions. The 1974 Indian nuclear test raised fears in the West that as many as ten Third World nations were close to deploying nuclear weapons.[42] Consequently, the possibility of regional nuclear conflict became a serious concern in and of itself (along with the still-worrisome but distinct possibility of the superpowers becoming inextricably entangled in a regional nuclear crisis). Most proliferation specialists argued that mutual nuclear deterrence between the United States and the Soviet Union rested on important "geopolitical and technical supports" that are absent in much of the Third World.[43] David C. Gompert expressed the typical view that "a global network of deterrence relationships, stable in theory, would be fraught with ambiguities and doubts in practice."[44] As a theory about the consequences of the spread of nuclear weapons, the logic of nonproliferation rests on four main foundations:

(1) *Third World security environments are qualitatively different from the central balance* or, indeed, from new nuclear balances that might arise in the advanced industrialized world. According to this view, unique geopolitical conditions make Third World nuclear weapons destabilizing. Some observers believe that the stakes are higher in Third World conflicts relative to those in the cold war superpower competition. According to this argument,

countries' territorial integrity, independence, even national survival may ride on the outcome of international conflicts in areas like the Middle East and South Asia. Because they "perceive the stakes to be so high, some of these countries' leaderships may be ready to risk nuclear confrontation, if not even to accept a surprisingly high level of nuclear damage, in pursuit of their objectives."[45]

Others believe nuclear weapons would be destabilizing in the Third World because of political instability in many developing countries. Michael Mandelbaum says: "the farther they proliferate, the greater the risk that nuclear weapons will enter areas troubled by political quarrels."[46] From this perspective, not only would stakes be high, but the opportunities for conflict to erupt and escalate to the nuclear level would be frequent. Common is Steve Fetter's view that states "such as India and Pakistan, North and South Korea, and Israel and various Arab states, have deep religious, ideological, or cultural animosities, often combined with active border disputes, that weaken deterrence."[47] The timing of proliferation in the Third World is also said to enhance nuclear instability: some analysts feel that regional powers may have so little time to assimilate nuclear weapons into their military forces and doctrines before conflict erupts that "nuclear learning" will be impossible.[48]

(2) *Third World leaders may be irrational in their handling of nuclear weapons* In this view, nuclear weapons may induce less caution in the leaders of Third World states than they have in the leaders of established nuclear weapon states. As early as 1957 Henry A. Kissinger warned that "just beyond the horizon lies the prospect of a world in which not only two superpowers will possess nuclear weapons, but also many weaker, and perhaps more irresponsible nations, with less to lose."[49] Typical of this line of reasoning is Mandelbaum's claim that "the farther the bomb spreads from the industrial circumference, the greater are the chances it could find its way into the hands of persons who would not show the prudence that the guardians of existing nuclear stockpiles have so far displayed."[50]

Arguments about irrational governments take two broad forms. In the first, future Third World proliferants may be led by rulers whose goals transcend material political interests, and whose decisions thus are not based on rational calculation. As Lewis A. Dunn writes, these leaders may not be "coldly calculating, cautious, and fully rational." Some, "motivated by their higher mission and destiny ... are less likely to weigh carefully the costs and gains of military action and are more likely to take high-risk gambles to serve their causes."[51] The fear here is of "states prone to autocratic rule by charismatic leaders who might not share the usual aversion to nuclear use or be sufficiently restrained by such institutional mechanisms as exist elsewhere."[52]

The second form of this argument is that so-called "crazy Third World statesmen"[53] are undeterrable.[54] For George H. Quester, the "most powerful argument of all" against nuclear proliferation is that sooner or later nuclear weapons "will fall into the hands of what history would remember as a madman – someone who did not sufficiently dislike nuclear war and thus plunged the world into it."[55] Kathleen C. Bailey also takes this position:

> While some have argued that possession of nuclear weapons might make a ruler or state behave more responsibly, evidence indicates otherwise. Ruthlessness or lack of sanity has led some Middle Eastern leaders to use chemical weapons on their own populations or to engage in terrorism. This lends credibility to the argument that some leaders might be irresponsible with a weapon of mass destruction, if one were at hand.[56]

(3) *Third World nuclear arsenals will be technologically unsophisticated, and therefore dangerous* This argument stresses that future proliferants will lack the resources and technological capabilities that have allowed the five original nuclear weapon states to avoid the use of nuclear weapons since 1945. Analysts stress three dangers: (1) accidents involving nuclear weapons; (2) the unauthorized use of nuclear weapons; and (3) the preemptive use of nuclear weapons during a crisis. Dunn summarizes the threat of nuclear accidents:

> Some of those countries that may soon acquire nuclear weapons ... probably do not have the scientific and engineering manpower or the financial resources to design and fabricate reliable, advanced accident-proofing systems integral to the nuclear weapon itself. Their limited resources will be expended in their struggle simply to join the nuclear club. Fearful of being caught unprepared, however, these countries may be unwilling to use such a simple, if less sophisticated, accident-proofing measure as not fully assembling their nuclear weapons until they are needed. And even stockpiling disassembled weapons would not reduce the risk of an accident once those weapons have to be assembled. In fact, hasty assembly of crude nuclear weapons would augment the chances of just such an accidental detonation in the most dangerous political context – a continuing crisis or even a low-level conflict.[57]

A second technical danger is said to be unauthorized nuclear weapon use. Dunn and others argue that nuclear weapons in Third World nations may not be adequately secured against theft and/or use by people outside the official chain of command. These might include terrorists, who could use stolen nuclear weapons for blackmail or anonymous nuclear strikes. Another possibility is said to be the nuclear coup, where one group in a power struggle uses its control of nuclear weapons to bolster its position. Dunn even speculates that renegade officers might launch a desperation blow against the enemy; their "commitment to a self-ordained higher mission would overwhelm any fear of the adverse personal or national consequences."[58]

A third danger analysts point to is that the nuclear arsenals of future proliferants may be so technologically backward as to promote crisis instability. According to this logic, Third World nuclear powers will have a limited number of weapons; during a crisis, what Thomas C. Schelling calls the "reciprocal fear of surprise attack" could foster a climate where striking second is seen as the worst option and striking first only the second-worst choice.[59] Fetter writes: if "one side believes that war is inevitable, it may try to preemptively destroy the other side's vulnerable but valuable weapons of mass destruction. Even if both sides prefer not to preempt, each may fear that the other side will; consequently, both may decide to launch at the first (perhaps false) indication of an attack."[60]

(4) *Pre-meditated nuclear strikes, while unlikely, are still possible in a more proliferated world* There are also those who take seriously the possibility that future nuclear weapon states could launch pre-meditated nuclear attacks. Michael D. Intriligator and Dagobert L. Brito suggest that the "circumstances when nuclear weapons are probably most likely to be used is when one nation in a conflict or war has a nuclear weapons monopoly."[61] Even with rough nuclear parity, others believe the prospects for nuclear war may be real. Dunn says that, contrary to the "mainstream of Western strategic studies," we should not assume that future proliferants will embrace nuclear weapons just for deterrent purposes: "some possible proliferators ... appear more likely to look upon nuclear weapons as potentially useful battlefield weapons." He argues that future nuclear weapon states could follow offensive or defensive, rather than deterrent, nuclear strategies. From this perspective, a country like Pakistan "could decide to escalate to the employment of nuclear weapons rather than accept a conventional military defeat." Or, if "Pakistan relies on aircraft for the delivery of a handful of nuclear weapons while India has available nuclear armed missiles, an Indian nuclear first strike might virtually destroy Pakistan's nuclear force, thereby greatly reducing the threat of nuclear retaliation."[62]

The Logic of Nuclear Deterrence

A few analysts take issue with the logic of nonproliferation outlined above. Although proponents of the logic of nuclear deterrence generally do not advocate the purposeful diffusion of nuclear weapon technology,[63] they argue that the impact of proliferation will be less menacing than commonly believed. Foremost among those who minimize the negative consequences of nuclear proliferation is Kenneth N. Waltz, who argues that predictions based on the nonproliferation logic "point less to likelihoods and more to dangers that we can all imagine. They identify some possibilities among many, and identifying more of the possibilities would not enable one to say how they are likely to unfold in a world made different by the slow spread of nuclear weapons." Waltz says flatly: "nuclear weapons have reduced the chances of war between the United States and the Soviet Union and between the Soviet Union and China. One may expect them to have similar results elsewhere." For Waltz, "much of the writing about the spread of nuclear weapons has this unusual trait: It tells us that what did *not* happen in the past is likely to happen in the future, that tomorrow's nuclear states are likely to do to one another what today's nuclear states have not done."[64]

The logic of nuclear deterrence stresses that nuclear weapons promote deterrent, not offensive, strategies and thus decrease the chances of war. Bruce Bueno de Mesquita and William H. Riker argue, for example, that the "presence of an explicit or underlying nuclear threat constrains conflict by reducing its likelihood of escalating into war."[65] This general effect should prevail in all nuclear balances, regional ones included. As John J. Weltman writes, "nuclearization of regional conflict patterns in the third world will tend to be self-regulating. Given enough time, mutual deterrent postures will tend to develop."[66] Discussing one conflict-prone region, Steven J. Rosen maintains that "apocalyptic visions and doomsday images of an openly nuclear Middle East have been accepted too readily .... A stable system of mutual deterrence may be entirely possible ...."[67] Waltz argues that the chief impact of nuclear weapons is to induce caution: "In a conventional world, one is uncertain about winning or losing. In a nuclear world, one is uncertain about surviving or being annihilated."[68] By removing uncertainty about outcomes, these analysts say, nuclear weapons will inhibit conflict even in the Third World.

Supporters of the logic of nuclear deterrence deny that Third World security environments are inherently more prone to nuclear weapon use than other situations. As Waltz writes, "bitterness among some potential nuclear states, so it is said, exceeds that experienced by the old ones. Playing down the bitterness sometimes felt by the United States, the Soviet Union, and China requires a creative reading of history."[69]

These analysts also dispute the thesis that Third World leaders may be irrational with nuclear weapons. As Weltman writes: "any assertion that regional nuclear balances would, as a general rule, be unstable for political reasons – the absence of those patterns of behavior and modes of thought that produced prudence in the Soviet-American relationship – must be regarded to be of questionable validity." In his view, "such an assertion would assume that the capacity for political rationality is a narrow, culturally based attribute," which contradicts the fact that "nothing in our experience with the political behavior of the new states gives evidence of a congenital inability to recognize dangers or to draw inferences of prudence therefrom." Like Schelling before him, Weltman notes that we "should not automatically take bizarre behavior patterns and wild rhetorical posturing at face value. Cultivation of an image of inflexibility, monomaniacal commitment, or even sheer madness may at times prove to be a useful bargaining device."[70]

Weltman further argues that nuclear weapons cannot be expected to "moderate the *intentions* of fanatical individuals with their hands upon the instruments of mass destruction; but equally we should not expect them to *use* those means to achieve their ends unless they conclude that they can use them without dire consequences to themselves."[71] Waltz points out that many Third World countries have a handful of densely populated cities that would be vulnerable to attack: "What government would risk sudden losses of such proportion or indeed of much lesser proportion? Rulers want to have a country that they can continue to rule."[72]

On issues of technical sophistication, critics of the logic of nonproliferation contend that its fears are overblown. Waltz argues that preemption is viable "only if the would-be attacker knows that the intended victim's warheads are few in number, knows their exact number and locations, and knows that they will not be moved or fired before they are struck. To know all of these things, and to know that you know them for sure, is exceedingly difficult."[73] Weltman agrees, arguing that a "failure to eliminate even a single deliverable weapon would thus be to risk catastrophe; and short distances mean that no great sophistication in means of delivery is required for a successful counter-value response."[74] These arguments also apply to the possibility of pre-meditated nuclear weapon use. For those who support the logic of nuclear deterrence, it is difficult to conceive of situations where the gains to be expected from a nuclear strike would outweigh the losses that a nuclear response would involve. Accordingly, they argue, since uncertainty about the efficacy of a first strike is easy to induce, nuclear deterrence should prevail.

In addition, these observers generally dismiss the dangers of unauthorized or accidental nuclear use as extremely unlikely. Nuclear terrorists, for example, would have to identify themselves or their constituents to achieve their political objectives, but in so doing could be countered or deterred just like any national government.[75] On the possibility that nuclear weapons could be used by factions in a domestic political dispute, Waltz writes, "what is hard to comprehend is why, in an internal struggle for power, any of the contenders should start using nuclear weapons. Who would they aim at? How would they use them as instruments for maintaining or gaining control?" From his perspective, "those who fear the worst have not shown with any plausibility how those expected events may lead to the use of nuclear weapons."[76] Finally, Waltz argues that new nuclear weapon states will have "every incentive" to "take good care of their weapons," making future nuclear accidents unlikely.[77]

Why a New Approach is Necessary

These opposing "logics" are ultimately inadequate, because neither yields compelling explanations of the consequences of nuclear proliferation. Analyses based on the experience of the declared proliferation universe are necessarily speculative, inasmuch as they try to predict the consequences of *future* proliferation based on political and technical assumptions that may or may not be useful in a world of opaque proliferants. By and large, they fail to recognize that nuclear proliferation has a history. In comparing the first book-length treatment of the spread of nuclear weapons[78] with later studies, one finds that the *process* of proliferation in the intervening decades is ignored. Rather than using earlier expectations as a baseline and examining the dynamics of proliferation in the interim, analysts generally extend old arguments and apply them to a future world of "N" nuclear weapon powers.[79]

Because these studies are future-oriented, their conclusions are logically indeterminate, a fact that has been recognized by a number of analysts. Bruce D. Berkowitz, for example, writes that neither of the general arguments about nuclear proliferation is "intrinsically more valid than the other. Rather, proliferation can increase or decrease the likelihood of nuclear war. The actual relationship between proliferation and the likelihood of nuclear war depends heavily on the assumptions one makes about the durability of deterrence."[80] Joseph S. Nye agrees: "Unfortunately, there can be no decisive answer in the debate over the effects of proliferation. Particular outcomes may differ. Some cases may start a disastrous chain of events; other may turn out to have benign effects."[81] Weltman writes:

> The introduction of nuclear weapons into a regional conflict may produce an impetus either to stability or instability, depending on the specifics of the technology introduced, the pace of its introduction, and the extent to which that introduction is made on a symmetrical or asymmetrical basis as between the parties to a regional conflict. There is no reason to assume, as a general proposition, that one or another outcome must eventuate.[82]

Rather than despair about the poor prospects for theoretically rigorous analysis of regional nuclear proliferation, it would be more profitable to shift the discourse from speculation about the future to analysis of the recent past.

Instead of guessing at the probable capabilities of future proliferants and inferring outcomes from those capabilities, we need honestly to acknowledge that the prevailing pattern of proliferation today does not lend itself very well to such "hardware"-oriented analysis. Instead of predicting the course of future nuclear proliferation in South Asia we can instead concentrate on the recent history of the regional nuclear competition, in the context of the theoretical debate outlined above, but in recognition of its limitations.

Benjamin Frankel and others have recognized that analyzing the opaque pattern of proliferation through the same lens with which we have viewed the declared proliferation of the five original nuclear weapon states is illogical. They have introduced a new literature on opaque proliferation that I believe is more useful in analyzing the consequences of proliferation in South Asia. Opaque nuclear proliferants "secretly" develop nuclear weapons while publicly denying their own nuclear capabilities. The prototype opaque proliferant is Israel, whose example has been emulated by other budding nuclear weapon states, including Pakistan and to a lesser degree India. The general features of an ideal type opaque nuclear weapon posture are: no nuclear test, official denial of weapon possession, no direct nuclear threats, no military doctrine, no military deployment, no open debate, and organizational insulation of nuclear weapon activities. Many cases will not exhibit one or two of these characteristics, but all of the de facto nuclear weapon states basically conform to this model.[83]

The implications of opaque proliferation for regional nuclear balances are intriguing: India and Pakistan are moving along the path of opaque proliferation in part because of the deterrent effect each hopes nuclear weapons will have on the other. As Cohen and Frankel write, however, "there is little in the literature to tell us how a country should plan to use its nuclear weapons to deter its adversaries while denying the possession of these weapons." The logic of nuclear deterrence has traditionally rested on evidence of tangible nuclear weapon capabilities. The opaque pattern of proliferation raises the question: is such evidence necessary for a country to achieve a deterrent effect from its nuclear weapon capabilities? Cohen and Frankel argue that it is "worth examining whether deterrence theory has the same validity in situations where this essential component of certainty is absent."[84]

Under conditions of opacity, where each side in a regional nuclear arms competition can muster little hard evidence concerning the precise force posture of the other side, any deterrent effect to be derived from nuclear weapon capabilities will logically be of an existential nature. That is to say, mutual calculations about the efficacy of nuclear deterrence in such a situation are based not on the details of relative nuclear weapon capabilities, which are beyond the national technical means of each side, but on the shared notion that each side is or can soon become nuclear-weapon capable, and thus that any outbreak of conflict might lead to a nuclear exchange.

Indeed, some analysts have suggested that even in the case of the cold war superpower balance, the existential nature of nuclear deterrence was more compelling than relative force postures for decisionmakers during crises. In 1969, McGeorge Bundy noted the huge difference between the complex technical calculations of nuclear planners and the much more simple equations that present themselves to political leaders in times of crisis. He argued that the threat of losing even one city to an opponent's nuclear strike is a disaster any political leader would do his utmost to avoid. Implicit in Bundy's argument is the notion that political leaders, unlike military planners, are deterred not by virtue of the fact that their nuclear forces are relatively inferior, but by the possibility that nuclear weapons exist and might be used *at all*.[85] Marc Trachtenberg's research on the Cuban missile crisis supports Bundy's contention that it was the absolute possibility of a nuclear exchange, not any relative calculation of advantage that conditioned the behavior of President Kennedy and his advisers. He writes: "It was as though all the key concepts associated with the administration's formal nuclear strategy ... in the final analysis counted for very little." As the President said during the crisis, "What difference does it make? They've got enough to blow us up now anyway."[86]

The main difference between the central balance and regional balances like the one found in South Asia, is, of course, the fact that the United States and the Soviet Union had unambiguous assured second-strike capabilities for most of the cold war. To the extent that each side had no doubt about the other's capabilities, these capabilities were "visible" and thus made mutual deterrence credible. Two questions arise in an analysis of opaque nuclear postures: (1) Can deterrence be achieved with much lesser nuclear weapon capabilities? (2) How can these lesser capabilities be made credible?

The first question can be answered by returning to the previous discussion of the conventional wisdom regarding nuclear proliferation. In my view, the only compelling scenario for the purposeful use of nuclear weapons stems from the "reciprocal fear of surprise attack;" that is, a situation where one side in a crisis views a pre-emptive strike to be the second-worst option and passivity the worst option. As Waltz argues, however, any nuclear weapon state contemplating a pre-emptive first strike against another nuclear weapon state must be sure that it knows the exact number and location of the opponent's nuclear warheads, and that it knows that these weapons will not be moved or fired before being struck. One need only think back to the 1991 Persian Gulf War and the allies' inability to locate and destroy all of Saddam Hussein's Scud missile launchers to realize the virtually unlimited possibilities for deception in concealing even a tiny number of nuclear warheads from an opponent. Moreover, the allied coalition in the Gulf war had at its disposal the best weaponry and intelligence-gathering capabilities in the world, and faced none of the time constraints that would guide behavior in a regional crisis. In the event of such a crisis, all that is necessary to deter the launching of a pre-emptive strike is "first-strike uncertainty," or the planting of a seed of doubt in the minds of the potential attacker's leaders about whether it would be possible to destroy all of the opponent's nuclear weapons pre-emptively. In short, it is unthinkable that an Indian or Pakistani military planner could convince his political leadership that a pre-emptive nuclear strike would *definitely* succeed in destroying *all* of the other side's nuclear warheads in a first strike.

The second question is how opaque nuclear weapon states can achieve deterrent credibility while denying that they deploy nuclear weapons. As Avery Goldstein writes, "the credibility of threats rests on the capabilities at one's disposal and the probability they will be used."[87] Even existential deterrence depends on demonstrated nuclear weapon *capabilities* and the resolve or intentions of the nuclear weapon state, which are a measure of the *probability* that the weapons will be used. Intentions can be signalled by bargaining between the two sides in the deterrent balance. As Joeck speculates, adversary opaque proliferants are forced by their public postures of denial into a nuclear dialogue built mainly on tacit communication. Following Schelling's ground-breaking work,[88] Joeck writes:

> in competition marked by tacit bargaining, the negotiating partners are seeking to coordinate for a common gain. In the subcontinent, some evidence suggests that India and Pakistan may agree far more than they disagree on nuclear problems. There may be sufficient agreement to allow for cooperative bargaining rather than overt competition that could result in an arms race or war.[89]

India and Pakistan "are trying to attain certain goals (security and prestige) that put them on a collision course, while at the same time trying to avoid an outcome (an unrestrained nuclear arms race that could more fundamentally jeopardize their security) that puts them on a cooperative course."[90] As Weltman explains this process, "states will implicitly or explicitly exploit their presumed capabilities to produce the Bomb, as well as a calculated ambiguity about quietly having actually done so. It constitutes a political device of multifarious utility in international political and strategic bargaining."[91]

If intentions can be signalled via tacit bargaining, capabilities must be *demonstrated* in some credible way. How is this possible, given the fact that opaque nuclear proliferants deny that they have nuclear weapons? Paradoxically, the international nuclear nonproliferation community has solved this problem: India and Pakistan do not have to demonstrate their nuclear weapon capabilities because the global nonproliferation regime does it for them. Every time a national government with extensive intelligence-gathering capabilities, such as the United States, or a reputable private organization, such as the Stockholm International Peace Research Institute or the Carnegie Endowment for International Peace calls attention to the development of nuclear weapon capabilities in India and Pakistan, they inadvertently bolster the credibility of those capabilities in the eyes of the adversary. The better the reputation of the government or private organization, the more credible are its observations about the nuclear weapon capabilities of the opaque proliferant.

When a private organization documents the strides Pakistan has made in its nuclear weapon program, the credibility of Pakistan's nuclear deterrent is enhanced where it matters most – in New Delhi. Similarly, when officials of the U.S. government discuss India's nuclear weapon capabilities, they enhance the credibility of the Indian deterrent in Islamabad. What better source of credibility could India and Pakistan want than the following public statement made by U.S. Central Intelligence Agency director Robert M. Gates in 1992: "We have no reason to believe that either country maintains assembled nuclear bombs, much less that either has deployed them. But such weapons could be assembled quickly, and both countries have combat aircraft that could be modified to deliver them during a crisis."?[92]

Adding to this irony, official U.S. pressure on Pakistan to *curtail* its nuclear weapon activities also has the inevitable effect of enhancing the credibility of Pakistan's program in the eyes of the Indian government. After years of certifying that Pakistan does not "possess a nuclear explosive device," as a statutory condition for continued U.S. aid to Islamabad, the U.S. government refused for the first time in 1990 to make such a certification.[93] This declaration was naturally front-page news in India, and it likely helped to make Pakistan's nuclear weapon capability an established fact in the eyes of Indian decisionmakers. The piling up of such revelations over the years surely has the effect of making an adversary opaque proliferant no less of a nuclear power than a declared nuclear weapon state. It also serves to galvanize public opinion in favor of an equally ambitious nuclear weapon stance as the opponent's. In the face of political pressures generated by pro-nuclear weapon lobbies in each country, no South Asian leader who wishes to remain in power can now reverse his nation's nuclear weapon progress.

South Asian Nuclear Dynamics

The simultaneous movement of two bitter South Asian enemies into a competition in opaque nuclear weaponization has produced a complex set of international dynamics. At one level, India and Pakistan continue to develop their nuclear weapon capabilities on the worst-case assumption that the other side is doing likewise. Both Islamabad and New Delhi claim it is necessary to maintain their nuclear option, in view of the other's nuclear weapon activity. At another level, the two governments are involved in a diplomatic dance that is intended to bolster their respective positions in the eyes of the world. Islamabad in particular has pushed what it says is a good faith effort to prevent a nuclear arms race in South Asia. Pakistan has offered repeatedly over the years to sign the Nuclear Non-Proliferation Treaty if India will. New Delhi has refused, in part because the treaty enshrines China but not India as a legitimate nuclear weapon state. More recently, Islamabad has put forth a U.S.-supported scheme to convene a conference on nuclear non-proliferation in South Asia that would include the United States, Russia, China, India, and Pakistan.[94] New Delhi's response has been negative, with Indian officials suggesting that Iran and Kazakhstan be brought into any regional discussion of nuclear weapons.[95]

For each government, domestic politics make it impossible to turn away from nuclear weapons, while international politics make it inadvisable to move ahead into actual deployment. Each side hovers on the nuclear threshold, ever refining its capability to do immense damage to the other. Both seem to believe that, on the whole, nuclear weapon capabilities enhance rather than threaten their security, and national elites increasingly speak of their deterrent value.[96] Pakistan's ambassador to the United States said at a 1992 forum in Washington, D.C., for example, that the "Indian and Pakistani capabilities deter war in South Asia just as the U.S. and Soviet nuclear threats did in Europe after World War II."[97] Stephen P. Cohen relates Pakistani President Zia ul-Haq's view in 1988 that "India and Pakistan had achieved deterrence stability because of uncertainty as to whether each possessed nuclear weapons and how many they might have."[98] Brahma Chellaney writes that "subcontinental policies of nuclear ambiguity are already acting as a nuclear deterrent, since each side believes its rival might have 'bombs in the basement.'"[99] Cohen, too, finds "strong evidence" that nuclear deterrence influenced both sides during a 1987 crisis precipitated by a large Indian military exercise called "Operation Brass Tacks." He also concludes that "proliferation may have a calming, even stabilizing regional effect. It probably had this effect in 1990, during the worst crisis in India-Pakistan relations in twenty years."[100]

While it is easy for national leaders and scholars to *say* that mutual nuclear deterrence is dampening tendencies toward conflict in South Asia, it is more difficult to find *evidence* that such an effect is influencing Indo-Pakistani relations, especially given deterrence theory's dilemma of proving non-events. Still, two recent crises involving Kashmir, including the one mentioned by Cohen above, offer an impressionistic picture that makes a strong case for the plausibility of the "existential deterrence" hypothesis. In 1989, long-simmering discontent in the Indian state of Jammu and Kashmir exploded into an insurgency that threatens the integrity of the Indian union.[101] New Delhi charges that Islamabad is arming and training the Islamic Kashmiri militants who advocate secession from India. Islamabad admits only to providing moral support to the militants, and maintains that Kashmiri separatism was instigated not by Pakistan, but by decades of Indian misrule in Kashmir. There are substantial elements of truth to the claims of both sides.

By early 1990, Indo-Pakistani relations had reached a boiling point. Media reports in both countries and around the world were filled with talk of a fourth Indo-Pakistani war. The U.S. government was sufficiently concerned to dispatch a high-level delegation to Islamabad and New Delhi to defuse tensions that might lead to war. According to one account, a senior American official called the situation "potentially the biggest threat of nuclear war since the Cuban missile crisis."[102] In the midst of this sustained Indo-Pakistani tension, two events occurred which suggest that both sides may have backed away from the brink of war because of the existence of nuclear weapon capabilities on the Indian subcontinent. The first was a U.S. peace-making mission led by then-Deputy National Security Adviser Robert Gates. According to the *Far Eastern Economic Review*, Gates was told by Pakistani president Ghulam Ishaq Khan that "in the event of a war with India, Pakistan would use nuclear weapons at an early stage."[103] Gates reportedly passed this message to Indian officials in New Delhi. Influential Pakistanis, at least, viewed this exchange as having deterred India from starting a war with Pakistan. As one defense expert suggested, "Robert Gates told the Indians that we were mad enough to use the bomb and they believed him."[104]

An even more intriguing example of nuclear signalling emerged in a London *Sunday Times* story, which suggested that "both countries have been readying their nuclear arsenals" for war. According to this account:

> American spy satellites have photographed heavily armed convoys leaving the top-secret Pakistani nuclear weapons complex at Kahuta, near Islamabad, and heading for military airfields. They have also filmed what some analysts said were special racks designed to carry nuclear bombs being fitted to Pakistani F-16 aircraft.[105]

In addition, the U.S. government received "detailed intelligence reports from the Soviet Union about India's preparations for war.... India is also believed to have nuclear weapons and the Soviet Union has detected signs that these, too, are being readied for use." Ultimately, India and Pakistan managed to move away from military conflict. Although it is impossible at this point to show conclusively that the fear of a nuclear exchange was the primary reason that no war broke out in South Asia in 1990, it is plausible that the major difference between this crisis and other Indo-Pakistani crises that *have* led to war is that the two governments were much more reluctant to engage in military conflict given the existence of nuclear weapon capabilities on the Indian subcontinent.

Early 1992 saw a second example of India and Pakistan backing away from war over Kashmir. In this instance, Indo-Pakistani tension had risen dramatically in February due to a planned march by thousands of supporters of the Jammu and Kashmir Liberation Front (JKLF), a group of Islamic militants that champions Kashmiri independence from *both* India and Pakistan. The marchers were to cross the "line of actual control" from "Azad (Free) Kashmir," occupied by Pakistan, into the Indian state of Jammu and Kashmir. Once in India, they planned to continue their *jihad*, or holy war, against the "foreign" occupation of Kashmir. In the actual event, the *Pakistani* government brutally suppressed the marchers before they could reach the line of actual control and cross into India. It did so by firing directly into a crowd of 1000 marchers as they "tore through rolls of barbed wire on a mountain bridge just kilometres" from the line of actual control. Pakistani police had earlier "dynamited avalanches of boulders, trees and deep mud" in order to block roads leading to the frontier. At least 12 JKLF supporters were reportedly killed, and another 150 injured.[106]

Again, without interviewing the actual participants, it is impossible to determine if existential nuclear deterrence contributed to peace in South Asia in 1992. It is interesting to note, however, that the Pakistani foreign secretary made his government's first-ever admission that it has a nuclear weapon capability just a week before the climax of the 1992 Kashmir crisis (see n. 5). This suggests that Islamabad wanted to deter India from hostility through a clearer signal about Pakistani nuclear capabilities than it had ever sent before. Moreover, the idea of a staunchly Islamic regime in Islamabad killing *Muslims* in Kashmir lends powerful support to the conclusion that India and Pakistan today are much more concerned about the possibility of war than in pre-nuclear South Asia. Also, Indian and Pakistani leaders worked *together* to defuse the crisis, suggesting, again, that neither side wanted it to escalate to the nuclear level. Indian and Pakistani military leaders were reportedly in close touch during the crisis via a "hot line," and the Indian army went so far as to praise Pakistani officials for "taking proper steps to prevent the JKLF activists from crossing the line of control."[107] At a terse news conference after the march had been halted, Pakistani prime minister Nawaz Sharif was blunt about his government's motivation for preventing the Muslim militants from crossing into India: "We've had three wars with India. We don't want to have a fourth war."[108]

These admittedly impressionistic accounts indicate the possibility of a new pattern of decision making behavior in South Asia today that is more circumspect about allowing crises to spin out of control. I believe it is plausible that the existence of nuclear weapons is the primary cause of this new behavior. Based on the evidence available to me now, I cannot show this conclusively. Nor can I rule out other factors that may also be deterring the two sides from war over the Kashmir problem. These might include: (1) conventional deterrence, that is, the fact that both sides have much more advanced conventional weaponry at their disposal than during previous wars, and that both sides would thus suffer a far higher level of casualties and destruction than before; and (2) the possibility that India and Pakistan have learned from past mistakes and are better able now to avoid military conflict. These questions await further research. In the meantime, it is certainly plausible that existential nuclear deterrence is dampening tendencies toward war in South Asia.

Notes

For their helpful comments, I would like to thank the participants in a conference on Social Science Theory and Nuclear Weapons Proliferation, Stanford University, 5–6 February 1993, especially Daniel Deudney, Ben Frankel, Avery Goldstein, Neil Joeck, and Peter Lavoy. Special thanks also to Jim Hentz and Rey Koslowski for their thoughtful suggestions on earlier drafts.

1. "The Next Nuclear War," *Washington Post*, Editorial, 13 October 1989, p. A13.
2. For a comprehensive analysis of the causes of war in South Asia, see Sumit Ganguly, *The Origins of War in South Asia: Indo-Pakistani Conflicts Since 1947* (Boulder, Colo.: Westview, 1986); for two accounts of the recent insurgency in Kashmir, see Sumit Ganguly, "Avoiding War in Kashmir," *Foreign Affairs* 74, no. 1 (Winter 1990/91): 57–73; and Ashutosh Varshney, "India, Pakistan, and Kashmir: Antinomies of Nationalism," *Asian Survey* 31, no. 11 (November 1991): 997–1019.
3. For succinct accounts of the development of the Pakistani and Indian nuclear weapon programs, see David Albright and Mark Hibbs, "Pakistan's Bomb: Out of the Closet," *Bulletin of the Atomic Scientists* (July/August 1992): 38–43; and David Albright and Mark Hibbs, "India's Silent Bomb," *Bulletin of the Atomic Scientists* (September 1992): 27–31.
4. Leonard S. Spector, *Nuclear Ambitions: The Spread of Nuclear Weapons, 1989–90* (Boulder, Colo.: Westview, 1990).
5. R. Jeffrey Smith, "Pakistan Can Build One Nuclear Device, Foreign Official Says," *Washington Post*, 7 February 1992, p. A18; Paul Lewis, "Pakistan Tells of Its A-Bomb Capacity," *New York Times*, 8 February 1992, p. 5.
6. Aaron Karp, "Ballistic Missile Proliferation," *SIPRI Yearbook 1991: World Armaments and Disarmament* (Stockholm: Stockholm International Peace Research Institute, 1991): 317–43.
7. The following sources discuss the nuclear "state of play" in South Asia: Brahma Chellaney, "South Asia's Passage to Nuclear Power," *International Security* 16, no. 1 (Summer 1991): 43–72; Stephen Philip Cohen, "Nuclear Neighbors," in

Stephen Philip Cohen, ed., *Nuclear Proliferation in South Asia: The Prospects for Arms Control* (Boulder, Colo.: Westview, 1991): 1–22; Spector, *Nuclear Ambitions*, 63–112.

8. Leonard S. Spector, "India-Pakistan War: It Could Be Nuclear," *New York Times*, 7 June 1990, p. A23.
9. Comprehensive statements of this majority position span the decades: Kathleen C. Bailey, *Doomsday Weapons in the Hands of Many: The Arms Control Challenge of the '90s* (Urbana: University of Illinois Press, 1991); Lewis A. Dunn, *Controlling the Bomb* (New Haven: Yale University Press, 1982); Ted Greenwood, Harold Feiveson, and Theodore B. Taylor, *Nuclear Proliferation: Motivations, Capabilities, and Strategies for Control* (New York: McGraw-Hill, 1977); Leonard Beaton and John Maddox, *The Spread of Nuclear Weapons* (New York: Praeger, 1962).
10. The seminal statement of this minority position is Kenneth N. Waltz, *The Spread of Nuclear Weapons: More May Be Better*, Adelphi Paper no. 171 (London: International Institute for Strategic Studies, 1981). Major works that support *parts* of Waltz's argument include: Bruce Bueno de Mequita and William H. Riker, "An Assessment of the Merits of Selective Nuclear Proliferation," *Journal of Conflict Resolution* 26, no. 2 (June 1982): 283–306; Steven J. Rosen, "A Stable System of Mutual Nuclear Deterrence in the Middle East," *American Political Science Review* 71, no. 4 (December 1977): 1367–83; John J. Weltman, "Managing Nuclear Multipolarity," *International Security* 6, no. 3 (Winter 1981/82): 182–94; John J. Weltman, "Nuclear Devolution and World Order," *World Politics* 32, no. 2 (January 1980): 169–93.
11. For several variations on this theme, see John Lewis Gaddis, "The Long Peace: Elements of Stability in the Postwar International System," *International Security* 10, no. 4 (Spring, 1986): 99–142; Robert Jervis, *The Meaning of the Nuclear Revolution* (Ithaca: Cornell University Press, 1989); Kenneth N. Waltz, "Nuclear Myths and Political Realities," *American Political Science Review* 84, no. 3 (September 1990): 731–45.
12. For a possible South Asian nuclear scenario, see Hamish McDonald, "Destroyer of Worlds," *Far Eastern Economic Review*, 30 April 1992, p. 23.
13. The term was first introduced by Frankel in Benjamin Frankel, "Notes on the Nuclear Underground," *The National Interest*, no. 9 (Fall 1987). It was elaborated in Avner Cohen and Benjamin Frankel, "Opaque Nuclear Proliferation," in Benjamin Frankel, ed. "Opaque Nuclear Proliferation: Methodological and Policy Implications," special issue of *Journal of Strategic Studies* 13, no. 3 (September 1990). Other essays in the collection discuss different aspects of the phenomenon.
14. See, for example, John Kerry King, ed., *International Political Effects of the Spread of Nuclear Weapons* (Washington, D.C.: U.S. Government Printing Office, 1979).
15. Bruce D. Berkowitz, "Proliferation, Deterrence, and the Likelihood of Nuclear War," *Journal of Conflict Resolution* 29, no. 1 (March 1985): 112–36; Joseph S. Nye, "Maintaining a Nonproliferation Regime," *International Organization* 35, no. 1 (Winter 1981): 33; Weltman, "Nuclear Devolution," 189.
16. Neil Joeck, "Tacit Bargaining and Stable Proliferation in South Asia," *Journal of Strategic Studies* 13, no. 3 (September 1990): 77–91.
17. Cohen and Frankel, "Opaque Nuclear Proliferation," 31–32.
18. See McGeorge Bundy, "Existential Deterrence and its Consequences," in Douglas MacLean, ed., *The Security Gamble: Deterrence Dilemmas in the Nuclear Age* (Totowa, N.J.: Rowman & Allanheld, 1984): 3–13.
19. Marc Trachtenberg, "The Influence of Nuclear Weapons in the Cuban Missile Crisis," *International Security* 10, no. 1 (Summer 1985): 139.

20. Existential deterrence is also a factor at each of the other levels in this hierarchy. At higher levels of the hierarchy it is bolstered by other forms of deterrence, for example, mutual assured destruction, which are based in part on relative capabilities. At the opaque proliferation level, existential deterrence is the *only* logical form of mutual deterrence, since relative capabilities are not a factor in the deterrence equation.
21. Neil Joeck, "Pakistani Security and Nuclear Proliferation in South Asia," in Neil Joeck, ed., *Strategic Consequences of Nuclear Proliferation in South Asia* (London: Frank Cass, 1986): 86–87.
22. For detailed accounts of the Indian and Pakistani nuclear weapon programs, see Ashok Kapur, *India's Nuclear Option* (New York: Praeger, 1976); and Ashok Kapur, *Pakistan's Nuclear Development* (London: Croom Helm, 1987).
23. For an excellent overview of the positions taken in this debate, see Bhabani Sen Gupta, *Nuclear Weapons? Policy Options for India* (New Delhi: Sage, 1983).
24. Onkar Marwah, "India's Nuclear and Space Programs: Intent and Policy," *International Security* 2, no. 2 (Fall 1977): 98.
25. Raju G. C. Thomas, "India's Nuclear and Space Programs: Defense or Development?" *World Politics* 38, no. 2 (January 1986): 321.
26. Lorne J. Kavic, *India's Quest for Security: Defense Policies, 1947–1965* (Berkeley: University of California Press, 1967): 27–28, n. 19.
27. Mitchell Reiss, *Without the Bomb: The Politics of Nuclear Nonproliferation* (New York: Columbia University Press, 1988): 211.
28. Spector, *Nuclear Ambitions*, 64.
29. Kapur, *India's Nuclear Option*, 125.
30. This general discussion of India's motives in 1974 is derived from: Reiss, *Without the Bomb*; Spector, *Nuclear Ambitions*; and Raju G. C. Thomas, *Indian Security Policy* (Princeton: Princeton University Press, 1986).
31. The preceding two paragraphs are based on Kapur, *Pakistan's Nuclear Development*; S. P. Seth, "The Indo-Pak Nuclear Duet and the United States," *Asian Survey* 28, no. 7 (July 1988): 711–28; Gerard C. Smith and Helena Cobban, "A Blind Eye to Nuclear Proliferation," *Foreign Affairs* 68, no. 3 (Summer 1989): 53–70; Hedrick Smith, "A Bomb Ticks in Pakistan," *New York Times Magazine*, 6 March 1988, pp. 38, 76–77, 95, 104, 106; Spector, *Nuclear Ambitions*, 90; and Joeck, "Pakistani Security and Nuclear Proliferation in South Asia."
32. See, for example, Mohammed Ayoob, "India in South Asia: The Quest for Regional Predominance," *World Policy Journal* 7, no. 1 (Winter 1989–90): 107–33; and Chellaney, "South Asia's Passage to Nuclear Power," 43–72.
33. Richard P. Cronin, "Prospects for Nuclear Proliferation in South Asia," *Middle East Journal* 37, no. 4 (Autumn 1983): 606.
34. Spector, *Nuclear Ambitions*, 66; see also Sen Gupta, *Nuclear Weapons?*, 39.
35. "Nuclear Neighbors," 6.
36. Spector, *Nuclear Ambitions*; and Spector, "Nuclear Proliferation in the 1990s: The Storm After the Lull," in Aspen Strategy Group, *New Threats: Responding to the Proliferation of Nuclear, Chemical, and Delivery Capabilities in the Third World* (Lanham, Md.: University Press of America, 1990).
37. Ross H. Munro, "Knocking at the Nuclear Door," *Time*, 30 March 1987, pp. 42–43.
38. McDonald, "Destroyer of Worlds," 24.
39. Spector, *Nuclear Ambitions*; Spector, "Nuclear Proliferation in the 1990s."
40. R. N. Rosecrance, "Introduction," in R.N. Rosecrance, ed., *The Dispersion of Nuclear Weapons: Strategy and Politics* (New York: Columbia University Press, 1964): 22–23.
41. Albert Wohlstetter, "Nuclear Sharing: NATO and the N + 1 Country," in Rosecrance, ed., *The Dispersion of Nuclear Weapons*, 186, 204.

42. Ernest Lefever, *Nuclear Arms in the Third World* (Washington, D.C.: Brookings, 1979).
43. Dunn, *Controlling the Bomb*, 69.
44. David C. Gompert, "On the Choice of a Nuclear Future," in David C. Gompert, et al., *Nuclear Weapons and World Politics: Alternatives for the Future* (New York: McGraw–Hill, 1977): 318; in the same volume, see also: Michael Mandelbaum, "International Stability and Nuclear Order: The First Nuclear Regime," 67–68. On the same theme, see: Dunn, *Controlling the Bomb*, 75; Michael D. Intriligator and Dagobert L. Brito, "Nuclear Proliferation and Stability," *Journal of Peace Science* 3, no. 2 (Fall 1978): 380–81; and Joseph S. Nye, Jr., "Comments," on pt. 4, "Managing Proliferation," in Dagobert L. Brito, Michael D. Intriligator, and Adele E. Wick, eds., *Strategies for Managing Nuclear Proliferation* (Lexington, Mass.: Lexington Books, 1983): 281.
45. Dunn, *Controlling the Bomb*, 69–70. On this same theme, see: Ted Greenwood, "Discouraging Proliferation in the Next Decade and Beyond," in Greenwood, Feiveson, and Taylor, *Nuclear Proliferation*, 28; Intriligator and Brito, "Nuclear Proliferation and Stability," 361; and Beaton and Maddox, *Spread of Nuclear Weapons*, 202.
46. Mandelbaum, "International Stability and Nuclear Order," 66.
47. Steve Fetter, "Ballistic Missiles and Weapons of Mass Destruction: What is the Threat? What Should Be Done?" *International Security* 16, no. 1 (Summer 1991): 60.
48. On this point, see Rosecrance, "Introduction," 20–21; and Gompert, "Nuclear Future," 233.
49. Henry A. Kissinger, *Nuclear Weapons and Foreign Policy* (New York: Harper and Brothers, 1957), 20.
50. Mandelbaum, "International Stability," 66.
51. Dunn, *Controlling the Bomb*, 71.
52. Greenwood, "Discouraging Proliferation," 29.
53. Robert Art, "A Defensible Defense: America's Grand Strategy After the Cold War," *International Security* 15, no. 4 (Spring 1991): 9.
54. See the discussion in Les Aspin, "From Deterrence to Denuking: Dealing With Proliferation in the 1990s." Memorandum, U.S. House of Representatives, Committee on Armed Services, 18 February 1992.
55. George H. Quester, "Nuclear Proliferation and Stability," in Brito, Intriligator, and Wick, *Managing Nuclear Proliferation*, 115.
56. Bailey, *Doomsday Weapons*, 2.
57. Dunn, *Controlling the Bomb*, 72.
58. Ibid., 88, 74, 76.
59. Thomas C. Schelling, *The Strategy of Conflict* (New York: Oxford University Press): 207–29.
60. Fetter, "Ballistic Missiles," 29.
61. Michael D. Intriligator and Dagobert L. Brito, "A Possible Future for the Arms Race," in Nils Petter Gleditsch and Olav Njolstad, eds., *Arms Races: Technological and Political Dynamics* (London: Sage, for the International Peace Research Institute, Oslo, 1990): 381.
62. Lewis A. Dunn, "Nuclear Proliferation and World Politics," *Annals of the American Academy of Political and Social Science*, no. 430 (March 1977): 98–99; Dunn, *Controlling the Bomb*, 73.
63. For exceptions, see Walter B. Wentz, *Nuclear Proliferation* (Washington, D.C.: Public Affairs Press, 1968); and Frederick C. Thayer, "Proliferation and the Future: Destruction or Transformation? *Annals of the American Academy of Political and Social Science*, no. 430 (March 1977): 133–46.
64. Waltz, *More May Be Better*, 1, 3; Kenneth N. Waltz, "What Will Nuclear Weapons Do to the World?" in King, *Political Effects*, 165.

65. Bueno de Mesquita and Riker, "Selective Nuclear Proliferation," 290–91.
66. Weltman, "Managing Nuclear Multipolarity," 190.
67. Steven J. Rosen, "Nuclearization and Stability in the Middle East," *Jerusalem Journal of International Relations* 1, no. 3 (Spring 1976): 2; see also Rosen, "Arab-Israeli Conflict."
68. Waltz, "More May Be Better," 5–6.
69. Ibid., 11.
70. Weltman, "Nuclear Devolution and World Order," 189–90.
71. Weltman, "Managing Nuclear Multipolarity," 191. For an interesting discussion of whether or not so-called Third World "crazies" can be deterred, see David Ignatius, "Madman's Bluff: Why Deterrence Still Works," *Washington Post*, 10 May 1992, p. c1.
72. Waltz, *More May Be Better*, 11–12.
73. Ibid., 16.
74. Weltman, "Managing Nuclear Multipolarity," 190.
75. Art, "Defensible Defense," 27; Weltman, "Managing Nuclear Multipolarity," 187.
76. Waltz, "More May Be Better," 10–11.
77. Ibid., 16.
78. Beaton and Maddox, *The Spread of Nuclear Weapons*.
79. Estimates about the magnitude of future nuclear proliferation have been routinely inflated. Lewis A. Dunn and Herman Kahn wrote about the "emergence of a world of as many as forty nuclear weapon states by 1995." *Trends in Nuclear Proliferation, 1975–1995: Projections, Problems, and Policy Options* (Croton-on-Hudson, N.Y.: Hudson Institute, 1976): 132. In 1977, David C. Gompert described a 1990 world of twenty-five de facto nuclear weapon states. "On the Choice of a Nuclear Future," in Gompert, et al., *Nuclear Weapons and World Politics*, 231. Even conservative estimates have proved to be high: Kenneth N. Waltz based his 1981 analysis on a future world of 10–18 nuclear powers. *The Spread of Nuclear Weapons*, 1. Since then, Pakistan and South Africa have crossed the line into de facto nuclear weapon status, but South Africa has apparently backed away from the threshold. In addition, three of the ex-Soviet republics' exact nuclear weapon status is uncertain. Thus the number of nuclear weapon states today is somewhere between eight and eleven, even with the "instant proliferation" caused by the break-up of the Soviet Union.
80. "Proliferation, Deterrence, and the Likelihood of Nuclear War," 113.
81. Nye, "Maintaining a Nonproliferation Regime," 33.
82. Weltman, "Nuclear Devolution and World Order," 189.
83. Cohen and Frankel, "Opaque Nuclear Proliferation," 14, 21–23.
84. Ibid., 31–32.
85. Bundy writes:

> There is an enormous gulf between what political leaders really think about nuclear weapons and what is assumed in complex calculations of relative 'advantage' in simulated strategic warfare. Think-tank analysts can set levels of 'acceptable' damage well up in the tens of millions of lives. They can assume that the loss of dozens of great cities is somehow a real choice for sane men. They are in an unreal world. In the real world of real political leaders – whether here or in the Soviet Union – a decision that would bring even one hydrogen bomb on one city of one's own country would be recognized in advance as a catastrophic blunder …
>
> … it is one thing for military men to maintain our deterrent force with vigilant skill, and it is quite another for anyone to assume that their necessary contingency plans have any serious interest for political leaders. The object of political men – quite rightly – is that these weapons should never be used. I

> have watched two Presidents working on strategic contingency plans, and what interested them most was simply to make sure that none of these awful events would occur.

McGeorge Bundy, "To Cap the Volcano," *Foreign Affairs* 48, no. 1 (October 1969): 9–12.

86. Trachtenberg, "Nuclear Weapons and the Cuban Missile Crisis," 148.
87. Avery Goldstein, "Deterrence and Security in a Changing World: Some Lessons from the Second-Ranking Powers," *Journal of Strategic Studies* 15, no. 4 (December 1992): 485.
88. *Strategy of Conflict*, 53.
89. Neil Joeck, "Tacit Bargaining and Stable Proliferation in South Asia," 80–81.
90. Ibid., 83.
91. Weltman, "Nuclear Devolution and World Order," 179.
92. Remarks by CIA Director Robert Gates to the Nixon Library Conference, Washington, D.C., 12 March 1992.
93. Kathy Evans, "U.S. Nuclear Concerns Endanger Pakistan Aid," *Christian Science Monitor*, 4 October 1990, p. 4.
94. Prime Minister Nawaz Sharif, "Address to the National Defence College," Rawalpindi, Pakistan, 6 June 1991, pp. 7–10. See also, Edward A. Gargan, "The Chastened Pakistanis," *New York Times*, 19 February 1992.
95. McDonald, "Destroyer of Worlds," p. 23.
96. Steve Coll, "India and Pakistan Seem to Seek Nuclear Deterrence to Keep Peace," *Philadelphia Inquirer*, 2 January 1991, p. 3-A.
97. "Pakistan Stands Firm on Nuclear Threshold," *Washington Times*, 11 February 1992, p. A2.
98. Cohen, "Nuclear Neighbors," 12.
99. Chellaney, "South Asia's Passage to Nuclear Power," 68–69.
100. Cohen, "Nuclear Neighbors," 12.
101. See Ganguly, "Avoiding War in Kashmir," and Varshney, "Antinomies of Nationalism."
102. James Adams, "Pakistan 'Nuclear War Threat'," *Sunday Times* (London), 27 May 1990.
103. McDonald, "Destroyer of Worlds," 24.
104. Charles Smith, "Atomic Absurdity," *Far Eastern Economic Review*, 30 April 1992, p. 25.
105. Adams, "Pakistan 'Nuclear War Threat'."
106. "'Suicide March' in Kashmir Leaves 12 Dead," *Toronto Star*, 13 February 1992, p. A3.
107. Mushahid Hussain, "South Asia: Back from the Brink of War in Kashmir," *Inter Press Service*, 14 February 1992.
108. "Pakistan Fires on Kashmiri Marchers; 12 Die," *Chicago Tribune*, 13 February 1992, p. C2.

III
Formal Models

The Economic and Political Incentives to Acquire Nuclear Weapons

Dagobert L. Brito and Michael D. Intriligator

New Risks of Nuclear Proliferation

The problem of nuclear proliferation has come back to haunt the world. It is an old problem, dating back to the first nuclear weapon in 1945, and it has resurfaced with each subsequent new nuclear power: the Soviet Union in 1949, the United Kingdom in 1952, France in 1962, China in 1964, and India in 1974. However, the new problem of proliferation in the post-cold war world is different in several respects from the old one, making it much more important.[1] This new problem of proliferation can be interpreted in terms of the basic economic factors of demand and supply. On the demand side, the political or economic value to many nations of possessing nuclear weapons has increased due to changes in the world situation, while on the supply side, technological and political changes have reduced the cost of acquiring such weapons.

Much of microeconomic theory is devoted to the calculus of the efficient allocation of scarce resources subject to technological constraints. The power of economic reasoning lies in its being able to predict the implications of changes in exogenous variables. This paper uses economic theory to analyze the economic and political incentives for a nation to acquire nuclear weapons and to consider the effects of recent changes in the world situation on nuclear proliferation and their implications for policy.

A major development that has changed both the value and the cost of acquiring nuclear weapons is the dissolution of the Soviet Union. The emergence of Soviet successor nations has left strategic nuclear weapons on the territory of four economically and politically unstable states and has increased the chance of accidental or inadvertent use of these weapons.[2] While Russia controls over 70 percent of the Soviet strategic nuclear weapons and all of its tactical nuclear weapons, there are large numbers of strategic nuclear weapons, including over one thousand warheads, in both Ukraine and Kazakhstan and a small number in Belarus. In fact, the number of nuclear warheads in either Ukraine or Kazakhstan exceeds those in the United Kingdom and France combined. Although Russia has stated its intention to keep its nuclear weapons and the other three former republics with nuclear weapons have agreed eventually to give up these weapons, they do not seem interested in giving them up quickly. In the interim, if they succeed in gaining technical control over the use of these weapons, then the breakup of the Soviet Union would have led to the creation of three major nuclear powers.

Another major development changing the value of acquiring nuclear weapons has been the end of the bipolar order, which brought to an end the stability and predictability that bipolarity had imposed on international affairs.[3] Now the situation is more precarious, suggesting the possibility of new threats to security. Economic theory suggests that this additional uncertainty would lead nations to attempt to reduce the risk, making the acquisition of independent nuclear capabilities more attractive. Adding to this uncertainty is covert (or "opaque") nuclear proliferation, which appears to be the approach of choice for nuclear weapons development in several nations.[4] Such nations would be following the Israeli rather than the U.S. model for nuclear weapons development. The traditional U.S. model of secret development, test, and announcement was followed by the Soviet Union, United Kingdom, France, China, and India. More recently, nations have tended to follow the Israeli model, with secret development, no test, and a policy of continued ambiguity and denial. This model appears to have been adopted by Pakistan and perhaps other nations.[5] There is also the example of Iraq, a member in good standing of the Nuclear Non-Proliferation Treaty (NPT), subject to regular inspections of the International Atomic Energy Agency (IAEA), but which, nevertheless, as shown in the aftermath of the 1991 Gulf War, was trying to build nuclear weapons covertly.

At the same time that the international order is making the acquisition of nuclear weapons more attractive from the viewpoint of the demand side, there are also political and economic forces that are making these weapons less expensive to acquire from the viewpoint of the supply side. First, the breakup of the Soviet Union has created the possibility that part of its enormous stockpile of fissile material could be diverted, and there are persistent rumors of fissile material from the former Soviet Union reaching the black market and moving to other nations.[6] In this regard, it should be recalled that substantial amounts of Western fissile material are unaccounted for, with perhaps some having been diverted to Israel, despite various national and international controls on such material. Second, today there is a greater reliance on plutonium in the production of energy in nuclear power plants, especially in breeder reactors. As a result, there has been a move to a global plutonium economy, with its increased transportation in international commerce, including large planned shipments to Japan from France to be used in a breeder reactor. Plutonium can be used to manufacture nuclear weapons, so its greater use nationally and increased shipments internationally make it easier for a potential nuclear power to acquire a key ingredient in nuclear weapons development. Third, the breakup of the Soviet Union has created a pool of scientists, engineers, and technical workers who could assist nations trying to build nuclear weapons. Fourth, technology is evolving in a fashion such that much technology which has legitimate civilian uses can also be used to manufacture nuclear weapons, and there is substantial international trade in such dual-purpose systems.[7] Exporters today are typically no longer motivated by military or diplomatic considerations but rather by the desire to earn foreign exchange and to keep production units operating in the absence of domestic sales, especially in the light of the difficulties of converting enterprises and workers from military to civilian production.[8]

This paper concludes that, as a result of these recent political and economic developments there has been an increase in the value and a decrease in the cost of acquiring nuclear weapons, making proliferation a more attractive option for many nations and presenting a major threat to global security.[9] It ends by suggesting several policy initiatives to reduce this threat, which would either decrease the political or economic value of nuclear weapons or increase their cost.

The Role of Nuclear Weapons

One of the major difficulties in using economic analysis in studying nuclear weapons is that there is no well defined or commonly accepted technology of warfare in which nuclear weapons play a role. If viewed from a strictly military viewpoint, nuclear weapons may be an anomaly, doomed to extinction by advances in technology, like crossbows and mounted knights. To put this argument in historical perspective, some have argued that the doctrine of strategic bombing evolved to justify the existence of an independent Royal Air Force.[10] The doctrine said that air power could be used as a means of direct attack on the enemy state with the object of depriving it of the means or will to continue the war.[11] Attempts to use aircraft as strategic weapons in the Second World War were expensive and not very effective. In April 1941, the theoretical error of drop was assumed to be 1,000 yards, but the Butt report in 1941 stated that during raids on the Ruhr only one-tenth of the bombs fell within five miles of their targets. In May 1942, 1,046 bombers were used to destroy 600 acres of the city of Cologne, but at the cost of forty bombers. In June 1942, 904 bombers attacked Bremen inflicting little damage and losing five percent of the bomber force. With loss rates like this, a bomber crew had one chance in eight of surviving forty missions.[12] Finally, in 1943, 200,000 tons of bombs were dropped on Germany; yet German production rose by fifty percent.[13]

It was not surprising, therefore, that in the context of the times the goal of bombing changed from destroying the enemy's ability to fight to destroying the enemy's morale. The result was the attacks on the populations of Hamburg, Dresden, and Tokyo. The decision to use the atomic bomb is still a subject of some controversy; however, it was in many ways a logical extension of existing practices.[14] The atomic bomb was seen as a means to destroy the enemy's cities, industries, and, especially, its will to fight. The bomb did what was already being done using conventional weapons by both sides in the war. In fact, more people were killed in the firebomb raids on Tokyo than by both atomic bombs. The special horrors and the threat to the human gene pool and the environment associated with radiation were not known or well understood at the time, as demonstrated by the exposure of American forces to radiation during postwar tests of nuclear weapons.

After the war, the scientific and military communities as well as moral and political philosophers pondered the implications of nuclear weapons. It became clear that nuclear weapons were not a very effective means for a nation to use in pursuit of its political and economic objectives, and it became necessary to develop complicated and sophisticated strategic doctrines so as to rationalize these weapons. Ultimately, it was seen that the role of nuclear weapons was primarily to deter their use by other nations.[15]

Although weapons designers eventually developed warheads that were "clean" and potentially useful on the battlefield as tactical nuclear weapons, the concept of a nuclear threshold that could not be crossed except at a high cost became ingrained in the strategic thinking of the mid-twentieth century. The nuclear threshold was so well defined and accepted that a small country like North Vietnam was able to inflict a political defeat on the United States without fear of the use of nuclear weapons. Nuclear weapons did not prove to be an effective instrument of policy, whether by the United States in Vietnam or by the Soviet Union in Afghanistan.

However, recent technological developments make it possible to use conventional warheads in a highly effective manner. In 1945 or even in 1965 the United States did not have the ability to cripple a country by precise nonnuclear bombing, but by 1990 this technology was available, delivering warheads to within a short distance of the target. This capability, based on computer and microelectronic technology, was demonstrated during the 1991 Gulf War. The United States experience in this war may foreshadow a new era in which nonnuclear weapons using advanced technologies may become the dominant instruments of power, being both effective and economically and politically inexpensive.

The effectiveness of a weapon is proportional to the explosive power and is inversely proportional to the accuracy of the delivery system raised to the third power.[16] Thus, doubling the accuracy of a weapon system reduces the necessary explosive power by a factor of eight. The payload of a B-29 in the Second World War was twenty tons, with an accuracy of around 2,000 yards. A twenty kiloton atomic bomb increased the effectiveness of a B-29 by a

Figure 1

Iso-Effectiveness Curves of Weapon Accuracy

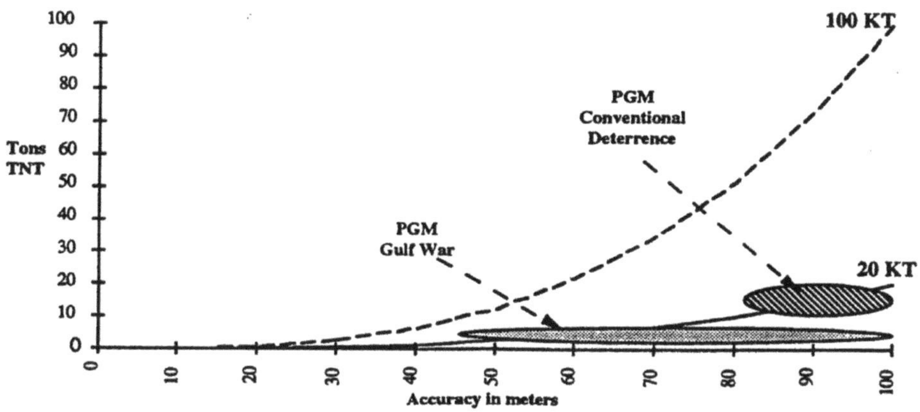

factor of about a thousand. If instead, the accuracy had been improved from 2,000 yards to 100 yards, then the effectiveness of the B-29 would have improved by a factor of eight thousand. The increase in the effectiveness of weapons that has resulted from precision guidance is eight to ten times greater than the increase that resulted from the introduction of nuclear weapons. Since precision guided weapons do not have drawbacks in terms of radiation and collateral damage associated with nuclear weapons, the role of nuclear weapons in future conflicts may be questioned.

Figure 1 illustrates the iso-effectiveness curves for a 100 kiloton bomb at 1,000 meters and a 20 kiloton bomb at 1,000 meters. The iso-effectiveness curve is analogous to the isoquant in production theory; it is the combination of inputs that yield the same output. Figure 1 shows that 100 tons at 100 meters is as effective as 100 kilotons at 1,000 meters. The shaded area is the reported feasible region for precision guided munitions in the Gulf War. It is clear that precision guided munitions can be used to accomplish tasks that would have formerly required nuclear weapons but without the collateral damage.

This is important because although general deterrence theory is premised on the proposition that outside powers will punish a nuclear aggressor, in practice such actions may prove to be very difficult for democracies. It would be very difficult to use nuclear weapons with potential collateral casualties in the tens of thousands to punish a military dictator like Saddam Hussein. One reason that massive bombing of population centers by the Allied Powers was feasible in the Second World War was that the enemy became identified as Germany and Japan. A distinction was not made between Hitler and the German people or Tojo and the Japanese people. The dilemma faced by the

United States and the European Community countries in deterring Serbian aggression in Bosina and Herzegovina suggest that the use of conventional forces against third countries may also be politically difficult unless there are clear national interests at stake. Guarantees offered to countries that do not acquire nuclear weapons may not be credible.

Developments in precision guided munitions may address the problem of nonnuclear deterrence. Modern economies are vulnerable to disruptions in fuel, transportation, and communications. Developing economies, inasmuch as they tend to be more centralized, are even more so. A tragic example was the near paralysis of communications in central Mexico after the 1985 earthquake destroyed the telephone exchange in Mexico City. Cruise missiles were used to attack targets in the Gulf War, but a more promising weapon may be a conventionally armed ballistic missile.

A conventionally armed ballistic missile that weighs 80,000 pounds and can carry a payload in the neighborhood of 12,000 to 18,000 pounds with a range of 1,000 miles with an accuracy of 100 meters or less is probably feasible.[17] (See Figure 1) The existence of such a weapon would reduce the temptation of nonnuclear nations to acquire nuclear weapons. Threats to use it would be credible. The nuclear weapons program of a country that was tempted to make political use of its nuclear weapons would always be at risk of a nonnuclear first strike, so not acquiring them could well be a more attractive and less expensive alternative.

Ballistic missiles also have substantial political and psychological value. Although the German Second World War rocket attacks on England did not cause as much actual damage as the Blitz, the blow to British morale was devastating. Being attacked without being able to respond may be psychologically cataclysmic. This is surely obvious to anyone who watched or experienced the Iraqi missile attacks on Jerusalem and Tel Aviv. Had it not been for the perceived effectiveness of the Patriot batteries in attacking the Iraqi Scuds, the political damage that these attacks would have inflicted on the coalition would have been much greater.

Nuclear Weapons in a Formal Model of Bargaining

This section presents a formal model of the use of nuclear weapons in bargaining between nations using the concept of the Nash bargaining problem.[18]

The "Nash bargaining problem" is the game-theoretic problem of how to divide a good between two players when either player can impose a threat unilaterally. Two nations, for example, may be concerned with certain economic rights in a model in which force can be used to redistribute these rights.[19] For example, they may be in contention over a territory that yields a

Figure 2
NASH BARGAINING PROBLEM

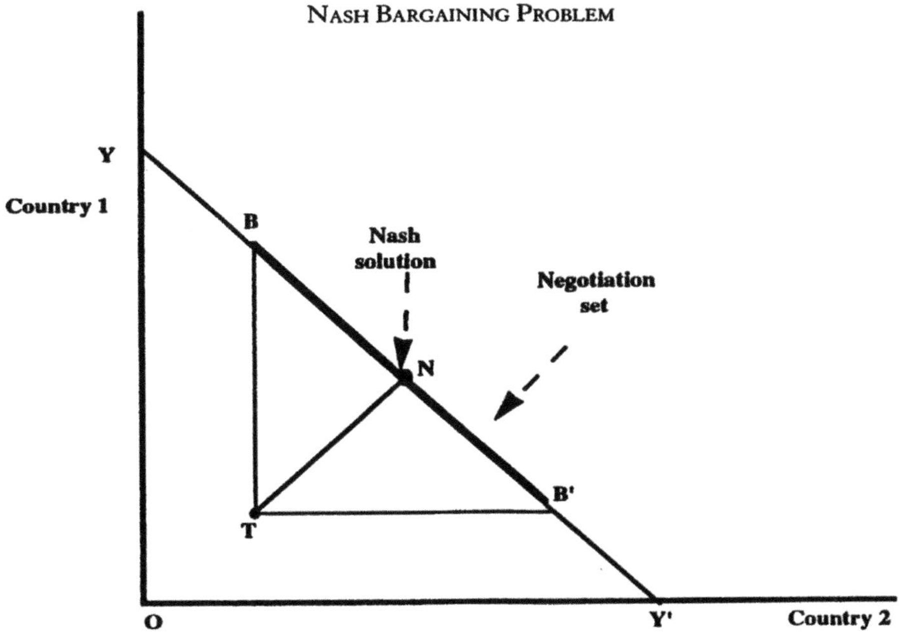

fungible resource such as oil (as in the case of the Iraq invasion of Kuwait) or the right to a market that yields income. Nations choose between consumption and investment in arms, and they can use force or the threat of force in their bargaining in order to reallocate resources. The expected outcome of a possible war depends on the weapons stocks each country possesses, which are determined by choices each has made earlier, during an arms race.[20] The weapons, whether conventional or nuclear, are used to define the "threat point," the point reached if the nations fail to reach an agreement over the division of the good. Here, the potential outcome of a possible war defines the threat point. The threat point generally influences the outcome of the negotiations between the nations over the division of the good in question.

Consider a situation in which the two nations, denoted 1 and 2, are bargaining over the rights to some resource and are behaving rationally and in a non-altruistic fashion. The total amount of the resource is given by Y, so all points that satisfy this constraint are given by the line YY' in Figure 1, the feasible set being the triangle YOY' in the figure. The threat point is an allocation in the interior of the feasible set, denoted by T in the figure, which can be imposed unilaterally by either party. Classical bargaining theory assumes *individual rationality*, in that a rational individual would not agree to an allocation that is dominated by the threat point. Thus, given the threat

Figure 3
CONVENTIONAL WEAPONS IN NASH BARGAINING

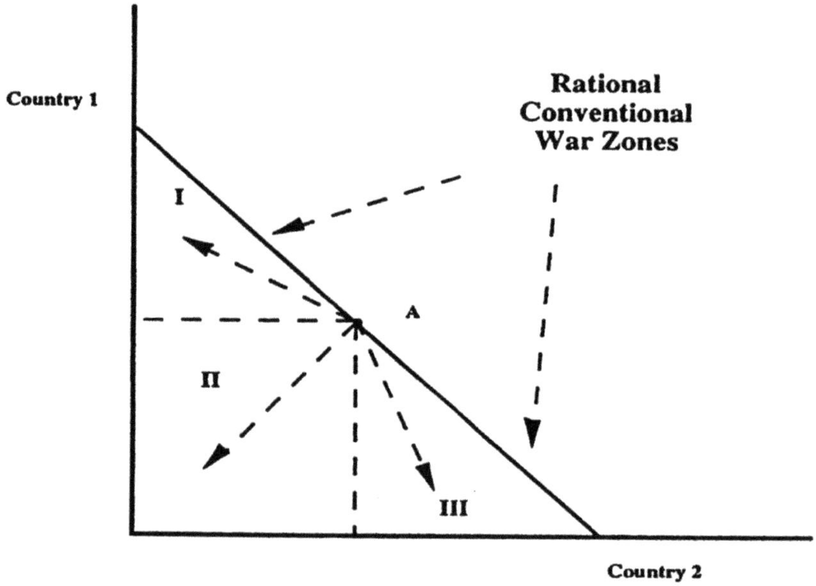

point T, the final allocation must occur in the triangle BTB' in the figure. Classical bargaining theory also assumes *joint rationality*, whereby two rational individuals would not agree to an allocation if there exists an alternative feasible outcome that will make them both better off. Since any outcome in the interior of the feasible set is Pareto dominated, the final outcome must lie on the boundary of the feasible set and consist of points that dominate the threat point. This is the line defined by BB', which represents the *negotiation set*. This solution of the bargaining game leaves open the question of why the threat point is ever implemented given that it is Pareto dominated by the negotiation set. This is a question that has not been completely solved, but one that is crucial to understanding the use of nuclear weapons in negotiations.

The "Nash solution" is an outcome for the bargaining problem defined by a set of reasonable axioms, and, under some additional technical assumptions, this Nash solution exists and is unique.[21] In the current context, the Nash solution is the midpoint of the negotiation set, denoted by the point N in Figure 2. While the analysis here uses the Nash solution, the results do not depend on any particular solution of the bargaining problem as long as the outcome is defined by the threat point and the negotiation set.

Consider the introduction of conventional weapons in this bargaining

Figure 4
CONVENTIONAL WEAPONS IN NEGOTIATION

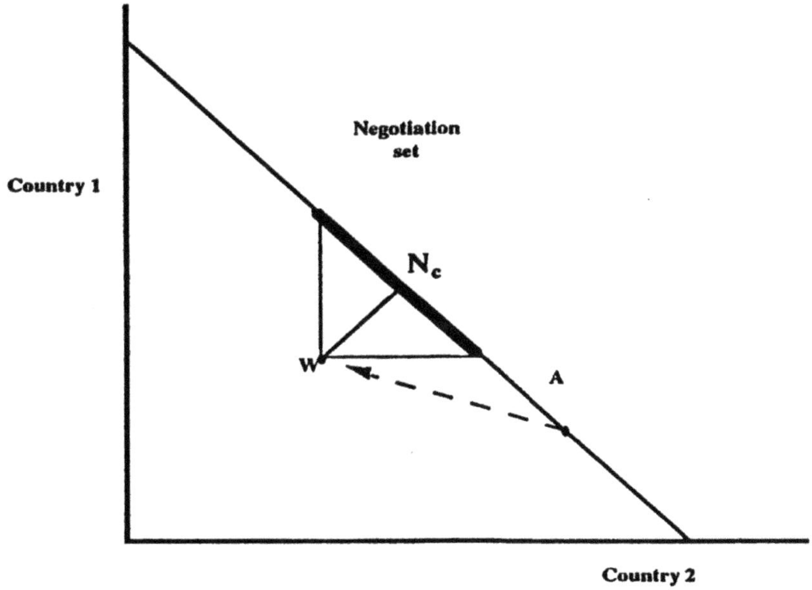

context. In a conflict over the distribution of resources, the use of conventional weapons would result in an outcome that has a net gain to the winner and a net loss to the loser, where the loss is greater than the gain. Thus, they can result in an allocation that is not zero sum, but where one party gains as in Regions I and III in Figure 3. Thus in Region I, country 1 could rationally initiate a war while in Region III, country 2 could rationally initiate a war. Whether a war would occur when the threat point is in Region II is a more complicated problem.

An example of the use of conventional weapons in negotiation is shown in Figure 4. The initial allocation is at the point A, showing the distribution to each country prior to the conflict. If country 1 has superiority in conventional weapons, then it has the capacity to force the reallocation given by point W, which is the expected outcome of a potential conventional war. The allocation given by W is inside the feasible set because the gains to country 1 are less than the losses to country 2. This net loss outcome of a potential conventional war reflects the fact that such a war is usually accompanied by destruction of infrastructure and loss of population.

There are two ways in which rights to the resource can be reallocated

Figure 5
STRATEGIC NUCLEAR WEAPONS IN NEGOTIATION

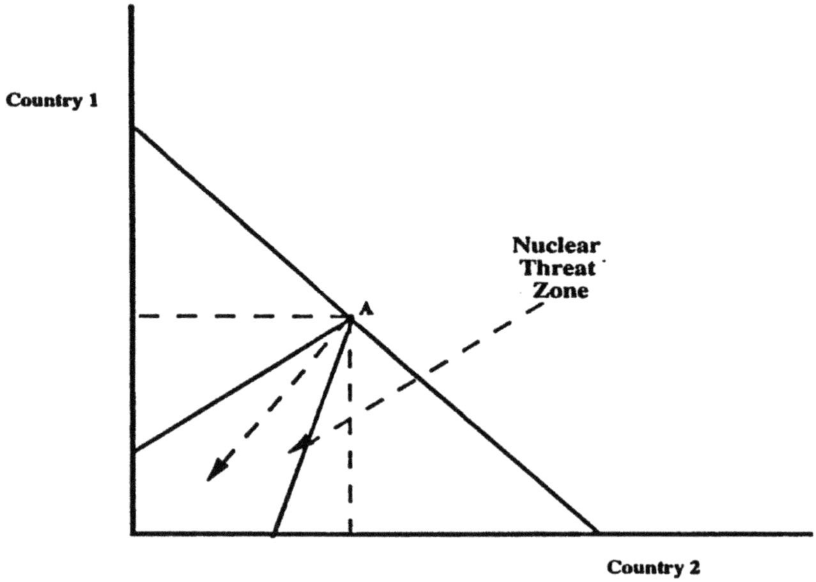

between the two nations: they can go to war or they can bargain. Point w becomes the threat point in the Nash bargaining game if they bargain over the allocation of the resource in the presence of conventional weapons. Since points in the bargaining set dominate the war outcome, both sides, if they are rational, may very well decide to choose to negotiate a solution in the bargaining set rather than fight a war. However, if they choose not to bargain, or if they do not agree, then the stronger party can impose the allocation it prefers.

Now consider in this context the introduction of strategic nuclear weapons. By their nature, such weapons can be used to impose a substantial cost on the other side, but not to redistribute by force, as in the case of conventional weapons. This is, of course, a simplification in that one country can use nuclear weapons to destroy the conventional forces of the other country and be able to limit externalities (damages to itself) so that it may be better off after the use of nuclear weapons. Nevertheless, this simplification is a good first-order approximation. It is unlikely that any new nuclear power will have nuclear weapons in sufficient number and sophistication that they will be able to use them effectively in combination with their other forces. The only exception may be Israel.

Figure 6
UNEQUAL NUCLEAR BARGAINING

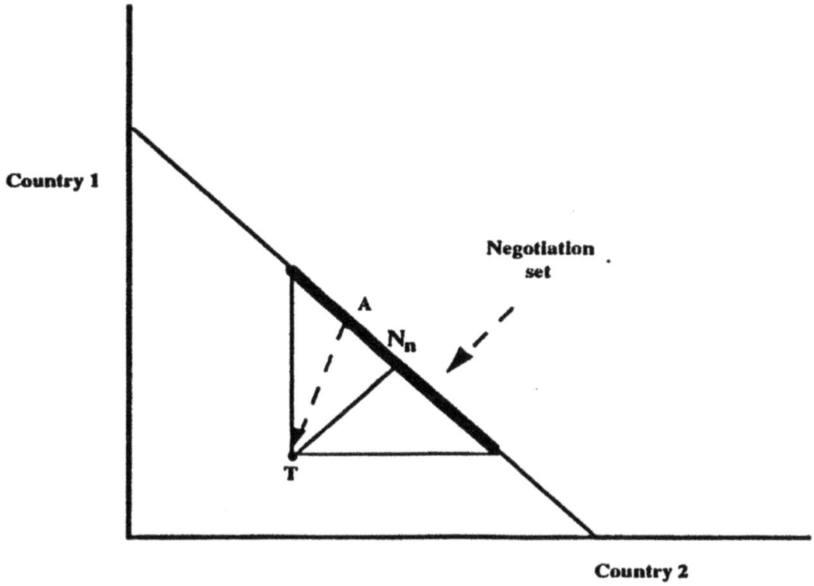

Figure 6 represents a situation in which country 2 has superiority in nuclear weapons over country 1. Using these weapons, country 2 can impose a larger cost on country 1 than country 1 can impose on it, but country 2 cannot use these weapons to appropriate resources directly from country 1. The threat point at T represents a loss to country 1 without a corresponding gain to country 2. In fact, both nations lose at T. However, if T is the threat point in a bargaining game then it defines a negotiation set which contains the initial allocation, point A. If the threat point defines a bargaining game, then the Nash solution, given by the point N_n, will result in a gain to country 2.

The fundamental question is whether a threat on the part of country 2 to inflict a cost on country 1, even at a cost to itself, is credible."[22] Without a credible threat it is hard to imagine that nuclear weapons could be used in an explicit fashion to impose a redistribution over the rights to the resource. Bargaining with threats is really a two-stage process. The first stage is explicit bargaining. The second stage, that in which the threat point will be implemented should the bargaining process fail, is implicit. The logical difficulty is that such games are usually not *subgame perfect*.[23] In the bargaining process it may be optimal to threaten to use nuclear weapons should a player's demands not be met. However, it is unlikely that the player

Figure 7
NUCLEAR WEAPONS AND CONVENTIONAL FORCES

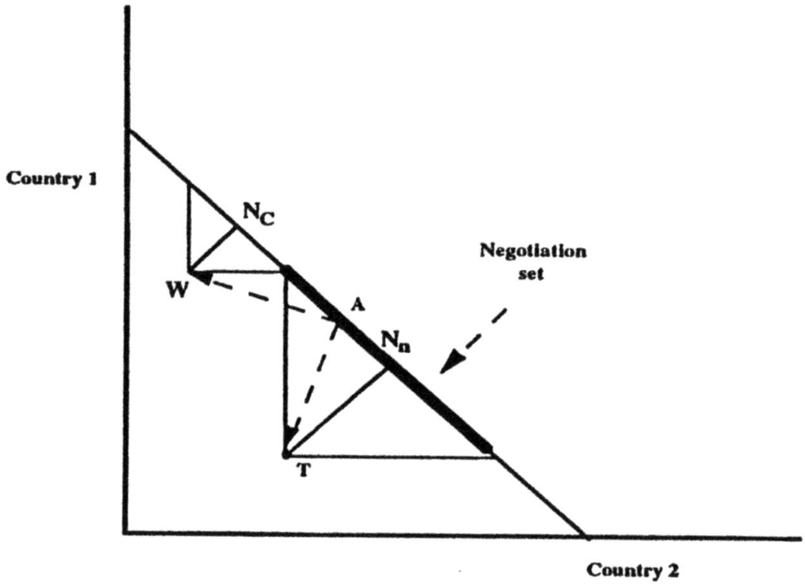

will want to implement its threats. One way to address this problem is through a commitment mechanism. If a player can restrict its freedom of action so that it has no alternative but to implement the threat, then the threat is credible.

Nuclear weapons can be used to deter or to limit the use of conventional forces to reallocate wealth. In Figure 7, country 1 has superiority in conventional forces and can force the allocation W if nuclear weapons were not used. However, country 2 can force the point T using nuclear weapons, so the Nash point associated with W, given by N_c, is not in the negotiation set. If country 2 can precommit to use nuclear weapons in the event of an attack then country 1 will be deterred from attacking. Precommitment mechanisms can be subtle. One example would be to transfer control of nuclear weapons to battlefield commanders. This creates the possibility that in the "fog of war" the weapons will be used. The aggressor is forced to try to evaluate the probability that the other side will use its nuclear weapons. Given the enormous damage nuclear weapons can inflict, their expected loss can be large even though the subjective probability of their use is small.

Since the Nash point N_n requires that the nuclear threat be credible, the status quo is likely to be stable. Nuclear weapons put the burden of accepting the risk of the threat being implemented on the party that wants to change the status quo. The potential use of nuclear weapons can also change the probability distribution of potential gains and losses from conventional war so as to reduce its expected gain. In fact, a very small probability of enormous loss due to the use of nuclear weapons can change the expected value of a conventional war from positive to negative. As a result, the presence of nuclear weapons can function so as to preserve the status quo.[24]

This model may characterize both the earlier East-West relationships and the Arab-Israeli conflict. In the East-West case, the presence of nuclear weapons may have limited the military initiatives of both sides during the cold war. For example, the danger of a nuclear war with the Soviet Union may have inhibited the use of force in Cuba and Vietnam, while the American nuclear umbrella may have inhibited the Soviet use of force in Europe despite Soviet superiority in conventional forces. More generally, the fact that the United States and the Soviet Union never engaged in a direct conventional war despite global rivalry, recurrent crises, and substantial military machines may be attributed, at least in part, to the presence of nuclear weapons.

The analysis of the Arab-Israeli conflict using the model is more complicated.[25] Israel has conventional as well as nuclear superiority in the region. Given the population and wealth of the Arab world, however, the Arab states have the potential of acquiring superiority in conventional forces. The model suggests that Israeli nuclear capability has neutralized this potential, offsetting any possible Arab superiority in conventional weapons. An interpretation of the Camp David agreement that is consistent with this analysis suggests that the agreement may have stemmed, in part, from the recognition of the implications of Israeli access to nuclear weapons by the Egyptians under President Anwar Sadat.[26]

An alternative motivation for the acquisition of nuclear weapons is as a means to validate conventional superiority, thereby also preserving the status quo. In Figure 8 country 2 has both conventional and nuclear superiority. It can use the threat of conventional weapons to achieve the allocation given by N_C and use its nuclear weapons to validate this reallocation. This threat would deter third parties from intervening in the regional conflict. Suppose, for example, that Iraq had been a nuclear power in the summer of 1990. Having taken Kuwait, Iraq would have forced the allies to choose between accepting the new allocation or risking that its threat of using nuclear weapons would be implemented.

The use of nuclear weapons in strategic bargaining is difficult because it may not be rational to carry out the threat and use these weapons. This difficulty does not occur in what may be called "nonstrategic bargaining." Consider a country bargaining with its neighbors over trade concessions or other forms of aid that is also facing an economic crisis with the potential of

Figure 8
CONVENTIONAL AND NUCLEAR SUPERIORITY

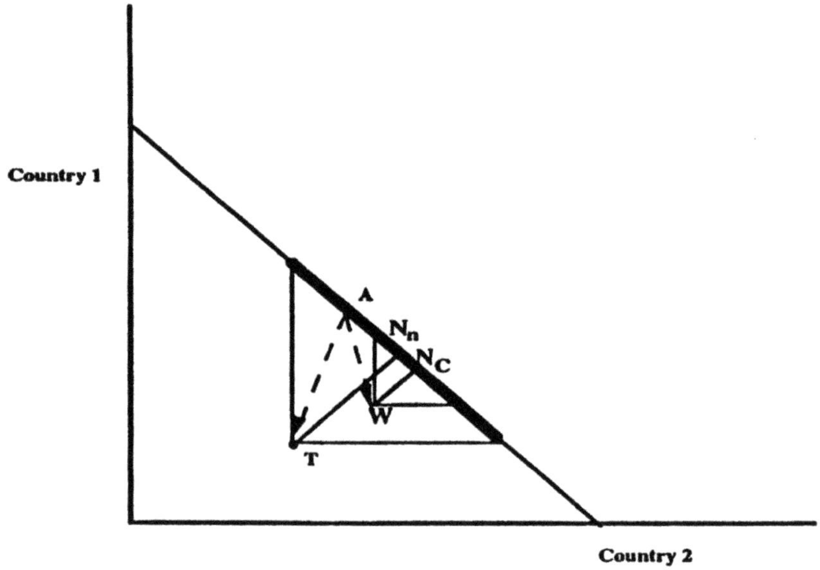

generating political instability. If it is a nuclear power it can invoke the threat point by using the argument in its negotiations that political instability may create the danger that the nuclear weapons may fall into irresponsible hands. Thus, it may be seen as in the interest of its neighbors to make concessions that will reduce this danger. The potential for political instability is itself a form of a precommitment mechanism. A nonnuclear power does not have the advantage of this type of leverage.

By changing the threat point in a bargaining game nuclear weapons can improve the status of the new nuclear nation. Many nations can develop nuclear weapons. The technology is no longer exotic, and the principal constraint for many of them has been economic, with the cost of acquiring these weapons exceeding any possible benefits these weapons may produce. These costs are changing, however. One major constraint, that of access to fissile material, would be substantially reduced if only a fraction of the vast stockpile of fissile material held by the nations of the former Soviet Union were to enter the stream of commerce or if plutonium, there or elsewhere, were diverted from its use as a fuel for power reactors. Another major constraint has been access to the technology, but dual-purpose systems are becoming more common and available in international commerce. Restrict-

Figure 9
PROLIFERATION AND PROBABILITY OF NUCLEAR WAR

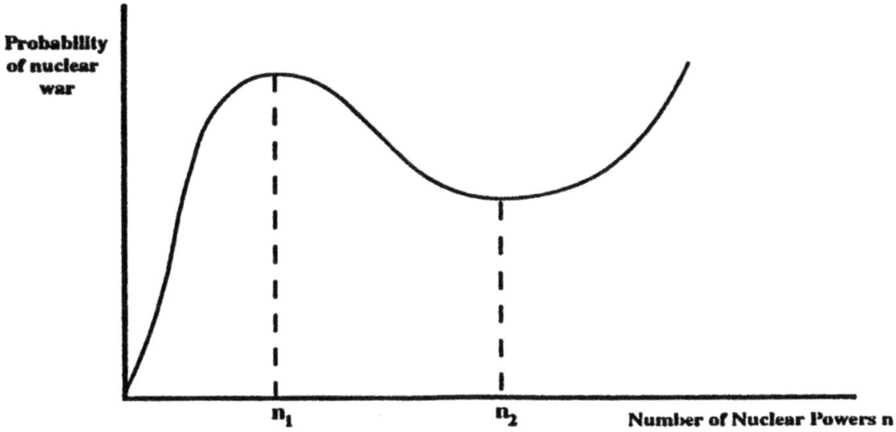

ing this technology may prove difficult, since there is no presumption that it will be used to produce weapons. As a result of these changes, nuclear weapons are likely to become less expensive in the future. A major proposition of economic theory states that if the price of a good falls, other things being equal, then its demand will increase. As the cost of nuclear weapons falls, this proposition suggests that there will be new nuclear states in the future unless new policies increase this cost.

NUCLEAR WEAPONS AND THE STABILITY OF THE WORLD ORDER

Nuclear proliferation may be the greatest danger to world stability at this time. If nuclear weapons are used to bargain in the manner described, there are three analytically different influences of proliferation on the probability of nuclear war. An added nuclear weapons state means that: (1) there is another potential initiator of a nuclear war and an additional potential target; (2) there is another potential retaliator against an initiator, augmenting general deterrence by restraining the previously existing nuclear-weapons states either individually or as part of a potential countercoalition; and (3) there is a greater chance of an accidental or inadvertent nuclear war. Effects 1 and 3 result in an increase in the chance of nuclear war outbreak, but effect 2 results in a decrease in this chance. The result is that in certain special circumstances proliferation could *reduce* the chance of nuclear war, as illustrated in Figure 9. In Figure 9 the probability of nuclear war is increasing when the number of nuclear nations is less than n_1 due to the first effect; it then drops between n_1 and n_2 due to the second effect; and then starts to increase due to the third effect. In general, however, proliferation is destabilizing in increasing the chance of intentional or accidental nuclear war, particularly after the number of countries is large enough so that the chance of a calculated attack is small due to general deterrence.[27]

In the future, the proliferation of nuclear weapons may lead to an environment in which new nuclear nations might attempt to change the distribution of wealth in the world. The model suggests two possible ways in which nuclear weapons can be used: first, as a shield for conventional aggression, and second, to support an implicit threat of political instability in bargaining over the terms of trade, patterns of aid, and other determinants of the global distribution of wealth. In both cases, an increase in the number of nuclear powers results in a higher probability of war because the number of nuclear powers is increasing the probability of war because of accidents or political instability.

New nuclear powers might feel confident to engage in regional adventures under the shield of their nuclear umbrella. They could attempt to extort economic or territorial concessions from their neighbors in the belief that their nuclear weapons would serve to deter any intervention by the major powers. For example, it would have been difficult for the United States to have forged an alliance against Iraq after its invasion of Kuwait if Saddam Hussein had been able to make a credible nuclear threat against his neighbors. Saudi Arabia might have been unwilling to have foreign troops in its country if its cities had been threatened by Iraqi nuclear weapons, despite the large inventories of nuclear weapons held by the United States and its allies, including Israel. Thus, Iraqi nuclear weapons might have made it feasible for Iraq to have succeeded in annexing Kuwait.

The presence of nuclear weapons could also have changed the outcome of other recent crises. For example, Britain might not have risked sending a task force to the South Atlantic in the 1982 Falklands/Malvinas War if Argentina had nuclear weapons. Similarly, a small Libyan nuclear force might have deterred the United States from using force against it in April 1986 since then the United States would have had to destroy all the Libyan nuclear weapons, which would have been a risky venture.

Nuclear weapons may also lead to a redistribution of world wealth without their use as a strategic threat but rather simply through the threat of political instability in the new nuclear powers. The major powers may be forced to make political and economic concessions to new nuclear nations to ensure the political stability of these states. A possible explanation for the reluctance of Ukraine and Kazakhstan quickly to give up the strategic nuclear warheads in their possession is their belief that the Western powers would be more likely to give security guarantees and aid to prevent social unrest to a nation with nuclear weapons than to one without.[28] Furthermore, because of the end of the bipolar order, states in the Third World have become increasingly marginalized, ignored by the major nations. There are only a few ways in which states can gain the attention of the world: one is economic importance, including ownership of oil reserves; another is geostrategic significance; while a third is the possession of nuclear weapons. Some nations may conclude that, in the absence of the others, the only way to be taken seriously is through the acquisition of nuclear weapons. A case in point is the difference in international treatment of the successor nations of the Soviet Union between those with and those without nuclear weapons. While the republics with nuclear warheads are also the most important ones economically, demographically, etc., other nations, including other successor republics of the Soviet Union, may perceive, rightly or wrongly, that an important difference that has led to greater visibility and world attention is precisely their possession of nuclear weapons.

Conclusions and Implications for Nonproliferation Policy

The coming decades may witness the further spread of nuclear weapons, followed by threats to use such weapons or even their actual use, particularly in unstable regions. The result of such proliferation would likely be the creation of instabilities, which threaten global stability. Nations may attempt to become regional or even global powers by acquiring nuclear weapons. Some national leaders may perceive that the only way for their nation to be taken seriously is by the acquisition of nuclear weapons. Such weapons have perceived military or security value to the nations acquiring them, and they may also be seen as valuable in diplomatic, political, and economic terms, as indicated by the above analysis. The structural changes in the international system have made Russia, as the principal successor state to the Soviet Union, and also, to some extent, the United States, less capable of controlling their traditional client states, in particular, less capable of limiting their access to nuclear technology and nuclear weapons. Thus, there are many incentives for proliferation and few disincentives.

Once a nation has nuclear weapons, however, there is an added incentive for its rivals or neighbors to develop their own, leading to *chains* of proliferation.[29] The result of these chains of proliferation is that the proliferating nation may ultimately be less secure in the proliferated region than if the region had no such weapons. Thus, while the short-term effects of developing nuclear weapons may, in some cases, be positive for the proliferating nation, the long-term effects might, in fact, be negative. Similarly, while it may appear to be in the interest of a single nation to develop nuclear weapons capability, from a wider perspective, the effects on global security are generally negative in view of the chance of the actual use of these weapons, whether intentionally or by accidental use.

What can be done to reduce the threat to global security stemming from these newer proliferation threats? Economics and common sense suggest that the answer must be a new set of policies, adopted on a global basis, that either increase the cost of acquiring nuclear weapons or decrease their economic and political value. The following set of policy initiatives could be effective in preventing the acquisition of nuclear weapons that could destabilize the international system:

First, the enhanced value of nuclear weapons resulting from the end of the bipolar order could be offset if some type of new order would restore stability and predictability in international affairs. As one possibility, a new concert of major nations, possibly working through the United Nations, might be able to provide security guarantees to nations facing military and other threats to their security or to help them establish regional cooperative security arrangements so as to defuse their possible interest in nuclear weapons development. In particular, the nuclear-weapons states could provide security guarantees and formally commit themselves to non-production and to no-first-use arrangements through new agreements or through amendment of the Non-Proliferation Treaty (NPT) in conjunction with its extension in 1995.

Second, it is necessary to counter the move to a worldwide plutonium economy. There must be collaborative international agreements, reached on the basis of both nonproliferation and environmental concerns, involving the renunciation of breeder reactor technology and of massive shipments of plutonium, all of which must be done on a worldwide basis. Restrictions might also be imposed on the production, transport, and storage of plutonium, using the new Chemical Weapons Treaty as a precedent, and moving toward the internationalization of plutonium storage.

Third, there must be some effort to discourage the migration of scientists, engineers, and technical workers, particularly those from the former Soviet Union, who could assist nations trying to build nuclear weapons. This is a delicate question, as there are serious questions involving human rights that make it difficult to attempt to restrict the voluntary immigration of individuals to a country that is willing to receive them. Positive incentives must be given to these weapons experts to prevent their assisting other nations in building nuclear weapons.

Fourth, covert nuclear proliferation must be dealt with directly by recognizing its existence and by using new technologies for intelligence gathering to reveal those nations using this route to nuclear weapons. Currently, many of the proliferation disincentives are reduced when nations follow the Israeli model of covert development of nuclear weapons, with a stated policy of ambiguity or denial. Nations following this course of covert proliferation can have most of the benefits of nuclear weapons with few of their costs. By developing weapons covertly nations can retain many of the supposed advantages of nuclear-weapons states, including having a weapon of last resort, deterring regional rivals, achieving some degree of hegemony in a region, and enhancing international status and prestige, without incurring certain of the costs of overt proliferation, including antagonizing allies, losing technical assistance, suffering international opprobrium, and perhaps even attracting a surgical strike on its nuclear facilities.[30] To stem covert proliferation all states should be given inducements to comply with the restrictions of the nonproliferation regime, including the NPT and the Suppliers' Group. The example of Iraq shows that it is also necessary to tighten this regime, possibly in connection with 1995 extension of the NPT.[31] Specific initiatives should include challenge inspections, full-scope safeguards, and return of spent fuel rods. It would also be valuable, as part of the nonproliferation regime, to have agreements ahead of time among the nuclear-weapons states on what actions they would take in various possible contingencies involving accidental, inadvertent, or intentional use of nuclear weapons.

Fifth, international agreements must be established so as to enforce limits on the transfer of nuclear technology, especially dual-purpose systems. There should be criminal prosecution of individuals and companies supplying sensitive technologies. In addition, exporting nations should establish a new regime limiting exports of sensitive equipment and technology to unstable regions, such as the Middle East, or to nations intent on building nuclear weapons.[32]

Sixth, developments in precision guided munitions can be used to reduce the political value of nuclear weapons and to reduce the temptation to acquire them. At present, guarantees offered to countries that do not acquire nuclear weapons may not be credible. It would be very difficult politically to use nuclear weapons with potential collateral casualties in the tens of thousands to punish a military dictator like Saddam Hussein. The use of conventional forces to protect third countries may also be politically difficult unless there are clear national interests at stake. A possible solution to this dilemma is to develop a conventional deterrence using precision guided munitions.

Finally, the successor states of the Soviet Union constitute the most important specific problem since they already have enormous stockpiles of nuclear weapons, huge inventories of fissile material, and the human and physical capital needed to produce these weapons even if their present stockpiles were depleted. A new set of policies must be established for these successor states, especially for Ukraine and Kazakhstan. Various incentives, including security guarantees, should be used to encourage these new nations to give up their strategic nuclear weapons and to become non-weapons state members of the NPT, with commitments not to build such weapons in the future. In any case, all the fissile material in the former Soviet Union must be safeguarded, perhaps using international cooperation to ensure that no material falls outside a safeguards regime. Going beyond such a regime, an international consortium might be formed to purchase the fissile material in the Soviet successor states, possibly for use as a fuel source for nuclear power plants or simply for dilution and storage. Such a step would be valuable in gaining control of this material as rapidly as possible. Additionally, a possible side benefit of this proposal is that dismantling the existing inventories of weapons could provide employment at Western wages to the scientists, engineers, and technical workers of the former Soviet Union.

These policy initiatives for nonproliferation would significantly contribute to global stability by reducing the incentives for nuclear proliferation, by raising the costs of such proliferation, and by limiting the instabilities caused by such proliferation.

Notes

Support from the Center for Institutions and Values at Rice University and the Center for International Relations at the University of California, Los Angeles is gratefully acknowledged. We would also like to acknowledge the very helpful suggestions of Bruce Bueno de Mesquita, Benjamin Frankel, Kurt Gaubatz, Ben Sanders, Richard Stoll, and Erica Worth.

1. Recent studies of nuclear proliferation include John Simpson, ed., *Nuclear Nonproliferation: An Agenda for the 1990s* (New York: Cambridge University Press, 1987); Mitchell Reiss, *Without the Bomb: The Politics of Nuclear Nonproliferation* (New York: Columbia University Press, 1988); Leonard S. Spector, *The Undeclared Bomb* (Cambridge: Ballinger, 1988); Leonard S. Spector and Jacqueline R. Smith, *Nuclear Ambitions: The Spread of Nuclear Weapons, 1989–1990* (Boulder, Colo.: Westview, 1990); Lewis A. Dunn, *Containing Nuclear Proliferation*, Adelphi Papers no. 263 (London: International Institute for Strategic Studies, 1991); and David Fischer, *Stopping the Spread of Nuclear Weapons: The Past and the Prospects* (London: Routledge, 1992). None of these studies, however, takes specific account of the various changes that have occurred in the international system since the end of the Cold War and their implications for proliferation, as discussed here.
2. See Kurt Campbell, Ashton Carter, Steven Miller, and Charles Zraket, *Soviet Nuclear Fission: Control of the Nuclear Arsenal in a Disintegrating Soviet Union*, CSIA Studies in International Security, no. 1 (Cambridge: Center for Science and International Affairs, Harvard University, November 1991); "Nuclear Weapons in the Former Soviet Union," entire issue of *Arms Control Today* 22, no. 1

(January–February 1992); and Carin Atterling Wedar, Michael D. Intriligator, and Peeter Vares, eds., *Implications of the Dissolution of the Soviet Union for Accidental/Inadvertent Use of Weapons of Mass Destruction* (Tallinn: Estonian Academy of Sciences, 1992) for discussions of the numbers of warheads in the Soviet successor states, problems regarding their control, and the potential for their accidental or inadvertent use. On the general issue of accidental or inadvertent nuclear war, see Derek Paul, Michael D. Intriligator, and Paul Smoker, eds., *Accidental Nuclear War* (Toronto: Science for Peace, 1990) and Hakan Wiberg, Ib Damgaard Peterson, and Paul Smoker, eds., *Inadvertent Nuclear War: The Implications of the Changing Global Order* (Oxford: Pergamon Press, 1992).

3. While international relations theorists have differered on whether bipolarity or multipolarity is more stable, it would appear that the end of the bipolar East–West order has led to greater instability in international affairs, including the 1991 Gulf War, the largest war in Europe since 1945 in the former Yugoslavia, and various local civil wars in the former Soviet Union.
4. On "opaque" nuclear proliferation, particularly as applied to Israel, see Benjamin Frankel, ed., *Opaque Nuclear Proliferation: Methodological and Policy Implications* (London: Frank Cass, 1991) and Shlomo Aronson, *The Politics and Strategy of Nuclear Weapons in the Middle East* (Albany: State University of New York Press, 1992).
5. India appears to have started out following the U.S. model and then shifted to the Israeli model of ambiguity or denial following its one test in 1974, labeled a "peaceful" nuclear explosion. For discussions of nuclear proliferation in India and Pakistan, see Stephen P. Cohen, ed., *Nuclear Proliferation in South Area* (Boulder, Colo.: Westview, 1991). For discussions of covert nuclear development in other nations, see Spector and Smith, *Nuclear Ambitions*.
6. See Wedar et al., eds., *Implications*, for discussions of the possible leakage of fissile material from the former Soviet Union. For discussions of material unaccounted for (MUF) see the references in n. 1.
7. On exports of nuclear-related items see William C. Potter, ed., *International Nuclear Trade and Nonproliferation: The Challenge of the Emerging Suppliers* (Lexington: Lexington Books, 1989).
8. On the problem of conversion from military to civilian production, see the UNIDIR Research Report, *Economic Aspects of Disarmament: Disarmament as an Investment Process* (Geneva: United Nations Institute for Disarmament Research, June 1992).
9. See Michael D. Intriligator, "Defining Global Security," *Disarmament* 14, no. 4 (October 1991): 59–72, where it is suggested that "global security" be defined as the absence of threats to the vital interests of the planet. While this paper stresses areas where newer problems of proliferation have developed, possibly jeopardizing global security, there have also been some positive developments in recent years with regard to the traditional nonproliferation agenda, particularly in Brazil and Argentina, North Korea, and South Africa.
10. See B. H. Liddell-Hart, *History of the Second World War* (New York: Putnam, 1970).
11. See Bernard Brodie, *Strategy in the Missile Age* (Princeton: Princeton University Press, 1959). See also Bernard Brodie, *War and Politics* (New York: Macmillan, 1973).
12. See the *United States Strategic Bombing Survey* (New York: Garland, 1976) and Albert Speer, *Inside the Third Reich* (New York: Macmillan, 1970).
13. Speer, *Inside the Third Reich*.
14. See L. Giovannitti and F. Freed, *The Decision to Drop the Bomb* (New York: Coward-McCann, 1965).

15. See Brodie, *Strategy in the Missile Age* and *War and Politics*. For more formal theories of deterrence see Michael D. Intriligator, "Strategic Considerations in the Richardson Model of Arms Races," *Journal of Political Economy* 83, no. 2 (April 1975): 339–53; Michael D. Intriligator and Dagobert L. Brito, "The Stability of Mutual Deterrence," in Jacek Kugler and Frank C. Zagare, eds., *Exploring the Stability of Deterrence*, Monograph Series in World Affairs (Denver: University of Denver School of International Studies, 1987); and Michael D. Intriligator, "On the Nature and Scope of Defence Economics," *Defence Economics* 1, no. 1 (October 1989): 3–11.
16. See Kosta Tsipis, *Arsenal, Understanding Weapons in the Nuclear Age* (New York: Simon and Schuster, 1983)
17. This calculation is based on missile parameters published in *Jane's Fighting Ships 1984–85*, (London: S. Low, Marston, 1985); and *New York Times*, 26 April 1986. There are not enough published parameters available to have sufficient boundary conditions to solve the orbital equations, so range/ payload approximations were computed by assuming conservation of energy of the payload outside the atmosphere.
18. See John F. Nash, "The Bargaining Problem," *Econometrica* 18, (1950): 155–62. The general discussion in this section follows that of John Harsanyi, "Bargaining," in James Eatwell, Murray Milgate, and Peter Newman, eds., *The New Palgrave: A Dictionary of Economic Theory and Doctrine* (London: Macmillan, 1987), while the analysis makes use of Dagobert L. Brito, A. M. Buoncristianai, and Michael D. Intriligator, "A New Approach to the Nash Bargaining Problem," *Econometrica* 45, no. 5 (September 1977): 1163–72, and is based on Dagobert L. Brito and Michael D. Intriligator, "Strategic Weapons and the Allocation of International Rights," in John V. Gillespie and Dina A. Zinnes, eds., *Mathematical Systems in International Relations Research* (New York: Praeger, 1977).
19. For another model in which the threat of force can be used to explain redistribution, see Dagobert L. Brito and Michael D. Intriligator, "Conflict, War, and Distribution," *American Political Science Review* 79, no. 4 (December 1985): 943–57.
20. For an economic model of an arms race see Dagobert L. Brito, "A Dynamic Model of an Armaments Race," *International Economic Review* 13 (June 1972): 359–75. For surveys of formal models of the dynamics of the arms race, see Michael D. Intriligator and Dagobert L. Brito, "Arms Race Modeling: A Reconsideration," in Nils Petter Gleditsch and Olov Njolstad, eds., *Arms Races: Technological and Political Dynamics* (London: Sage Publications, 1990); and Michael D. Intriligator and Dagobert L. Brito, "Formal Models of Arms Races," *Journal of Peace Science* 2, no. 1 (Spring 1976): 77–88. See also Michael D. Intriligator and Dagobert L. Brito, "A Possible Future for the Arms Race," in the Gleditsch and Njolstad volume, where it is argued that, unlike the earlier East-West arms race, the current regional arms races do not lead to a stable situation of mutual deterrence.
21. See Nash, "The Bargaining Problem."
22. This is a very complicated question that has not been completely solved. For its treatment in the case of asymmetric information see Brito and Intriligator, "Conflict, War, and Distribution."
23. Games are *subgame perfect* if the optimal strategy in period 1 for period 2 play is in fact the optimal strategy in period 2 for period 2 play. See Drew Fudenberg and Jean Tirole, *Game Theory* (Cambridge: MIT Press, 1991).
24. There is no reason why an aggressor cannot precommit prior to bargaining. In 1519, Cortez burned his ships after landing in Mexico prior to attacking the Aztec Empire, precommitting his men to battle. In Shakespeare's play, Henry V

threatened Harfleur with a looting army out of his control if they do not yield: "If I began the battery once again, ... We may as bootless spend our vain command Upon th' enraged soldiers in their spoil... Therefore, you men of Harfleur, Take pity of your town and your people, Whiles yet my soldiers are in my command."

25. See Frankel, *Opaque Nuclear Proliferation*; and Aronson, *The Politics and Strategy of Nuclear Weapons in the Middle East*.
26. President Sadat may have concluded that there was no reason to waste resources trying to gain military superiority over Israel, given that Israeli nuclear capability would have ensured that any triumph would be a Pyrrhic victory. For example, one scenario for the use of Israeli nuclear weapons in the so-called "Samson option" is an airburst over Lake Nasser, creating a wave that would destroy the Aswan High Dam, resulting in a massive flood that would wipe out most of Egyptian civilization.
27. See Michael D. Intriligator and Dagobert L. Brito, "Nuclear Proliferation and the Probability of Nuclear War," *Public Choice* 37, no. 2 (1981): 247–60. For other (and very different) views on the subject of proliferation and stability, see Kenneth N. Waltz, *The Spread of Nuclear Weapons: More May be Better*, Adelphi Papers no. 171 (London: International Institute for Strategic Studies, 1981); Bruce Bueno de Mesquita and William Riker, "An Assessment of the Merits of Selective Nuclear Proliferation," *Journal of Conflict Resolution* 26, no. 2 (June 1982): 283–306; and George H. Quester, "Nuclear Proliferation and Stability," Kenneth N. Waltz, "Toward Nuclear Peace," Dagobert L. Brito and Michael D. Intriligator, "Proliferation and the Probability of Nuclear War: Global and Regional Issues," Bruce Russett, "Away from Nuclear Mythology," Thomas C. Schelling, "Comments," and Dina A. Zinnes, "Comments," all in Dagobert L. Brito, Michael D. Intriligator, and Adele Wick, eds., *Strategies for Managing Nuclear Proliferation: Economic and Political Issues* (Lexington: Lexington Books, 1983).
28. See Wedar et al., eds., *Implications*. See also Reginald Bartholomew [Undersecretary of State for International Security Affairs], "U.S. Effort to Halt Weapons Proliferation in Former Soviet Republics," *U.S. Department of State Dispatch* (Washington, D.C., 10 February 1982).
29. The main historical chain of proliferation was from fear of Nazi Germany to the United States to the Soviet Union to China to India to Pakistan, with side influences on the United Kingdom, France and other nations and with separate chains involving yet other nations.
30. Consider the acquisition of nuclear weapons in the first five decades of the nuclear age, from 1945 to 1995. The overt nuclear proliferators include three in the first decade, 1945–55: the United States (1945), the Soviet Union (1949), and the United Kingdom (1952); two in the second decade, 1955–65: France (1962) and China (1964); one in the third decade, 1965–75: India (1974); and none in the last two decades 1975–95. From this alone, it would appear that the problem of nuclear proliferation has been decelerating. But consider the covert proliferators. By the very nature of covert proliferation the numbers are speculative, but there were probably none in the first two decades; perhaps one in the third decade, 1965–75: Israel; possibly two in the fourth decade, 1975–85: South Africa and Pakistan; and there could be some in the fifth decade, 1985–95. Obviously, because of the nature of covert proliferation, this is highly speculative and cannot be validated, but this way of looking at nuclear weapons acquisition suggests that, while the problem of overt nuclear weapons acquisition has been decreasing, the problem of covert proliferation may be increasing as various nations follow the Israeli model.
31. On NPT and its extension, see Jozef Goldblat, *Twenty Years of the Non-Proliferation Treaty: Implementations and Prospects* (Oslo: PRIO, 1990); and Joseph

F. Pilat and Robert E. Pendley, eds., *Beyond 1995: The Future of the NPT Regime* (New York: Plenum, 1990).

32. A precedent for limiting arms sales to the Middle East was the Tripartite Agreement, among the United Kingdom, France, and the United States, which worked successfully up to 1958 when the Soviet Union began major arms sales to the region.

Forecasting the Risks of Nuclear Proliferation:
Taiwan as an Illustration of the Method

Bruce Bueno de Mesquita, James D. Morrow, and Samuel S. G. Wu

THE BREAKDOWN of the bipolar world may create strong incentives for nations to seek nuclear weapons. As neorealists argue, the insecurities inherent in a multipolar world provide a reason for states to seek the ultimate guarantee of their security.[1] However, predictions of proliferation do not necessarily follow from the systemic incentives; different states will respond differently to those incentives.[2] States may differ in how they interpret threats to their security. Some may see the threats of a multipolar world as minor, while others could see the same threats as vital. Even when states share a perception of threat, there are many possible responses other than the acquisition of nuclear weapons.[3] Why would a state choose proliferation over another source of security, such as additional conventional capabilities? Finally, even when a state has decided to acquire nuclear weapons, it must create a domestic coalition to support the expense of a nuclear weapons program.[4] Nations which cannot build such a coalition will be unable to build nuclear weapons. Determining which nations will choose to proliferate requires answering these questions. In this paper, we analyze a specific possible proliferant, Taiwan, using a general predictive model of foreign policy.

The study of national security generally focuses either on bureaucratic infighting or on structural models of the international system. We present an approach that combines elements of each, taking seriously international constraints and domestic politics. The formal model we discuss provides accurate and precise predictions of future policies, detailed descriptions of relevant internal political coalitions, their stability and fluidity, and a means of evaluating the perceptions and anticipated actions of key political actors.[5] A computerized version of this model, in conjunction with issue or area expert inputs, evaluates the predicted consequences of alternative policy scenarios. We use these scenarios to anticipate the likely outcome of almost any policy debate on specific, well-defined issues.

Many will be appropriately skeptical about the possibility of using formal models to predict and explain politics. Models, however, provide certain insurance against ad hoc arguments or reliance on ex post knowledge. The implications derived from the model we use are replicable regardless of the personal beliefs of the analyst. All assumptions are explicit and open to scrutiny so that those who disagree with them can evaluate the consequences that follow from alternative assumptions. In this way, the researcher can better judge exactly how his or her argument takes shape.

Others will wonder at the possibility of rigorously analyzing domestic politics as it bears on national security. Research over the past several decades into spatial models of voting and assessments of structure induced equilibria as applied to American politics, however, have proven quite successful in shedding empirical as well as theoretical light on the intricacies of American politics.[6] Of course, much of domestic politics around the world does not involve voting so that international relations researchers have generally dismissed these insights as irrelevant to their problems. We contend, however, that the central realist notion of the exertion of power is, in non-democratic settings, analogous to the casting of votes. With this simple intuition, we analyze domestic politics as a struggle for security and power over policy.

Stake holders for any policy issue are presumed to have three fundamental attributes: *power* to influence decisions; *preferences* over policies; and *salience* or willingness to put their power at risk for this or that outcome. Political choices are viewed as always involving some tradeoff between security goals and policy goals. Those trying to influence decisions want to promote their desired outcome, but they also want to be perceived as consequential players who are part of the winning coalition. Sometimes they cannot achieve both simultaneously and so must strike a balance between the two that is satisfactory to them.

The method we are using assesses specific, well-defined issues. Inferences can often be drawn about over-arching questions of national or domestic strategy, but this method is geared toward evaluating such strategies as an aggregation of individual policy choices. We illustrate the application of this method with three issues related to nuclear weapons policy in the Republic of China/Taiwan (ROC).

Nuclear proliferation is one area of foreign policy to which this model can be applied. The model allows us to analyze how internal and external political pressures could influence Taiwanese policy on developing their own nuclear capability. Such a policy includes several issues beyond just the question of whether to acquire such a capability. Nuclear technology could be acquired abroad or developed domestically. Other security policies could substitute for a Taiwanese nuclear program and so discourage a nuclear program. This paper begins with a brief review of the history of Taiwanese nuclear development. We then describe the model used in the analysis. The analysis of three issues follows. For each issue, we present a brief description of the relevant political actors, their power, position, and salience for that issue. We then discuss the results of the analysis, using a different issue to highlight the ability of the model to predict different perceptions, strategic moves by actors in response to political pressure, and the outcome of the issue. We conclude with a brief discussion of the value of the method for security studies.

BACKGROUND ON TAIWANESE NUCLEAR DEVELOPMENT

On 16 October 1964 the People's Republic of China (PRC) successfully tested its first atomic bomb. Less than three years later, on 17 June 1967 the PRC joined the set of nations with a demonstrated nuclear capability. These developments, during a period of significant hostility and friction in the relations of the United States and Taiwan with China, prompted new strategic thinking in Taiwan about its future security. In 1968, presumably in response to China's nuclear test, Chiang Ching-Kuo, Taiwan's defense minister (and son of Chiang Kai-shek), established the National Chung Shan Institute of Science and Technology to conduct research on advanced weapons. Almost certainly, the Institute focused its efforts on the development of an independent Taiwanese nuclear capability.

As is typical of such efforts, the Taiwanese kept the project top secret, admitting only that they were doing research on peaceful uses of nuclear energy. It was not until 1988 that any admission of broader purposes was made. On 21 September 1991, the director of the Institute, Dr. Liau, admitted that more than ten years before, the Institute had conducted research on the development of long range missiles.[7]

Taiwan's nuclear research program is monitored closely by the United Nations and the United States. The objective behind the close monitoring is to prevent the development of nuclear weapons. This has, of course, made it difficult for the Taiwanese to continue to pursue nuclear weapons research.

In the past decade and a half, Taiwan has turned to Israel and South Africa for assistance. According to the *Journalist* (a weekly magazine published in Taipei), it appears that Taiwan imported plutonium from South Africa in 1977. Some reports suggest that Taiwan was very close to building a nuclear weapon in 1988.[8]

As Taiwan approached the moment of proliferation, the Vice Director of the Laboratory of Nuclear Energy of the National Chung Shan Institute suddenly disappeared. Several days later a team of U.S. experts from the Nuclear Regulatory Commission (NRC) arrived in Taiwan to disassemble the experimental nuclear reactor. Dr. Chang reappeared to testify before the United States Congress about Taiwan's nuclear weapons development program. It appears that Dr. Chang may have been a United States intelligence source from the start. The incident involving Dr. Chang seems to have ended this particular program.

Although many believe that Taiwan has the technical know how to build a nuclear weapon, Taiwanese officials deny that there is any attempt to build strategic weapons. Despite repeated assurances from high level sources, still we cannot overlook the statement made by the defense minister in testimony before the Legislative Yuan. He is quoted as having argued that Taiwan "should not rely on other countries [to deter the PRC from attacking Taiwan], ... we are evaluating the appropriate missile systems for the purpose [of deterrence]".[9] A leading member of the main opposition party, the Democratic Progressive Party (DPP), supported the defense minister's position. President Lee seems also to have supported this view indirectly. He has announced that Taiwan should move into the area of advanced aerospace technology. One Taiwan-based industrial group is currently negotiating with the PRC for a joint effort to establish a satellite launching business. Naturally, the associated technology is also relevant to launching ballistic missiles.

The Method and Model

We describe briefly a model more fully discussed elsewhere that focuses on the application of Black's median voter theorem and a theorem about the monotonicity between certain expectations and the escalation of political disputes.[10] The conjunction of these theorems along with concepts from bargaining theory foster the development of a quasi-dynamic political model that includes detailed expectations about the agreements or compromises that various "players" are willing to make over time and the implications of those compromises for the ultimate resolution of the issues in question.

Two constraints are assumed to facilitate prediction and explanation: that issues are unidimensional, so that preferences can be represented on a line segment, and that preferences (and associated utilities) for potential outcomes diminish steadily the farther in Euclidean distance a possible settlement is from one's preferred outcome. These two constraints are requirements of the median voter theorem. Black's theorem demonstrates that the outcome desired by the median voter is the winning position under the constraints just assumed provided a simple majority is required for victory. Of course, we do not assume that all interesting political problems involve voting. Rather, we assume that power is the non-voting analog of votes in most political interactions.

The monotonicity theorem provides a basis for predicting when policy debates are expected to produce negotiated settlements or are expected to lead to an escalation of friction between competing interests. Banks's monotonicity theorem highlights an important feature of all politics. It tells us that the more one expects to gain from challenging a rival perspective, the more likely one is to undertake the challenge. This simple statement turns out to have interesting and sometimes surprising implications for political intercourse.

Let $N = \{1, 2, 3, \ldots, n\}$ be the set of actors or stake holders trying to influence a multilateral decision. An actor might be a government representative, an official from a faction within a political party or a bureaucracy, a leader of some interest group, an influential private citizen, and so forth.

Let $M = \{a, b, c, \ldots, m\}$ be the set of issues in a multilateral negotiation and let R_a be the line segment, bounded for convenience between 0 and 100, that describes the unidimensional policy continuum for any individual issue *a* selected from among the larger set of issues M. Let each actor *i*, $i \in N$ have its own *preferred* resolution of issue a, with that preferred resolution denoted as x^*_{ia}, such that on the issue continuum R_a, $0 \leqslant x^*_{ia} \leqslant 100$.[11] For any feasible proposed outcome on issue a, say k's proposal, x_k, *i*'s utility for x_k, $u^i x_k$, is a decreasing function of the distance between the proposal and *i*'s preferred resolution, so that $u^i x_k = f|x_k - x^*_i|$. Exactly how fast an actor's utility falls off with the distance between the proposal and its preferred resolution depends on the actor's attitude towards risk. This risk attitude captures what tradeoffs an actor will accept between being in the winning coalition and achieving its preferred outcome. Some actors are willing to accept an outcome further from their preferred resolution of the issue in order to be in the winning coalition. Other actors are willing to take great risks to achieve an outcome close to their preferred resolution, even at the cost of likely defeat.[12]

Of the infinitely many possible proposals to resolve some issue a, how are we to predict which will be chosen? To answer this question, let us first learn a little more about each actor *i*. In this analysis, each decisionmaker is endowed with three characteristics. Each player attaches some *utility* to each possible outcome on issue a, as already noted. Each participant in the bargaining process is also endowed with the *power* to exert some influence on decisions. Let c_{ia} be the capabilities (or power) actor *i* could bring to bear on issue a, such that the sum of the capabilities of the participants in a multilateral decisionmaking setting is 1.[13] c_i is, then, actor *i*'s share of the total *potential* influence that could be brought to bear in the negotiations over some issue a. Each participant has its own agenda of priorities or *salience* that it attaches to the issues that must be confronted. Thus, *i* may attach considerable importance to issue a and considerably less importance to issue b. Denote the salience of issue a for actor *i* as s_{ia}, with $0 \leqslant s_{ia} \leqslant 1$. Each actor is described by the values of $u^i x_k$ for all $i, k \in N$, c_i, and s_i on each issue. A group, player or actor is any aggregation of individuals with identical values on all three of these variables.

When alternative courses of action are pitted against each other, the array of forces on either side often determines victory. Of course, this array depends on more than the relative power of the competing interests. It depends also on the willingness to spend influence on the issue in question [s_i] and the intensity with which each actor prefers one proposed settlement, say *j*'s proposal, u^ix_j), to another proposal, say k's (u^ix_k). Each group has a total number of *potential* "votes." The "votes" presumed to be cast by actor *i* in a comparison of alternatives x_j and x_k are said to equal v^{jk}_{ia} where:

$$v^{jk}_{ia} |x_j, x_k) = (c_i)\ (s_i)\ (u^ix_j - u^ix_k) \qquad (1)$$

Equation (1) states that the "vote" or power mobilized by actor *i* in a comparison of two alternatives (x_j and x_k) is equal to the potential capabilities of *i* discounted by how important the issue is to *i* (that is, s_i) and by how much *i* prefers one proposal to the other ($u^ix_j - u^ix_k$). The "voting" scheme reflects, if you like, what takes place "inside the smoke-filled room" before any formal, visible decisionmaking process occurs. It assumes that any formal process echoes the agreements reached beforehand.

The prospect that a proposal will succeed is assumed to depend on how much support can be mustered in favor of the proposal as compared to the feasible alternatives. In the model this is calculated as the sum of "votes" across all actors in a comparison between x_j and x_k. This sum equals v^{jk} with

$$v^{jk} = \sum_{i=1}^{n} v^{jk}_i \qquad (2)$$

If v^{jk} is greater than zero that implies that x_j defeats x_k because the tacit coalition in favor of *j*'s proposal is more motivated and powerful than the coalition supporting *k*'s proposal. If v^{jk} is less than zero, x_j is expected to be defeated by x_k and if v^{jk} equals zero the competing interests are collectively indifferent between the two alternatives.

In any negotiation, there are likely to be many more than two proposed settlements. By pitting all alternatives against one another two at a time, the outcome preferred by the median voter (weighted by power, salience and intensity of preference) is found. Each policy issue has one outcome within the set of possible outcomes from 0 to 100. In this sense, policies resemble winner-take-all elections. Of course, any policy has many different parts, but we represent these parts as separate issues. In our analysis of Taiwan, we separate whether Taiwan chooses to build a nuclear capability from whether they choose to seek external aid at building such a capability.

Barring perceptions or beliefs that lead decisionmakers to switch their position, the median voter position is the predicted outcome.[14] In practice, of course, perceptions or beliefs often lead decisionmakers to grant concessions or to give in to a rival's point of view, sometimes even needlessly. Such concessions or capitulations can change the location of the median voter.

Consequently, it is crucial to provide an accounting of when such switches in position are expected to take place.

If an interest group is dissatisfied with the *expected* outcome of a negotiation, it can alter its own level of effort (that is, change s_i); shift its revealed position, selecting x_i such that $u^ix_i^* \neq u^ix_i$; influence those who are willing to make concessions to the focal group so that those other groups alter their level of effort (that is, s_k); or it can influence those who are willing to make concessions to the focal group so that those other groups alter their revealed position x_k so that $u^kx_k^* \neq u^kx_k$, leading to switches in positions. Although in the model we examine all possibilities, here we focus only on the latter one.

Decisionmakers interested in ascertaining what leverage they can exert could benefit from estimating the beliefs held by each other actor. To do so requires a focus on the three characteristics – u^ix_j for all $i,j \in N$, s_i, c_i – used to estimate each player's expected utility from challenging or not challenging the policy proposal backed by each potential rival and for approximating the expected utility each actor *i* believes its rival expects to derive from challenging or not challenging the policy goals of actor *i*. In the model envisioned here, decisionmakers are assumed to calculate the expected consequences of challenging and of not challenging alternative proposals. The expected utility for *i* from not challenging rival *j*'s position is denoted as $E^iu^i\Delta x_j \mid \bar{d}$, with $\bar{d}$ denoting the failure to challenge or make a proposal. This expected utility is estimated by projecting what the relevant decisionmaker believes is likely to happen in the absence of the exertion of pressure on a rival to persuade the opponent to alter its behavior. One of three contingencies may arise: actor *i* may anticipate that with some probability (Q^i) rival *j* will not alter its current policies over the time period of concern to *i*, giving group *i* whatever utility it receives from the preservation of the status quo between itself and *j* ($u^i\Delta x_j^\circ \mid \bar{d}$); *i* may anticipate that *j*'s position on the issues will change, in which case there is some chance (T^i) that, from *i*'s perspective, the policies of *j* are anticipated to get better (with $u^i\Delta x_j^+ \mid \bar{d}$ being the associated utility) or to get worse ($u^i\Delta x_j^- \mid \bar{d}$), so that $u^i\Delta x_j^+ \mid \bar{d} > u^i\Delta x_j^\circ \mid \bar{d} > u^i\Delta x_j^- \mid \bar{d}$. *i*'s expected utility if it leaves *j*'s proposal unchallenged is described as:

$$E^iu^i\,\Delta x_j \mid \bar{d} = Q^iu^i\Delta x_j^\circ \mid \bar{d} + (1-Q^i)\,[T^iu^i\Delta x_j^+ \mid \bar{d} + (1-T^i)u^i\Delta x_j^- \mid \bar{d}] \quad (3)$$

i can challenge *j*'s position on issue a by proposing a change in *j*'s position. In doing so, actor *i* presumably takes into account the probability that *j* does not care enough about the issue to resist the proposed settlement by *i* ($1-s_j$). *i* also considers the possibility that *j* will resist *i*'s proposal (s_j), in which case there is some likelihood that *i* will succeed in its efforts to enforce its wishes on *j* (P^i) and some probability that it will fail ($1-P^i$). Should *i* succeed, then *i* will derive the utility associated with convincing *j* to switch from its current policy stance to that supported by i. This is denoted by $u^i\Delta x_j^+ \mid$ d, which equals $u^i(x_i - x_j)$. Should *i* fail, then it confronts the prospect of having to abandon its objectives in favor of those pursued by *j*, denoted by $u^i\Delta x_j^- \mid d = u^i(x_j - x_i)$. The expected utility for challenging *j*'s proposed resolution of the multilateral dispute ($E^iu^i\Delta x_j \mid d$) is:

$$E^iu^i\Delta x_j \mid d = s_j\{P^i[u^i\Delta x_j^+ \mid d] + (1-P^i)[u^i\Delta x_j^- \mid d]\} + [1-s_j][u^i\Delta x_j^+ \mid d] \quad (4)$$

so that the overall expected utility of *i* with respect to *j*'s outlook on issue a is:

$$E^iu^i\Delta x_j = E^iu^i\Delta x_j \mid d - E^iu^i\Delta x_j \mid \bar{d} \quad (5)$$

If equation (5) is greater than zero, then *i* believes that challenging *j*'s position is superior to not challenging it and so *i* is assumed to make a proposal of its own. If equation (5) is less than zero, then not challenging is preferred and *i* is said to be deterred. If (5) equals zero, then *i* is indifferent between challenging and not challenging *j*'s proposed settlement. Each actor evaluates equation (5) vis-à-vis each other actor. In doing so, actors take the expected actions of third parties into account. The estimates of P^i include calculations of how *i* expects all other parties to respond to a dispute over policy settlements between *i* and *j*. In particular, P^i places each other actor in *i*'s coalition, *j*'s coalition, or in a neutral position as indicated by each third party's preference for *i*'s policy proposal or *j*'s. *j* makes a comparable calculation (as does each $k \in N$). Because equation (5) includes such subjective elements as utilities and subjective probabilities, it is possible to estimate a complete matrix of expected utilities that capture all possible confrontations, compromises, and capitulations among all the participants in the relevant political arena.

The various components of equation (5) must each be measured if the model proposed here has practical value.[15] Equation (5) is estimated from four perspectives, with relevant superscripts on equation (5) indicating from whose perspective the calculation is being viewed:

(1) *i*'s expected utility vis-à-vis each rival *j*'s proposal;
(2) *i*'s perception of each *j*'s expected utility vis-à-vis *i*'s proposal;
(3) *j*'s expected utility vis-à-vis each *i*'s proposal; and
(4) *j*'s perception of each *i*'s expected utility vis-à-vis *j*'s proposal.

The expected utility values summarized in (1) and (2) and in (3) and (4) respectively describe each actor's perception of its relationship vis-à-vis each other actor. With Banks's monotonicity of escalation theorem in mind, these relationships can be described in continuous form. According to Banks's theorem, the probability with which an actor anticipates confronting a given rival increases with its expected utility for challenging the rival's proposal, so that the higher some actor *i*'s expected utility is with regard to persuading some other actor *j* to accept *i*'s position, the higher the likelihood that *i* will confront *j*.

The likelihood with which confrontation or concessions occur can be easily displayed in a polar coordinate space. For ease of presentation, we divide such a space into six sectors, with the boundary between each reflecting a fundamental turning point in the probability functions. Figure 1 displays such a polar coordinate space, along with relevant labels for each of the six sectors, reflecting the general likelihood of alternative outcomes in accordance with Banks's monotonicity theorem.

Figure 1

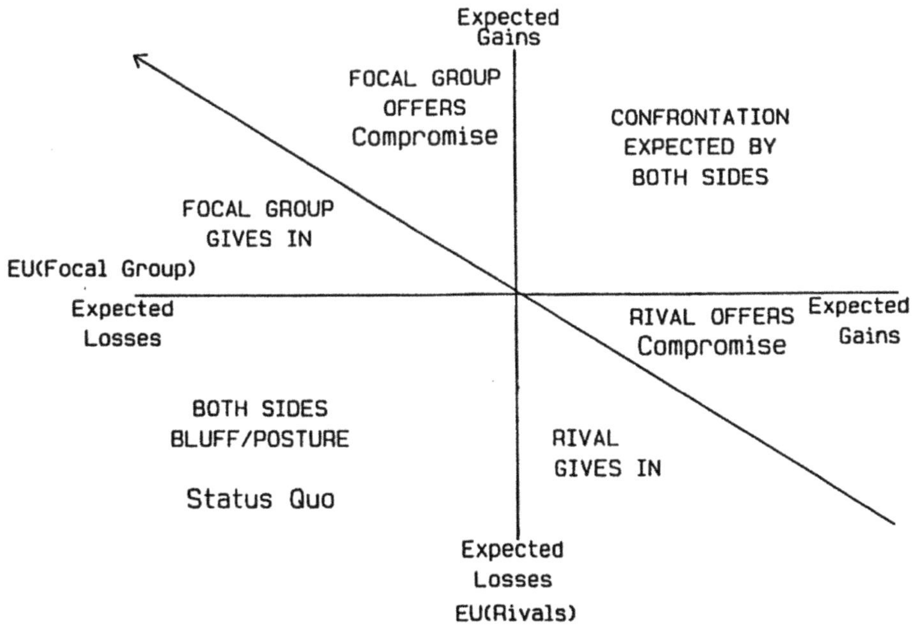

By examining the distribution of information in graphs like Figure 1, we attempt to approximate the private information available to each participant in a dispute or negotiation. From that information we estimate how each party will behave and what consequences are likely to ensue.

The perceptions of each actor as illustrated by figure 1 imply actions. Those actions, in the form of the extraction or granting of concessions over support for this or that specific position, lead to a re-evaluation of the situation by each decisionmaker. As stake holders respond to revised proposals, with their responses supported by their beliefs and expectations, the prospects for a favorable or unfavorable settlement change for many participants. Beliefs and expectations provide the foundation for a quasi-dynamic assessment of the evolution of issue positions and to recalculations of the location of the median voter.

When actors are persuaded or coerced into accepting a proposal different from their initial (or current) position on an issue the decision process enters a new phase. Coalitions change and the support or risks associated with alternative proposals vary. New proposals are brought forward as revised beliefs and expectations open new possibilities or foreclose old ones. Each such sequence of revised stances on an issue is called an iteration. The model computes as many iterations as it takes for the policy issue to resolve itself by reaching a stable outcome; an outcome from which there is not a meaningful possibility of change given the estimated expectations of the actors.

The model portrays a process of decisionmaking during each iteration. To gain an intuitive sense of what happens within the logic of the model during each iteration, think of the decisionmakers as being engaged in a game of cards. At the outset, each player is dealt a hand. The quality of the hand dealt to each player depends on the commonly known characteristics of each player. Stronger players (or those with strong backing from others) generally draw better cards than weaker players. Because of variations in salience, some players pay closer attention to their cards than do others and so form different perceptions of the situation. Based on the cards they hold and the known characteristics of other players, each decisionmaker forms perceptions about how good the hand is of each rival relative to the hand dealt the particular decisionmaker. With that information, each player decides on proposals or bids to make to the other decision makers.

If a player believes his or her hand is very weak compared to a specific rival, then no proposal is made to that actor. If *i* expects to lose to *j*, for instance, then *i* does not make a proposal to *j*. If, however, *i* thinks it holds a good hand relative to *j* then *i* makes a proposal in the form of a suggested change in position by *j* on the issue at hand.[16] If *i* thinks *j* stands to lose quite a lot, then *i* will propose that *j* accept *i*'s current position.[17] If *i* thinks it has a good enough hand to shift *j*'s position, but not so good that *j* will give in to what *i* wants, then *i* proposes a compromise somewhere between *i*'s position and *j*'s.[18]

After all the players have submitted their secret proposals to one another, each player now reviews the new cards – the proposals – that it holds. Of course, some proposals are better for the recipient than others. Indeed, some proposals turn out to be frivolous in that the proposer cannot enforce the proposal, something that the proposer might only learn at the end of the round of proposal making. Other proposals received by a decision maker are potentially enforceable, but fall by the wayside because a superior, enforceable proposal was made by a different player. Each player would like to choose the best offer made to it and each proposer enforces its bids to the extent that it can. Those better able to enforce their wishes than others can make their proposals stick. Given equally enforceable proposals, players move the least that they can. Each actor selects from among the bids it made and the bids it receives. The bid that is chosen is the proposal that is the optimal choice for the player given the constraints under which it operates. These constraints include its own perceptions and the reality of which proposals turn out to be enforceable and which turn out to be beaten back by rivals or rejected outright as unenforceable by the recipient.

At the end of a round of proposal making, players learn new information about their opponents. If, for instance, a player finds that some proposal it thought of as enforceable is successfully rejected then it learns the proposal was unenforceable (that is, the player has less support than it thought). By monitoring responses to its proposals a player learns how much leverage it has with other decisionmakers. If a proposal is accepted, then a player learns that it made the best offer among all the proposals made to the recipient of its accepted bid.

When the players finish sorting out their choices among proposals, each shifts to the position contained in the proposal it accepted (if any). These shifts in position can be dramatic. If the outcome moves dramatically, actors who previously did not form a coalition may coalesce to move the outcome back. This internal dynamic allows the model to capture the dramatic changes in politics caused by other actors' actions. We then re-solve the model using the new positions as the inputs. These iterations of the game allow us to capture the dynamics of bargaining among the groups as the issue evolves. Each iteration is assumed to represent a short period of time, like one week to one month. The game ends when no player believes it has a remaining credible proposal or the value of remaining proposals is sufficiently small that the cost of continuing to bargain outweighs the value for each player of the expected improvement in the outcome. Typically, the model converges to an end after only a few iterations. The median voter of the final stage of the game is the predicted policy outcome.[19]

The model considers each issue separately and does not investigate the possibility that groups may be able to trade concessions across different issues. There may be situations where two groups can exchange concessions in their positions across two issues in such a way that both are better off after the trade. Models based on network analysis attempt to capture this logic of linkage. In competitive testing between this model and a model based on network analysis, this model has been more successful in predicting the outcomes of policy decisions in the European Community.[20] We continue to use the model because it has demonstrated greater ability to predict policy outcomes than competing models, but we remain open to the possibility that trades across issues could improve the performance of the model.

How Can the Model Be Estimated: A General Approach to Developing the Data

Political outcomes, whether they involve intra- or inter-governmental relations or negotiations between public and private organizations or even within a single organization, can be and have been predicted using the model delineated above. To do so, however, requires converting theoretical concepts into practical application. Although this can be an extremely difficult task, there is, fortunately, a body of knowledge that can be called upon to estimate the critical variables. By combining the perspective of this rational-actor model with the knowledge and expertise of area or issue *experts* it is possible to estimate the variables of interest and to solve the perceptual and "voting" components of the model discussed here.

The forecasting and perceptual models require the identification of the groups or actors interested in trying to influence a policy outcome on the issues in question. For each actor, data must then be estimated on three variables: capabilities, preferred outcome and salience. Sometimes, in institutionally structured settings, it is also important to take into account structural constraints that operate to help shape outcomes. With just this minimal information in hand, and *without any other information*, regarding, for instance, the history of the situation, the history of relations between particular actors within the situation, other sunk costs, or without even interviewing the actors involved to assess their own judgment about their beliefs and expectations, it is possible to predict what the likely outcome will be.

The full details of the estimation techniques are too lengthy to enumerate here.[21] However, it should be noted that it is possible to achieve a very high level of accuracy and cross-expert agreement on the essential information, even when the experts in question disagree markedly with one another on the likely outcome or evolution of discussion on an issue. The greatest power and insight from such an approach can be attained by combining the analytic consistency of models such as those explained here with the nuanced and rich insights of area or issue experts.

When issue experts combine their skills with the analytic strengths of these models then accurate and subtle predictions are most likely to be attained. To be sure, an abstract model such as the one proposed here is of no value without the information from experts needed to turn its abstractions into practical estimates. Expert knowledge, of course, is quite valuable even without a model to suggest what information is critical and how that information should be organized. But, in numerous controlled experiments it has been found that the predictions extracted from the model in conjunction with the information from area experts is substantially more reliable than the predictions made by the experts themselves. Furthermore, the predicted outcomes using input from different experts are quite close even when the exact nature of their information varies with the expert.

Analysis: The Possibility of Taiwanese Nuclear Proliferation

Nuclear proliferation poses a complex nexus of issues. Most immediate is whether a country will develop a nuclear weapons capability. How they choose to do so, through internal research or with external technology, is important for judging the appropriate strategies to prevent proliferation. Also of interest is the preparedness of some countries to accept substitutes for an independent nuclear capability, such as protection under a nuclear umbrella. Advanced conventional capabilities may also suffice for some as an alternative to nuclear forces. Although many other issues are germane to the development of a successful nonproliferation strategy, these three seem to be at the core.

Our analysis is centered on the following three specific issues:

(1) What is the preference of each interested party with regard to maintaining Taiwan's defense reliance on the United States or Taiwan's becoming self-reliant?

(2) What is the attitude toward seeking continued security through the development of an independent Taiwanese nuclear force versus maintaining a purely conventional force?

(3) If a nuclear capability is developed, what is the attitude toward relying solely on internal technology versus the importation of such technology and materials from external sources?

Data on the actors, their power, preference and salience, for these issues were developed by Samuel Wu in consultation with several other experts on Taiwanese policy. For the first issue, the actors of interest include two factions of the Kuomintang, the mainstream leadership (KMT1) versus those who are less confident about the reliability of the United States as a guarantor of Taiwanese security (KMT2). The DPP also falls into two factions, the radicals (DPP1) who are willing to trust the United States and Japan to provide for Taiwan's defense and the moderates (DPP2), who are hopeful but less confident about foreign protection. Within the United States, we identify two factions. The first (USA1) reflects the interests of those in the security elite in the United States who believe that the special relationship with Taiwan advances American interests. The second group (USA2) is less committed to Taiwan's defense and more interested in nurturing the growing relationship between the United States and the PRC. Finally, of course, Taiwanese officials must take into account the interests of and threat from the PRC. The data on power, preferences and salience of these actors are displayed in Table 1.

Table 1

GROUPS AND THEIR DATA ON TAIWANESE SELF-RELIANCE IN DEFENSE

Group	Power	Position	Salience
DPP1	20	80	70
DPP2	35	70	60
USA1	30	60	40
KMT1	100	55	70
KMT2	45	45	80
USA2	8	30	20
PRC	10	0	60

A position of 0 indicates a preference for complete self-reliance and 100 a preference for complete reliance on the United States for Taiwanese security.

Higher scores on power, of course, reflect greater potential influence over outcomes relative to other actors. Higher salience scores reflect greater interest and attention to the issue relative to other issues, with 100 meaning that the group dedicates all of its attention and resources to the issue during the period in which a decision is to be made. The preference scale runs from 0 to 100, with 0 meaning complete reliance on Taiwanese security forces and 100 implying complete reliance on the United States for Taiwanese security.

The prediction derived from the model indicates a mix of reliance, indicating that it is important for Taiwan to maintain an independent defense capability within the framework of the special arrangement with the United States (55 on the scale). That special arrangement should have a significant deterrent effect, especially on the PRC. The predicted outcome, then, is consistent with the wishes of the mainstream elements in the KMT (KMT1).

The issue is not without some controversy. Figure 2a and 2b show the difference between the PRC perception of its situation vis-à-vis the stake holders in Taiwan and the perceptions of those groups about the credibility of threats from the PRC.

Of particular note in comparing Figure 2a to 2b is that the PRC believes it can negotiate a tacit understanding with the Taiwanese domestic groups. Yet, from the perspective of those groups, they do not believe the PRC has any credible leverage. Rather, they expect the PRC to tolerate reluctantly the policy that Taiwan intends to adopt. At the same time, the figures reveal that tensions will run high between the United States and the People's Republic of China over the continuation of American security provisions for Taiwan.

The mainland Chinese government expects a conflict with the United States. Those in the United States who wish to continue the existing security relationship with the ROC mistakenly believe that they have the political clout to extract concessions from the PRC. As implied in Figure 2a, this American belief is mistaken. The Chinese are not prepared to compromise on this

Figure 2a

Reliance on the US Nuclear Umbrella: PRC Perceptions

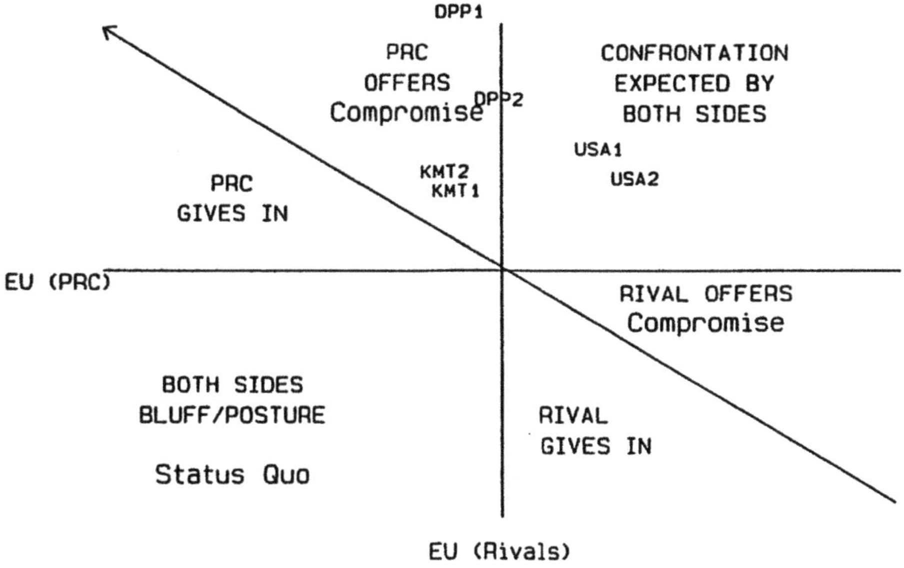

Figure 2b

Reliance on the US Nuclear Umbrella: Rival Perceptions

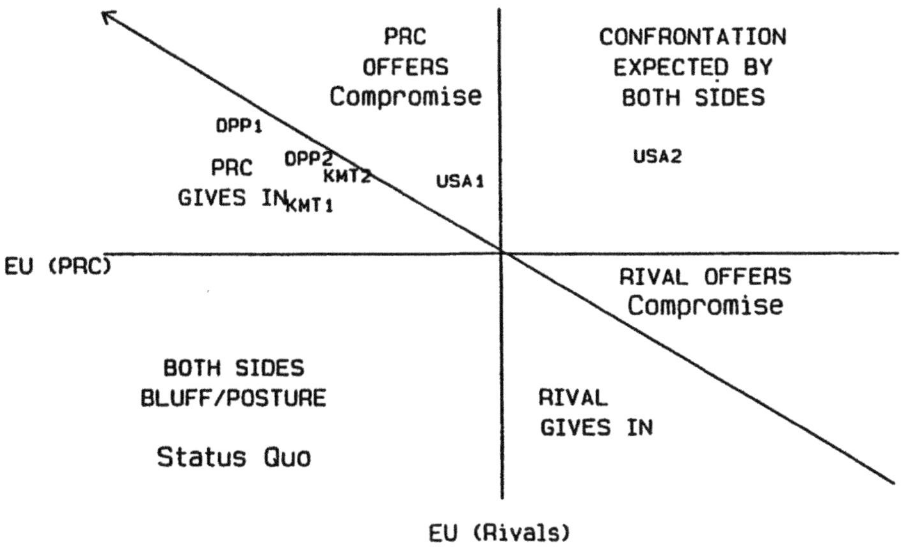

important issue and are willing to confront the United States diplomatically to assert their claims. The United States, then, can expect that its relationship with Taiwan will again loom large as a threat to the growing ties between the United States and the mainland. It is unlikely that Washington will have its cake and eat it too.

As a final observation regarding this issue, we note that our analysis indicates that there is little fluidity of positions. The main KMT faction gets what it wants with little credible objection from other domestic groups or from foreign actors. The KMT has positioned itself so as to create a wining coalition against challenges to its position from either direction. Notice, however, that the KMT *by itself* is not strong enough to enforce its will. Its positioning and coalitional strategy are both crucial to its success. It is quite likely, then, that the KMT leadership has sacrificed some underlying true preferences for the sake of establishing its winning posture.

The second issue we examine draws our attention to the prospects that Taiwan will rely upon its conventional forces or will develop strategic nuclear forces to insure its security. On the preference scale delineated in Table 2, zero refers to total reliance on conventional forces and 100 total reliance on strategic nuclear capabilities.

Table 2

GROUPS AND THEIR RELIANCE ON NUCLEAR CAPABILITIES

Group	Power	Position	Salience
GP2	20	70	60
TM1	40	67	75
SA	15	60	10
GP1	68	55	20
TM2	55	35	60
ECP	100	25	60
MAM	30	20	40
ENV	25	8	80
USA	87	4	20
UN	13	4	20
PRC	55	0	60

A position of 0 indicates a preference for total reliance on conventional capabilities and 100 a preference for total reliance on strategic nuclear capabilities.

GP1 and GP2 represent alternative views within the general public on Taiwan. GP2 tends to be sympathetic toward the development of a nuclear capability, while GP1 is more skeptical. GP1 finds the potential monetary savings from strategic weapons to be attractive, but is concerned about the environmental and international political consequences of a Taiwanese nuclear capability.

TM1 is a pro-nuclear faction of the Taiwanese military, while TM2 is more attentive to the prospective disadvantages of nuclear forces. ECP, the potentially most powerful actor, is a large coalition drawn from the KMT and the DPP and prominent think tanks. The ECP constituencies share the belief that economic power (hence ECP), not military might, lies at the root of Taiwan's security. This loosely knit group is fearful of the economic consequences of nuclear proliferation. MAM denotes the mass media on Taiwan. They favor building up conventional capabilities to maintain control over the sea routes into and out of Taiwan. Also of some significance are Taiwanese environmental groups (ENV) who are adamantly opposed to any nuclear development.

The key foreign actors – the United States, the United Nations and the People's Republic of China – share a common perspective in opposition to the development of a Taiwanese nuclear force. South Africa (SA), however, may be a willing provider of fissionable materials to Taiwan. As such, they are more sympathetic to the notion that Taiwan should develop its own strategic capabilities. South Africa's influence is somewhat enhanced by the fact that they are apparently the only developed nation in the world that currently recognizes Taiwan as the Republic of China. South Africa is, of course, a relatively weak participant on this issue.

Whether to emphasize conventional or nuclear weapons development is likely to be highly controversial. Indeed, we find that the actors themselves expect about two-thirds of all initial interactions on this issue to lead to conflict and political rancor. As the groups exchange views, however, there is some diminution in strife and some convergence in outlook. Table 3 shows the predicted changes in policy positions of each group in response to their negotiations and competition.

Table 3
MOVEMENT OF POSITIONS ON NUCLEAR CAPABILITIES ACROSS TWO ITERATIONS OF MODEL

Group	Initial	Intermediate	Final
GP2	70	31	31
TM1	67	31	27
SA	60	60	53
GP1	55	38	25
TM2	35	31	35
ECP	25	8	18
MAM	20	24	24
ENV	8	8	8
USA	4	10	10
UN	4	10	22
PRC	0	0	4
FORECAST	25	→ → → → → → →	24

Table 3 highlights the political dynamics behind this issue and also focuses our attention on the model's ability to sort out bargaining strategies. Most interesting is the movement of the moderate coalition, the ECP. At the outset they believe their position is strong enough to resist challenges from either side. What is more, they expect the strongly anti-nuclear international groups (USA, UN, and PRC) to make concessions to them. However, the model analysis suggests that they are overly optimistic. These groups not only do not offer concessions, but rather they react with virulent opposition. The ECP misperception leads them to over-react to hostile responses from international actors. As Table 3 indicates, the ECP at first moves sharply toward the position of the anti-nuclear groups. After the ECP moves from its initial position of 25 to 8, they find that their policy shift does not eliminate the potential for conflict with the anti-nuclear groups, but it certainly takes them out of the winning coalition. They have gone too far. To regain support, the ECP is predicted to move back toward their original position, apparently striking agreements with representatives of moderate public opinion (GP1), the pro-nuclear military (TM1), which moves dramatically from its initial position of 67 to 27, and the mass media (MAM). In short, recognizing that conflict with anti-nuclear forces is inevitable, the ECP seeks a middle ground that attracts substantial domestic political support for their outlook.

The third and final issue we assess concerns whether Taiwan should seek external aid in the event it pursues a nuclear weapons program. The issue continuum ranges from zero, indicating reliance on external assistance exclusively, to 100, in which case Taiwan would eschew foreign assistance, relying instead on its own technical know-how. Table 4 displays the initial data assumptions.

Table 4

GROUPS AND THEIR DATA ON SEEKING EXTERNAL AID FOR A TAIWANESE NUCLEAR PROGRAM

Group	Power	Position	Salience
PM	60	55	75
LEE	100	35	60
DEF	30	45	40
DPP	35	10	60
USA	70	5	30
JAPAN	20	15	35
PRC	25	95	75
GP1	55	50	40
GP2	45	20	80

A position of 0 indicates a preference for total reliance on external assistance in a Taiwanese nuclear program and 100 a preference for a completely internal program.

The Kuomintang coalesces into two factions, one surrounding Prime Minister Hau (PM) and one around President Lee (LEE). The former faction has advocated the development of an internal Taiwanese defense industry. President Lee and his followers are more oriented toward garnering international support on this issue. The defense minister (DEF) and his backers are not quite as committed to developing a domestic defense industry as is the prime minister.

The DPP, United States and Japan, to varying degrees, believe that Taiwan should be reliant on external assistance if they are to develop a nuclear program. Presumably the notion is that external reliance will create greater leverage for external actors to restrain a prospective Taiwanese nuclear program. Naturally, the PRC is adamantly opposed to any such assistance. GP1, the larger of the two public opinion sectors, favors some internal development to reduce Taiwan's security dependence on outside actors. GP2, representing other elements of the mass public, oppose domestic development of defense capabilities because they believe it is dangerous and unrealistic.

Initially, the prime minister appears to be in an excellent position to advance his policy agenda on this issue. However, the president, Mr. Lee, maneuvers by shifting his stance to be more in line with the preferences of the DPP, GP2, the United States and Japan. In this way he shores up his coalition, defeating prime minister Hau's ambition to establish a significant internal high technology defense industry.

The prime minister is not without coalitional support of his own and, indeed, on other issues he is able to force President Lee's hand. Whereas Mr. Hau appears to be outmaneuvered on the development of a high tech defense industry, Mr. Lee is vulnerable politically on his support for an Asian security alliance, an issue which we do not fully analyze here.

Conclusion

This paper has presented a model which can be applied to analyze security issues and employed that model to examine the possibility of nuclear proliferation by Taiwan. To summarize the results of our analysis of the three issues, Taiwan is unlikely to try to achieve a nuclear capability. Taiwanese security policy will continue to rely on conventional forces and the special relationship with the United States. Those groups that might have some interest in pushing towards nuclear development face a strong coalition of external and internal actors opposed to a nuclear program. Internally, important groups inside and outside the government are concerned about the likely economic and political consequences of a nuclear program. Externally, the United States and others are willing to continue to their current relations with Taiwan provided that nuclear development does not occur. This combination of internal and external actors would be difficult to defeat.

This paper has illustrated a new approach to security analysis. The model allows the formal analysis of how internal politics and external actors shape a state's security policy. States have many paths to security, and their chosen path depends in part on how politically active groups evaluate the available policies.[22] Political pressures from these groups and external pressures both matter. The model presented here provides a way to analyze this political struggle and forecast its outcome. The model provides detailed and reproducible results about the actors' perceptions of the situation, their strategic moves, and the outcome of the struggle over policy. It also allows the analyst to explore alternative scenarios and strategies for different actors, although we have not demonstrated this use of the model here. We can also test the reliability of the results in the face of different assumptions about the actors' capabilities and interests. We hope that others will find this method as interesting and valuable as we have.

Notes

The authors would like to thank Paul Chrzanowski, Peter Feaver, Kurt Gaubatz, Avery Goldstein, Michael Intriligator, and Ramon Myers for their comments on this paper.

1. For a good presentation of the neorealist position on nuclear proliferation, see Benjamin Frankel, "The Brooding Shadow: Systemic Incentives for Nuclear Weapons Proliferation," *Security Studies*, this volume.
2. Peter R. Lavoy lays out the possible responses in his "Nuclear Myths and the Causes of Nuclear Proliferation," *Security Studies*, this volume.
3. For a discussion of how nations choose between different sources of security, see James D. Morrow, "Arms Versus Allies: Tradeoffs in the Search for Security," *International Organization*, forthcoming. Benjamin A. Most and Harvey Starr discuss the problem of the substitutability of different foreign policies in their "International Relations Theory, Foreign Policy Substitutability, and 'Nice Laws,'" *World Politics* 36, no. 3 (April 1984): 383–406.
4. Steven Flank discusses the political and economic support needed for the large technological project of building nuclear weapons in his "Exploding the Black Box: The Historical Sociology of Nuclear Proliferation," *Security Studies* 3, no. 2 (Winter 1993), forthcoming.
5. Descriptions of the model can be found in Bruce Bueno de Mesquita, David Newman, and Alvin Rabushka, *Forecasting Political Events: The Future of Hong Kong* (New Haven: Yale University Press, 1985); and Bruce Bueno de Mesquita and Franz Stokman, *Twelve into One* (New Haven: Yale University Press, forthcoming). The model has been applied to predict the succession in Iran following Ayatollah Khomeini's death in Bruce Bueno de Mesquita, "Forecasting Policy Decisions: An Expected Utility Approach to Post-Khomeini Iran," *PS* (Spring 1984): 226–36 and the progress of peace negotiations in the Middle East in Bruce Bueno de Mesquita, "Multilateral Negotiations: A Spatial Analysis of the Arab-Israeli Dispute," *International Organization* 44, n. 3 (Summer 1990): 317–40.
6. The seminal source on spatial models of politics is Duncan Black, *The Theory of*

Committees and Elections (Cambridge: Cambridge University Press, 1958). On structure-induced equilibria, see Kenneth Shepsle, "Institutional Arrangements and Equilibrium in Multidimensional Voting Models," *American Journal of Political Science* 23 (1979): 27–59.

7. *World Journal*, 21 September 1991.
8. *The Journalist*, issue 222, 10–16 June 1991, p. 20.
9. Testimony before the Legislative Yuan on 11 March 1992.
10. For a complete discussion of the model, see Bueno de Mesquita, "Forecasting Policy Decisions"; Bueno de Mesquita, et al., *Forecasting Political Events*; Bueno de Mesquita and Stokman, *Twelve into One*. Black's median voter theorem can be found in his *The Theory of Committees and Elections*. The monotonicity theorem is from Jeffrey S. Banks, "Equilibrium Behavior in Crisis Bargaining Games," *American Journal of Political Science* 34 (1990): 599–614.
   11 x^{*}_{ia} is the outcome actor *i* has revealed to be preferred on issue a. It may or may not be *i*'s *true* ideal point. We generally do not know for sure what another actor's true ideal point is as there may be strategic incentives for an actor to misrepresent his or her ideal point. Because the model as applied here assesses policy decisions on one issue at a time, we drop the issue-denoting subscript (a, or b, etc.) from the notation so that henceforth x^{*}_{i} is the preferred position of actor i on the issue being evaluated at the moment.
12. Further discussion of risk attitudes and how they can be measured can be found in Bueno de Mesquita, "The War Trap Revisited"; and James D. Morrow, "On the Theoretical Basis of a Measure of National Risk Attitudes," *International Studies Quarterly* 31 (December 1987): 423–38.
13. Again we will drop the "a" subscript from the notation throughout, but the reader is alerted to the fact that the model does not assume that an actor's capabilities or potential power is the same on all issues.
14. Black, *The Theory of Committees and Elections*.
15. The measurement procedures for each term are explained in considerable detail in Bruce Bueno de Mesquita, "The War Trap Revisited." *American Political Science Review* 79 (1985): 156–77; Bruce Bueno de Mesquita and David Lalman. "Reason and War," *American Political Science Review* 80 (1986): 1113–30; Bueno de Mesquita, et al., *Forecasting Political Events*; Bueno de Mesquita, "Multilateral Negotiations;" and Bueno de Mesquita and Stokman, *Twelve into One*.
16. *i* makes a proposal if, in figure 1, the conjunction of *i*'s expected utility and *i*'s estimate of *j*'s expected utility falls between zero degrees from the horizontal axis and 45 degrees or falls between 270 degrees and 360 degrees from the horizontal axis. That is the domain within which *i* believes it has a comparative advantage over *j* and *i* expects more gains than losses from challenging *j*'s position.
17. A proposed capitulation by *j* to *i*'s wishes is made, *i* locates the conjunction of the respective expected utilities in the wedge that falls between 270 degrees and 315 degrees below the horizontal axis in figure 1 or in the wedge between zero degrees and 45 degrees. In the latter instance, *i* expects resistance from *j*, but *i* believes it can enforce its demand. In the former case, *i* expects no resistance from *j*.
18. A compromise is proposed if *i* believes the conjunction of the relevant expected utilities falls between 315 degrees and 360 degrees from the horizontal axis in figure 1.
19. For a fuller description of the sequential process in this game, we suggest seeing Bueno de Mesquita and Stokman, *Twelve into One*.
20. Bueno de Mesquita and Stokman, *Twelve into One*.
21. See Bueno de Mesquita and Stokman, *Twelve into One* for full details.
22. Morrow, "Arms Versus Allies."

Multilateral Cooperation and Nuclear Nonproliferation

Richard T. Cupitt and William J. Long

IN THE AFTERMATH of the 1991 Persian Gulf War, efforts to stem the proliferation of weapons of mass destruction gained new momentum. Above all, revelations about the extensive Iraqi nuclear program prompted calls for more efficient and effective multilateral controls on the transfer of nuclear material, equipment, and technology. Improving controls on these items involves strengthening two arrangements that form the bulwark of multilateral efforts to counter nuclear proliferation, the nuclear Non-Proliferation Treaty (NPT) and the Nuclear Suppliers Group (NSG).

In considering ways to improve multilateral cooperation through the NPT and the NSG, we offer a strategic perspective, drawing insights from game theory. In doing so, we do not focus on the determinants of proliferation as such. Instead, we review the obstacles to international cooperation from a strategic perspective and relate them to reforms in the NPT and the NSG. Using this approach, we intend to demonstrate that the NPT and NSG represent distinct types of cooperative arrangements in both theoretical and practical terms. Consequently, different, and sometimes contradictory, policies will be necessary to enhance their effectiveness.

A Strategic Perspective on International Cooperation

At its core, international cooperation requires a state to adjust its individual policies in ways that increase mutual benefits. To secure cooperation, states must coordinate their policies to fit "the actual or anticipated preferences" of other states.[1] The strategic perspective submits that states will engage in both cooperation and conflict by choice.[2] It subsumes both the realist and liberal approaches as representative of special classes of environmental conditions. Historically, scholars have used the strategic perspective to illustrate problems of deterrence, arms races, and arms control.[3] More recently, they have applied strategic models to understand international cooperation.[4]

The NPT presents a case of mutual adjustment. It is a bargain between nuclear weapons states and nonnuclear weapons states. Nuclear weapons states agree to assist the nonnuclear weapons states in developing peaceful applications of nuclear power and to reduce "vertical" proliferation (a growth in the size and sophistication of their nuclear arsenals). In return, the nonnuclear weapons states agree not to acquire nuclear weapons. Both sets of states constrain their policies in accordance with the preferences of other states, allowing both sets of states to reap benefits. The NSG also involves a process of policy coordination among member states. In the NSG, states that supply nuclear items develop guidelines defining exportable items and conditions for their transfer.

From a strategic perspective, enhancing cooperation in the NPT and the NSG requires an understanding of the nature and conditions surrounding each arrangement. As the authors will illustrate, the NPT and the NSG reflect different conditions and face different impediments to cooperation. Policies that enhance cooperation for one arrangement may fail to do so, or even decrease the prospects for cooperation in the other.

When Will States Cooperate?

The strategic perspective implies that the potential for cooperation for a group of states depends on the particular mix of policy preferences among the group's members. To illustrate points about international cooperation the strategic perspective most often relies on formulations of 2×2 games.[5] In most forms, two players can choose either to cooperate or defect (not cooperate). This stylized approach creates four potential outcomes for each player – cooperation, mutual defection, unilateral defection, and unreciprocated cooperation.

International relations involve situations where cooperation is possible, even preferable, but difficult and problematic to secure. In game theory terms, states have mixed motives, and their interactions have the potential for either a cooperative or a noncooperative outcome. Most research focuses on two classes of mixed-motive games: games where players face dilemmas of common interest and games where the players face dilemmas of common aversion.[6]

Dilemmas of common interest arise when choices made independently by each player lead to suboptimal results. In such situations, individual actors have a common interest in constraining their independent policies and achieving the optimal outcome through mutual adjustment.

Dilemmas of common interest exist in many games, including Prisoner's Dilemma, the game most commonly applied to international relations.[7] Prisoner's Dilemma begins by assuming two states each have two policy choices: cooperation or defection. Each state prefers, in order, unilateral defection, cooperation, mutual defection, and unreciprocated cooperation.

This situation closely resembles the conditions surrounding the NPT. For security reasons, a state might prefer to have nuclear weapons, especially if its potential opponents did not – a case of unilateral defection. A state's second preference would be to forgo nuclear weapons, if assured others would do the same (that is, cooperate). However, a state would prefer to adopt nuclear weapons, even if others do so (mutual defection), than to forego nuclear weapons when others adopt them (unreciprocated cooperation). The irony of this situation, as Robert Jervis notes, "is that even if each side prefers mutual cooperation to mutual defection (and each knows that this is the other's preference), the result can be non-cooperation because each party is driven by the hope of gaining its first choice – which would be to exploit the other – and its fear that, if it cooperates, the other will exploit it."[8]

The NPT is an agreement that helps states overcome the tendency toward noncooperative outcomes. Under the NPT, nuclear and nonnuclear weapons states are able to strike a bargain to curb the global proliferation of nuclear weapons. Since the signing of the NPT in 1968, surprisingly few nonnuclear weapons states have developed a nuclear weapons program. One can argue that the NPT helped avoid a potentially destabilizing and certainly debilitating global nuclear arms race.

Dilemmas of common aversion exist when states have a mutual interest in avoiding a particular outcome. In such cases, many strategies will lead to cooperation. Although individual actors may prefer different strategies, coordinating their actions permits them to choose one option from many that allows them to avoid an outcome neither wants.[9] Typically, players in common aversion games are equated with friends trying to pick a restaurant for lunch. When friends are indifferent to the choice of restaurants, so long as they lunch together, they only need to agree to a method of choice before picking a restaurant. When friends are not indifferent to the choice of restaurants, the players must agree on which outcome is fair, as well as a fair way of choosing.[10]

NSG members face a dilemma of common aversion. NSG members could adopt any of several regulatory frameworks that help them avoid diversion of nuclear items to weapons production. NSG states generally agree on the need to control some items to some destinations, although individual members may differ over the particular array of items to control or the destinations to proscribe. Unless member states coordinate their control lists and targets, however, all will suffer undesired consequences.

The theoretical distinction between these types of arrangements is of practical importance for efforts to enhance international cooperation against nuclear proliferation. Resolving a dilemma of common interest usually requires the creation of a contract that outlines the duties and responsibilities of all parties and ensures its enforcement. For common interest situations like the NPT, the issues are identifying and distributing the costs and benefits of cooperation, verifying compliance, and sanctioning noncompliance. The critical issue is usually verification – the more difficult an agreement is to verify, the less likely cooperation will occur. Recent problems with North Korea and Iraq dramatically illustrate this point.

Resolving a dilemma of common aversion usually involves setting standards and coordinating policies. For the NSG, the issues will be choosing a fair outcome and harmonizing procedures, rather than verification and enforcement. For example, spawned by the Indian nuclear test, the NSG met in London from 1975 to 1977 to standardize export controls on nuclear items and to avoid uncoordinated actions leading to further unwanted outcomes. Again in 1991, concerns about the role of dual-use nuclear items in the Iraqi nuclear program led the NSG to adopt a standard list of items to be regulated by all members.

Recognition of the fundamentally different nature of the two nonproliferation regimes has profound consequences for reforms designed to make them more effective. Techniques that strengthen one regime may be irrelevant or counterproductive for the other. For example, when states reap substantial benefits from cheating on an agreement when others cooperate (as in Prisoner's Dilemma), setting up collective enforcement procedures can make sanctions more effective and enhance cooperation. Under the NPT, collective enforcement is achieved by allowing signatories to deny other signatories access to nuclear technologies in retaliation for noncompliance. This sanction is an appropriate strategy for strengthening cooperation in a common interest arrangement.

In contrast, with few incentives to cheat, as in a common aversion situation, sanctions are redundant or counterproductive. The NSG does not impose sanctions against its member states for noncompliance with NSG guidelines. Given the preferences of NSG members, it is possible that when "leakage" (an illicit transfer of nuclear items) occurs, it is the result of accident, ignorance, or incapacity. In those instances, assistance in meeting the agreed standard, not punishment, will enhance cooperation. Even where willful neglect or outright connivance by a government occurs, the sense of scandal that emerges as members learn of lax enforcement or other questionable practices leads to reform (as many states have done since revelations of Western participation in Iraqi programs). Unilateral measures may backfire. For example, the extraterritorial application of the more restrictive U.S. export controls against other members of the Coordinating Committee for Multilateral Export Controls (COCOM) in the 1980s prompted some members to block U.S. efforts through national legislation rather than tighten their export controls.

Besides differences in the core problems facing the NPT and the NSG, factors external to the immediate strategic environment affect cooperation in the two types of arrangements in very different, even contradictory, ways and commend different policy responses. Below the authors examine some exogenous factors noted in the literature and consider their differential impact on the NPT and the NSG.

Exogenous Factors Affecting Cooperation

Certain exogenous factors have the same general impact on cooperation in either common interest or common aversion arrangements. For example, although the transaction costs (such as the cost of making or enforcing an agreement) in common interest and common aversion games may differ, high transaction costs impede cooperation in both.

Other exogenous factors may be relevant to one arrangement, but not both. By way of illustration, the relative distribution of gain affects the outcome of common aversion games very little. However, in common interest situations, if one player perceives that another will use a favorable relative gain to seek an advantage in their relationship, then cooperation may not occur.[11]

For a third set of exogenous factors, the consequences are random. These factors can either inhibit or enhance cooperation in both common interest or common aversion games. For example, the impact of sub-unit (domestic) players on the unitary (national) player may enhance or impede cooperation in both types of regimes. Rent-seeking behavior by domestic pressure groups can make cooperation less likely.[12] Pressure groups can induce governments to adopt noncooperative policy positions, forsaking benefits for the nation as a whole to secure gains for a special group. When Western firms abetted Iraqi nuclear weapons programs, they undercut national efforts to coordinate nonproliferation policies. Conversely, other domestic factors may make cooperation easier. The emergence of democratic forces in Brazil and Argentina made both countries more cooperative on issues of nuclear nonproliferation.

All of these exogenous variables are relevant to the operation of the NPT and the NSG. Our intent, however, is to illustrate the distinctions between the two arrangements and the need for policymakers to disaggregate their nonproliferation proposals when considering strategies for enhancing cooperation. For that purpose, the authors examine a fourth set of exogenous factors that consistently affect cooperation in common interest and common aversion games in opposing ways. In practical terms, NPT and NSG members need to respond differently to the same external forces to promote cooperation. Game theory suggests that at least three exogenous factors will have contrasting effects on the prospects for cooperation in common interest and common aversion games. These are the impact of time, the number of actors, and the asymmetric distribution of capacity among members.

The Impact of Time

Perhaps more than any other exogenous factor, time alters the prospects for cooperation. Indeed, the very foundation for cooperation in common interest games depends on the potential for the game (or turns within a game) to be played sequentially over time.

For common interest games, if the players believe there will be iterations to the game and that the stakes will remain regular, then the farther away the last game is from the present, the greater the potential for cooperation.[13] This result occurs because iterated play allows actors to use bargaining strategies that can lead to reciprocal cooperation, particularly variations of a Tit-for-Tat strategy.[14] If player A uses a Tit-for-Tat strategy, player A begins by cooperating. In each subsequent game, player A imitates the action of the other player or players from the previous game. This strategy allows a player to respond swiftly, in a reciprocal fashion, to punish defection and reward cooperation. Over repeated plays of Prisoner's Dilemma games, for example, theorists find this strategy elicits cooperation between rational players.[15]

The nearer a player believes the last game or last play is from the current round, the more difficult it is to achieve cooperation, as the incentive to cheat is greater at the end than at the beginning or middle of play. As the time horizon shrinks, the "shadow of the future" becomes less and less of an inducement to cooperation.[16]

The slow evolution of cooperation between the nuclear haves and have-nots under the NPT and its predecessors reflects this pattern of learned cooperation. Fifteen years of bilateral and regional cooperative efforts laid the foundation for the eventual creation of the treaty-based NPT regime in 1968. Nearly twenty-five years of relative success in slowing proliferation under the NPT also lengthened the shadow of the future and encouraged states to remain in compliance with their nonproliferation obligations.

In contrast, the shadow of the future has a somewhat perverse impact on common aversion games whenever the players are not indifferent about the particular outcome or mode of cooperation. Assume two players and two stable cooperative outcomes, each of which favors a different player (though either are preferable to any noncooperative outcome). Under common aversion conditions, the longer the shadow of the future, the more likely one or both players will delay cooperation and risk failure, especially if they disagree on what behavior they consider fair.[17] This outcome occurs because players understand that any perceived inequity will become institutionalized through standardization, which would undermine the value of the relationship over time.

The different impact of time horizons on the two types of arrangements suggests that crises, or other events that shorten the time horizon, affect cooperation within the NPT and the NSG systems very differently. The crisis presented by the detonation of a nuclear device by India in 1974 illustrates these diverse effects. By shortening the shadow of the future for a few key neighboring states, the detonation of the Indian nuclear device made globalizing the NPT much more difficult. In particular, Pakistan came to see the long-term benefits of the NPT as insufficient to counter the near-term benefits of acquiring a nuclear deterrent against India. The reverberations of a potential Indian-Pakistani nuclear arms race had an immediate impact on security policy for China, the Middle East, and the United States. In contrast, the same crisis induced cooperation through creation of what would become the NSG. The supplier states saw that without greater control over the export of nuclear materials and equipment, many more states could quickly gain the capacity to build nuclear weapons.

What action might states take based on the theoretical implications of the impact of time? Let us assume crises that shorten the time horizon for NSG and NPT states will occur periodically. If leading states in the NSG create and discuss plans for a radical strengthening of nuclear export controls or other programs in an informal manner, then states could take advantage of crises to garner the political support to make these plans operational. This semi-reactive strategy would avoid debilitating controversies over radical change when members do not face a crisis, while building consensus on the need for change when a crisis occurs. For the NPT, states should place greater emphasis on enforcement of existing arrangements and crisis management. If states are not convinced that swift and reciprocal punishment will be the response to cheating, then in times of crisis, when the incentives to cheat may be higher than normal, states will break the agreement. Because crises have negative consequences for the NPT, members should also give high priority to improving crisis management mechanisms.

The Number of Players

Game theorists suggest that increasing the number of players in a game often impedes cooperation.[18] Although the potential total gain from cooperation should increase as the number of players increases, additional players can affect the stability of cooperative efforts because of the diversity of national interests and practices that an arrangement must reconcile. Theory also tells us to anticipate that a large and growing number of actors will pose greater problems for common interest arrangements than for common aversion situations.

In common interest games, cooperative solutions usually require the players to strike a narrow agreement. Much like contracts in international trade, solutions must carefully define the responsibilities and benefits accruing to each party, the means for dispute settlement, and the penalties for noncompliance. When there are many states, bargains are likely to be delicately balanced and somewhat rigid.[19] Sustaining cooperation in the face of changing conditions – such as the addition of new players – becomes difficult.

Initially, the NPT was to be a bargain among the nuclear powers and the advanced, near-nuclear states in Europe along with Japan. By the time the NPT negotiations ended, the initial bargain was insufficient to secure the cooperation of certain emerging nuclear and near-nuclear powers, such as China, India, Pakistan, North Korea, and Brazil. Nor did the NPT bargain maintain uniform compliance among existing members, as the Iraqi programs proved. Thus, the growing number of potential nuclear powers strained and, at times, overwhelmed the original bargain between the nuclear weapons states with the nonnuclear weapon states. Increasing the number of players complicated efforts to engage cooperation.

In contrast, for common aversion games, once cooperation occurs, the participation of many players implies that a broad, extensive, and interdependent network of coordinated practices has come into being. Once coordination or harmonization takes place, cooperation is likely to persist, even with the introduction of new players.[20] This pattern develops because common aversion arrangements, with less precisely defined obligations and responsibilities for members, have the flexibility to incorporate new members that share the general consensus of the group, even if their application of its rules do not conform precisely. The ability of the NSG to add new members rapidly while creating more restrictive guidelines on coordination illustrates this characteristic.

In the future, theory tells one to expect that efforts to add signatories to the NPT are likely to be more complex, more problematic, and more regime-threatening than efforts to increase the membership of the NSG. Incorporating new signatories to the NPT entails amending a fragile and delicately-balanced bargain, whereas getting more nuclear supplier states to abide by NSG guidelines involves building a general consensus. In light of this difference, one practical approach would be for NPT members to adopt a negotiating framework similar to the accession practices used to encourage participation in the General Agreement on Tariffs and Trade (GATT). Under this format, states may become full-fledged contracting parties to the agreement only by undertaking the necessary tariff and trade concessions towards member states. At the same time, some states may participate in negotiations and reap some of the benefits of cooperation without incurring the costs of becoming a contracting party through a form of limited membership. While this approach would create a two-tier NPT, the tiers would not pit nuclear-weapons states against nonnuclear weapons states. The NSG, in turn, should aggressively recruit new members. Bringing states into the NSG will yield more deeply embedded standardization of nonproliferation policies.

The Issue of Asymetric Capacity

In practice, the many states issue is synonymous with an asymmetry of capacity among the members. Theory maintains that asymmetries can enhance cooperation in common interest games.[21] Powerful players have the resources to underwrite larger transaction costs, including the costs of creating appropriate international institutions. Powerful players have the resources to make side-payments to entice players to cooperate and not to cheat. Many scholars go so far as to argue that international arrangements only come into being through action by a hegemonic state.[22] Forming or transforming a cooperative arrangement may require substantial resources and leadership that only a few countries can provide.

While the grand bargain or contract in a common interest game reflects the priorities of the powerful, several scholars emphasize that the rules created by a hegemonic player or players often bring greater relative rewards to the weaker players, which persuades the weaker players to participate.[23] This effect creates a dilemma for the powerful. Cooperation brings substantial rewards to the powerful in an absolute sense, but the weak become somewhat more powerful. Noncooperation may sustain the gap between the powerful and the weak, but the powerful lose the gains reaped from cooperation.

In the NPT, for example, the promise of access to nuclear technology and assistance for civilian nuclear programs under the treaty was not sufficient to entice many nonnuclear weapon states to join. Outside the formal agreement, the nuclear weapon states – the United States, Great Britain, and the former Soviet Union – had to deposit documents with the United Nations Security Council outlining positive security assurances regarding their pledge not to use nuclear weapons to get weaker, nonnuclear states to join the NPT. The transfer of nuclear material and technology under the NPT also closed the technology gap between the nuclear have and have-not states.

In common aversion games, asymmetry of power also facilitates coordination, but it usually favors powerful players. Generally, this result is a minor problem in common aversion arrangements where disparities in relative gains are not a great concern for members. For example, the sixty-eight item dual-use list recently adopted by the NSG almost exactly parallels the seventy item Nuclear Referral List in U.S. law, and allegedly sixty of those items were identified by U.S. nuclear laboratories for the NSG. Moreover, the extent of U.S. intelligence capabilities guarantees that U.S. assessments of potential targets of the controls will dominate discussions.

The effects of asymmetrical power on benefits accruing to leaders in the two arrangements have implications for U.S. policy. In common interest arrangements, leadership often involves transferring resources to weaker members to induce cooperation. In an era of limited resources, this responsibility implies that a prudent course for the U.S. would be to mutilateralize any effort (including any resource transfers) necessary to induce states to join and comply with the NPT. In contrast, in the NSG, U.S. leadership can reap a greater share of the benefits of cooperation without resource transfers. The NSG is likely to provide a more efficient venue for U.S. leadership.

CONCLUSIONS

The policy-making community is currently awash in recommendations for strengthening international cooperation in nuclear nonproliferation. It is beyond the scope of this article and the expertise of the authors to critique thoroughly the many nonproliferation policy proposals afoot. Whatever the wisdom of these many proposals, however, this article cautions against sweeping statements that recommend undifferentiated strategies for strengthening the NPT and the NSG. Strategic theory suggests that recognizing the distinctive features of the two nonproliferation arrangements and tailoring policies that work within their respective strategic environments will more likely enhance multilateral cooperation in the fight against proliferation.

NOTES

1. Robert Keohane, *After Hegemony* (Princeton: Princeton University Press, 1984), 51.
2. See Duncan Snidal, "International Cooperation Among Relative Gain Maximizers," *International Studies Quarterly* 35, no. 4 (December 1991): 387-402.
3. See for example, Steven J. Brahms, *Superpower Games: Applying Game Theory to Superpower Conflict* (New Haven: Yale University Press, 1985).
4. See for example, Robert Alexrod, *The Evolution of Cooperation* (New York: Basic Books, 1984); Keohane, *After Hegemony*; Kenneth Oye, ed., *Cooperation Under Anarchy* (Princeton: Princeton University Press, 1986).
5. Anatol Rapoport, M. J. Guyer, and D. G. Gordon, *The 2x2 Game* (Ann Arbor: University of Michigan Press, 1976).

6. See Arthur Stein, "Coordination and Collaboration: Regimes in an Anarchic World," *International Organization* 36, no. 2 (Spring 1982): 299-324.
7. See for example, Alexrod, *The Evolution of Cooperation*.
8. Robert Jervis, "Realism, Game Theory, and Cooperation", *World Politics* 40, no. 3 (April 1988): 493.
9. Stein, "Coordination and Collaboration: Regimes in an Anarchic World," 309; Duncan Snidal, "Coordination Versus Prisoner's Dilemma: Implications for International Cooperation and Regimes," *American Political Science Review* 79, no. 4 (December 1985): 923–42.
10. In addition to the articles cited in n. 9, see also Beth V. Yarbrough and Robert M. Yarbrough, "Reciprocity, Bilateralism, and Economic 'Hostages': Self-Enforcing Agreements in International Trade," *International Studies Quarterly* 30, no. 1 (March 1986): 7–21.
11. Duncan Snidal, "Relative Gains and the Pattern of International Cooperation," *American Political Science Review* 85, no. 3 (September 1991): 701–26.
12. Rent-seeking exists when a resource owner demands a payment above the amount its resource could otherwise command in the economy.
13. Robert Axelrod and Robert Keohane, "Achieving Cooperation Under Anarchy: Strategic Institutions," *World Politics* 38, no. 1 (October 1985): 254.
14. Alexrod, *The Evolution of Cooperation*.
15. Ibid.
16. See Axelrod and Keohane, "Achieving Cooperation Under Anarchy: Strategies and Institutions."
17. Snidal, "Coordination Versus Prisoner's Dilemma: Implications for International Cooperation and Regimes."
18. Ibid., 929; Michael Nicholson, *Formal Theories of International Relations* (Cambridge: Cambridge University Press, 1990), 133.
19. Oran R. Young, "International Regimes: Toward a New Theory of Institutions," *World Politics* 39, no. 1 (October 1986): 104–22.
20. Snidal, "Coordination Versus Prisoner's Dilemma: Implications for International Cooperation and Regimes."
21. Alexrod and Keohane, "Achieving Cooperation Under Anarchy: Strategies and Institutions."
22. See Keohane, *After Hegemony*; Charles P. Kindleberger, *The World in Depression, 1929–1939* (Berkeley: University of California Press, 1986); Duncan Snidal, "The Limits of Hegemonic Stability Theory," *International Organization* 39, no. 4 (Autumn 1984): 579–614.
23. Arthur Stein, "The Hegemon's Dilemma," *International Organization* 38, no. 2 (Spring 1984): 355–86.

Appendix

Technical Barriers to Nuclear Proliferation

Peter D. Zimmerman

IT TOOK THE United States three and a half years and two billion (1940s) dollars to build its first nuclear explosive device.[1] The official start of the Manhattan Project was 6 December 1941, and the Trinity test took place on 16 July 1945. The Soviets began their nuclear project at a low level in 1943,[2] increased its tempo after the war's end, and detonated a first device in August 1949. It seems to take about four years of steady work coupled with massive investment for a nation with an industrial and technical base corresponding to that of the United States in 1941 to progress from a knowledge of the fundamental physics of nuclear weaponry to a working device.

The principles of fission weapons were well described in the 1943 lectures of Robert Serber which became known as *The Los Alamos Primer* or "LA-1," the first technical report issued by the Los Alamos laboratory.[3] LA-1, a secret document until it was declassified in 1965, remained a cult object among physics students until its recent publication by the University of California Press.[4] Except for the perfection of the implosion method of critical mass assembly, little which is fundamental has changed since 1943. The true secret of nuclear weaponry is that nature does not step in at the last instant and forbid the process; rather, the difficulties lie in the engineering and industrial tasks.

Requirements for Building a Weapon

A fission weapon consists simply of a core of fissile material – uranium containing a high percentage of the isotope ^{235}U, or plutonium – together with an explosive system to assemble rapidly a critical mass of the material, and an "initiator" which ensures that a supply of neutrons is available in the core at the appropriate moment during the assembly. In addition, a nuclear device includes the arming, fuzing, and detonating components that are needed to initiate the entire explosive chain at the right place and time. Each individual component is deceptively easy to describe qualitatively, but far more difficult to design, test, and construct.

Weaponizing Special Material

Uranium weapons may be assembled with a modified cannon. The bomb that destroyed Hiroshima was of this sort and so conservatively designed that it was used without a nuclear test. Plutonium, in contrast, has a high spontaneous fission background, which produces a large number of unwanted neutrons. This, according to the *Los Alamos Primer*, appears to make gun assembly of a plutonium weapon seem just barely feasible.[5]

To make things worse, however, it was soon discovered at Los Alamos that even the purest plutonium produced in a reactor has a small admixture of the isotope ^{240}Pu which decays extremely rapidly by spontaneous fission. The presence of ^{240}Pu produces such a high neutron background that the muzzle velocities of even the most powerful cannons are far too low to make a plutonium gun feasible.[6] Faced with that fact, Los Alamos scientists invented the second, and by now standard, method of assembling a critical mass: implosion driven by focused shock waves from the detonation of high explosives.

Design of a safe and reliable implosion system is also not simple. Careful study of the chronology of the Manhattan Project[7] indicates that the first tentative experiments began in late fall and early winter of 1943 when S. Neddermeyer first wrapped solid high explosive around metal cylinders. James Tuck made the critical suggestion to use shaped charges in April of 1944, and a "promising" test was achieved for the first time in December 1944. From that point progress was apparently made extremely rapidly, for laboratory director J. Robert Oppenheimer froze the final design of the explosive lenses on 5 March 1945.

A modern weaponization project should not take as long to achieve satisfactory implosions for several reasons. First, the principles of lens charges are now widely understood, so that several months need not be wasted using other charge geometries. Second, the limits on simultaneity of detonation and homogeneity and symmetry of the imploding system may not be quite as high as were assumed to be necessary during the Manhattan Project. Finally, the techniques of preparing and casting high explosives have improved generally so that what was considered nearly unreachable precision in 1945 is now not much better than routine.

Even so, the development work needed to obtain implosions of the quality of the Trinity and Nagasaki weapons is not trivial. X-Division (explosives) at Los Alamos used 100,000 pounds of high explosive per month at peak consumption,[8] although the final weapon probably did not contain more than 4,000 to 5,000 pounds of explosive.[9, 10] According to various sources, the Fat Man weapon weighed between 10,000 and 10,800 pounds including its aerodynamic case and fuzing mechanism.[11]

While the detonators used in an implosion system are special purpose, and the firing system must produce a fast rising pulse of high voltage and current in order to explode the bridge wires of the detonators with acceptable simultaneity (on the order of one microsecond according to the *Primer*), modern switching devices and capacitors available "off the shelf" are probably adequate. The principles of the electrical circuits involved are not secret; indeed, a visitor to the unclassified nuclear weapons museum at Kirtland Air Force Base in Albuquerque, New Mexico will find a display of early solid-state switching devices called "sprytrons" together with schematic diagrams of the circuits in which they are used. The fabled "krytron" often referred to in the news media is not required. Neither krytrons nor sprytrons were available in 1945. A designer with only modest skill in electrical engineering should be able to produce firing circuitry of perfectly acceptable performance.

The fuzing system which commands the detonation sequence to begin is also not technically difficult. A barometric fuze such as was used in the destruction of Pan American flight 103 over Lockerbie, Scotland, should be adequate, at least in principle, for an air-delivered weapon. A weapon delivered by a ballistic missile might be set off by a time fuze where the firing instant is established at the moment when the thrust of the rocket motor is terminated and the flight time to target has been set by Newton's laws.

A weapon is not complete and ready for delivery with the construction of its fissile core and tamper; in addition, an "initiator" is required to introduce a burst of neutrons into the supercritical mass at precisely the correct time in order to start the chain reaction. Initiators working on several different sets of physical principles have been described in the open literature in adequate detail to permit a competent scientist or engineer to fabricate an acceptable device without very much experimentation.

There are some areas of difficulty which, while not barriers to the production of nuclear weapons, must be addressed with care by a proliferant. The first of these is establishing a set of guidelines and process limitations which prevent a "criticality accident" in which sufficient fissile material is brought together to start a chain reaction. This is not the same thing as ensuring that a bare sphere critical mass of active material is never assembled or stored in one place at one time. Criticality accidents can happen if a subcritical mass of uranium is present in a water solution, because the water can moderate the neutrons so that the actual amount of nuclear material needed to go critical is greatly reduced. Moderated neutrons are too low in energy to produce a nuclear explosion, but a slow critical reaction can nevertheless release great amounts of heat and radioactive material in addition to producing a burst of neutron and gamma radiation. Most of the radiation fatalities in the American nuclear weapons program have been the result of criticality accidents.[12]

Another area of difficulty is "health physics," the protection of workers from radiation and the provision of protection from the highly toxic materials used in nuclear engineering and weapons work. Skilled technicians and scientists are the primary capital of a weapons project, and keeping them alive and healthy is a primary consideration. Iraq did not neglect this aspect of a successful nuclear weapons project, although even by Manhattan Project standards their precautions were primitive and inadequate to protect the majority of workers; top scientists were, however, better protected.[13]

The metallurgy and chemistry of both uranium and plutonium are unusual. Uranium is difficult to machine because it easily hardens.[14] Furthermore, it is pyrophoric,[15] as is plutonium. And plutonium is highly toxic. These difficulties are well documented today, and competent students of the nuclear fuel industry will have little difficulty learning how to handle the materials and to prevent radiation accidents.

Although computer-controlled machine tools are useful for manufacturing the intricate shapes of the "physics packages" of modern fission weapons and their high explosive components, such sophisticated tools are unnecessary. Sophisticated nuclear weapons were built in the era from 1945 through the middle 1970s, before computer-controlled tools became available.

Once high explosive assemblies and fission cores have been designed, they must be tested, although the testing can use inert material (^{238}U instead of ^{235}U or plutonium) with similar physical and chemical characteristics. The basic diagnostic tools for such tests have been widely publicized – streak and framing cameras, flash x-rays, and neutron flux as a function of time.[16] None of these go much beyond what has become standard industrial practice, and all of them were employed during the Manhattan Project.

Obtaining and operating the diagnostic tools for non-nuclear explosive testing pose no insurmountable barriers to any proliferant, even one trying to use as much indigenous technology as possible. Non-nuclear testing should be adequate to prove the functioning of weapons at least as sophisticated as the Trinity and Nagasaki devices, and conceivably somewhat more advanced designs.

The technical barriers to weaponization of fissile material come down to designing and proving the explosive sets needed to assemble the supercritical mass; producing a reliable initiator; coping with the physical and chemical properties of the fissile materials; and performing the necessary proof testing of the designs. In the era of the Manhattan Project it was necessary for brilliant scientists not only to produce the designs, but to measure the physical constants needed to perform the calculations, to learn the chemistry and metallurgy of uranium and plutonium, and to invent many of the techniques involved. In 1993 all of the relevant physical constants have been published, and the diagnostic and construction techniques are well understood and widely known. While a proliferant would have to assemble a competent team, it could be smaller than that assembled at Los Alamos and need not be as brilliant. Today there is little need for creative invention; a skillful retracing of a well-worn path will do.

OBTAINING SPECIAL MATERIAL

Despite the technical difficulties of constructing the first atomic bombs and the constellation of brilliant scientists associated with the laboratory at Los Alamos, the majority of the investment of the Manhattan Project was made in procuring the special materials, plutonium and enriched uranium. Of the $1.88 billion total cost of the Manhattan Engineer District through 31 December 1945, $951 million, or just more than half, went to Oak Ridge and the enrichment facilities there, an additional $390 million, or 21 percent, was spent at the Hanford Engineer Works (plutonium production reactors), and only $74 million dollars, or 3.9 percent of the entire project, went to research, design, and production of the weapons themselves at Los Alamos.[17] At each site except Los Alamos the funds used to construct the plant greatly exceeded the operating budget, sometimes by as much as a factor of 8.5 (the K-25 gaseous diffusion plant).[18] It is reasonable that a new entrant into the nuclear club would expend funds in roughly the same ratio on special materials as compared to weaponization – except that a new proliferant would be likely to select one route to the bomb, either plutonium or uranium, and not pursue both simultaneously. This was true in the case of Iraq which chose to go the route of uranium enrichment. Other examples include India and Israel which appear to have concentrated solely on plutonium.

Uranium Enrichment

Knowledge of the general methods of uranium enrichment is now widespread. Indeed, enrichment capability to the range of 3–5 percent ^{235}U now exists in many nations, for example at the URENCO plant in the Netherlands. Pakistan has built and is presumably operating a large enrichment facility using gaseous centrifuges; Iraq had plans to construct a 10,000 centrifuge cascade. South Africa invented its own "nozzle enrichment" process. Pakistan apparently enriches uranium to the 90 percent range, which is weapons grade; and Iraq clearly intended to produce equally useful material. Even gaseous diffusion plants, such as that built at the K-25 site at Oak Ridge during the Manhattan Project, have been built in many nations.

The simplest method for enriching uranium, one which is well documented in the open literature, relies on straight-forward application of the principles of a mass spectrograph, an idea familiar to most freshman physics students. This is called Electromagnetic Isotope Separation, or EMIS. The machines used for EMIS are generically termed calutrons, a word derived from the University of CALifornia CycloTRON Laboratory where the technique was invented. Iraq had constructed several calutrons before Operation Desert Storm began and was believed to have produced small quantities of highly enriched uranium.

Calutrons are no longer used for enrichment in advanced countries. This is not because EMIS is an unsatisfactory technique to produce high quality material at a reasonable rate, but because it uses an extraordinary amount of electric power. Only a country such as the United States in 1944, with the huge Tennessee Valley Authority hydroelectric infrastructure, or Iraq in the 1980s with a similar surplus of hydroelectric generation capacity, is likely to be able to afford the drain on its power grid. The design of a calutron, however, is straightforward and the requirements for precision machining and high vacuum are not beyond what a developing country can procure. The simplicity of design and construction makes a calutron enrichment system particularly attractive to a proliferant from the developing world.

Because of the enormous power requirements, the United States has considered electromagnetic separation unattractive in comparison to either gaseous diffusion or centrifuge enrichment. As a consequence, most technical information regarding the design and construction of calutrons, ion sources to produce the beams, and appropriate vacuum technology have all been declassified.

For many years gaseous diffusion was the mainstay of uranium enrichment. It also requires an enormous amount of electricity and a very large cascade of "porous" barriers and high pressure pumps. Operating a gaseous diffusion plant successfully is as difficult today as it was in the early 1940s: finding the proper diffusion barrier materials and constructing a sealed system in which uranium hexafluoride gas can be passed through the barriers. In terms of construction costs, a gaseous diffusion enrichment plant is probably the most expensive to build, but it is less expensive to operate than an electromagnetic isotope separator.

The technical base from which to construct laser isotope separators is probably beyond the reach of most potential proliferants. Gaseous and liquid thermal enrichment facilities are easy to build and adequate to step natural uranium to a level of enrichment at which calutrons operate efficiently, but they are not capable of enriching uranium to levels useful in weapons.

Plutonium

Plutonium is the fissile material of choice. It is more active than uranium and has a higher fission cross-section, and therefore requires less active material per weapon. The smaller quantities of plutonium required for a bomb make it easier to design an implosion weapon using plutonium than one using uranium. Plutonium, however, *does* require an implosion system; the relatively high neutron background even in pure ^{239}Pu makes the gun option impractical. When the additional neutron background from the unavoidable

^{240}Pu contamination is included, it becomes clear that gun assembly is not merely difficult but impossible.

Plutonium can only be produced in a nuclear reactor. The rule of thumb is that for every one megawatt(thermal)-day of reactor operation one gram of plutonium is produced.[19] If six kilograms of plutonium are needed to produce a bomb, the production reactor must be operated for 6,000 megawatt (thermal) days. A 100 MW(t) reactor would, therefore, have to operate around the clock for sixty days to produce the active material for one bomb.

In order to reduce the content of ^{240}Pu in the total plutonium inventory, one must frequently change the fuel rods in the reactor. As a result, the duty factor of a plutonium production reactor is probably only about 60–70 percent, extending the required number of calendar days to produce material for a bomb.

Reactor design is well understood and well described in textbooks on nuclear engineering. It is likely that any potential proliferant could send its students to study in universities with nuclear engineering departments in Western nations or in the republics of the former Soviet Union, where they could master the fundamentals of reactor design and construction. Plutonium production reactors date, technologically, back to 1944.

There are no insuperable barriers to the indigenous construction and operation of such facilities in most countries. This is particularly true if the potential proliferant opts for several relatively low power units rather than a single high capacity reactor. Smaller reactors are easier to cool, control, and construct than larger ones. Furthermore, a number of smaller reactors provides redundancy, an opportunity to have most reactors producing while one is down for maintenance or refuelling, and an opportunity to modify the design of newly-constructed reactors based on operating experience with the earlier ones.

Leonard S. Spector and Selig Harrison of the Carnegie Endowment, who visited North Korea, reported that the North Koreans were having difficulties commissioning and operating even relatively low capacity but indigenously designed and built reactors.[20] North Korea may, however, be the special case of an international outcast unable to obtain technical assistance from any other country – unlike Israel, which reportedly had assistance from France, and India, which received reactor technology from Canada. Few other nations may be as unable to trade for technology as the North Koreans, although an exchange of Korean Scud missiles, originally supplied by China, for Algerian nuclear knowledge, for example, might well be attractive to both countries.

After operating its production reactors long enough to obtain adequate stocks of plutonium for testing and to build a supply of weapons, a country must reprocess the spent nuclear fuel to extract the material. While reprocessing is a 1940s technology well described in many places, it is hazardous and difficult. The spent fuel is fiercely radioactive and a potential proliferant may not wish to wait months or years while the short half-life isotopes decay. Nonetheless, an out-of-print Oak Ridge National Laboratory report from the middle 1970s suggested that an adequate "hot cell" to produce one or two bombs a year could be safely built in a space about the size of a two-car garage.

The nature of the plutonium reprocessing technique was described early in the nuclear age by Henry DeWolf Smyth.[21] The trick is to find chemical reactions which are efficient at separating the small number of atoms of plutonium made in the reactor from the large number of unwanted uranium (mostly ^{238}U) atoms, and separating the plutonium from the chemically similar but highly radioactive fission products. Most of the useful chemical reactions are good at one or the other of these two tasks, but not both. Hence, the reprocessing facility needs to be capable of cycling the material through the various reactions several times.[22]

India, Iraq, North Korea, and Sweden, among many others, have all demonstrated the ability to extract plutonium from spent fuel despite the radioactivity and toxicity of the material. The barrier to proliferation posed by the need to reprocess is thus very low indeed.

Plutonium metallurgy, however, presents unique difficulties. The metal has five different allotropic phases, labelled alpha through epsilon in the order of the temperatures at which they are stable. The alpha phase is stable at room temperature (that is, 300 K), but the other phases, being produced at elevated temperatures, are not. The delta phase has a density of 15.9 gm/cm^3, and is malleable. The alpha phase is denser (about 19 gm/cm^3), but also brittle. A weapon designer would prefer to work with delta phase material, minimizing machining difficulties and also problems of criticality since more kilograms of active material could be assembled without exceeding a critical mass. This, in turn, might permit higher-yield weapons to be produced.

The trick is to find a way to stabilize delta phase plutonium at room temperature. The problem was solved in the Manhattan Project by alloying the plutonium with 3 percent gallium, producing material with the density of delta-phase plutonium but which was stable over a temperature range from −75 to +450 C.[23]

Plutonium also presents special hazards because it can quickly contaminate a laboratory by self-migration. That is, the alpha decay of plutonium is sufficiently energetic that a single decay can remove many atoms of plutonium from an exposed surface and propel them about the room. For this reason plutonium pieces are usually heavily clad or plated with such metals as nickel. Despite the difficulties of handling the material, no particular technical barriers arise because the problems are so well documented in the open literature.

Technical Hurdles, Not Barriers

The principal technological barrier to nuclear proliferation is the difficulty of obtaining a great enough quantity of plutonium or enriched uranium to use in testing and to fuel the weapons of an arsenal. The cost of obtaining the special materials, working from scratch, may exceed that of designing and building the first devices by a factor of ten, to judge from Manhattan Project experience. Converting chemical compounds rich in plutonium or ^{235}U to metal and then fabricating the nuclear components of a weapon are difficult tasks, but clearly within reach of many nations with only modest technical bases.

Materials signatures of a nuclear weapons project, as opposed to a nuclear power project, might include: a reprocessing capability in a nation with only a small installed nuclear power base; nuclear reactors designed to permit refuelling while the reactor is operating (as at Hanford); and any presence of metallic uranium or plutonium in the system. This last is important because nuclear weapons of any quality demand pure metallic fuels, but virtually all civilian reactors use only metal oxides. It is, therefore, difficult to "explain away" the possession of more than gram quantities of uranium or plutonium metal.[24]

Weapons fabrication can begin when supplies of fissile material have been assured. The design of a Trinity type implosion system has been described in detail in the open literature. While constructing such a device presents no barrier in principle, the development process probably requires the expenditure of several thousand kilograms of high explosive in a test program. Most experienced weapon designers are convinced that once a reliable implosion system has been designed and tested using inert material similar to plutonium or enriched uranium (for example, natural uranium), a nuclear test is probably unnecessary in order for the proliferant to have confidence in the functioning of the weapon.[25] Those same designers are divided on the question of whether nuclear tests would be necessary to certify somewhat more advanced designs of the "physics package" which use the active material more efficiently.

Manufacture of the explosive lenses for an implosion weapon is not as challenging as it was in 1945. Rough castings of the explosive can be machined using water as the coolant; RDX and HMX explosives can be pressed. The technology for the manufacture of precision explosive products is widespread.[26] An effective initiator is a necessity. Iraq developed its initiators indigenously, and they differ in design from those used elsewhere.[27]

Digitally controlled machine tools are not needed to shape either implosion lenses or the nuclear components of fission weapons. None were available until at least the late 1960s or early 1970s, by which time most of the nuclear powers of the world had already built their first devices and many had proceeded to sophisticated designs and to thermonuclear weapons.

A nuclear testing program is almost certainly unnecessary for a nation content to reproduce the first generation of fission weapons; high explosive tests should suffice. The most advanced pure fission devices, boosted fission weapons, and true thermonuclear weapons probably cannot be constructed without a certain amount of experience with what "works" and what does not, as the sophistication of the nuclear design increases. The required expertise can be obtained either by testing or by finding mercenary weapons experts who can replicate successful designs for well-to-do customers.

On 24 February 1993 the new director of Central Intelligence, R. James Woolsey, testified before the Senate Government Operations Committee that Iran was about eight years from possessing a nuclear capability. That may be a correct estimate if Iran maintains its project at its present rate and if it does not receive fissile material from outside. None of the five acknowledged nuclear weapons states, however, took anywhere near that long to proceed from a decision to build a nuclear weapon until their first test took place. Given the fact that the basics of nuclear weapon design and construction are now widely known, as is the technology for producing plutonium and enriched uranium, no nation with the technical capabilities of the United States in 1941 can ever be said to be more than five years from possessing a nuclear weapon, regardless of the state of its program at any given moment. There are no technical barriers to nuclear proliferation, only hurdles of greater or lesser size which can be leaped over at greater or lesser cost and on time scales commensurate with the budget for the project and the skill of the workers.

A final caution: so far as is known, no nation which has attempted to detonate a nuclear explosive has failed on its first attempt.

Notes

1. There is always a certain tension involved in any article which discusses in some detail the technical barriers to attaining a nuclear capability. On the one hand it is necessary to lay out for policy makers how few difficulties actually remain, other than assembling the very large project required, and to do this requires elucidation of how nuclear weapons are built with enough specificity to convince the reader of the ease with which the technical problems can be solved even by developing nations such as Iraq. On the other hand, prudence demands that critical details be omitted so that an article such as this does not become a primer for proliferants. When choices needed to be made, the author has erred on the side of prudence, usually by a large margin.
2. Robert Serber, *The Los Alamos Primer*, quoted by Richard Rhodes in his introduction to the 1992 University of California Press edition, p. xiv.

3. The *Primer* is a fascinating introduction to the fundamental nuclear physics of a nuclear device as seen from the Los Alamos mesa in 1943, but it leaves out the equally critical lessons needed from high explosives, hydrodynamics, enrichment, metallurgy, and chemistry. Because it was written in March 1943, the original version of the *Primer* has virtually no mention of implosion, the principal method of assembling a supercritical mass in a modern weapon. The use of implosion did not become a major feature of the Los Alamos project until late in 1943.
4. Serber, *The Los Alamos Primer*.
5. Serber, *The Los Alamos Primer*, xv, xvi, 55, and 56.
6. Neutron background is important because the chain reaction must not start too early in the assembly process. If it does, it can cause the explosion to start before a sufficiently super-critical mass has been produced; this predetonation results in an extreme reduction of yield – from kilotons to tons or even less.
7. See Richard Rhodes, *The Making of the Atomic Bomb*, searching the index for entries under "implosion," "Tuck, James," "Neddermeyer, S.", and "Kistiakowsky, G."
8. David Hawkins, *Project Y: The Los Alamos Story*, pt. 1, *Toward Trinity* (Los Angeles: Tomash Publishers, 1983), 211.
9. Rhodes, *The Making of the Atomic Bomb*, 660. The weight of the sphere of high explosives, including its metal casing, but less one lens charge, piece of tamper, and plutonium, is given as two tons.
10. Thomas P. Cochran, William M. Arkin, and Milton W. Hoenig, *The Nuclear Weapons Databook*, vol. 1, *U.S. Nuclear Forces and Capabilities* (Cambridge, Mass.: Ballinger, 1984), 31
11. Cochran et al., *The Nuclear Weapons Databook*, vol. 1, *U.S. Nuclear Forces and Capabilities*, 32; Richard G. Hewlett and Oscar E. Anderson, Jr., *A History of the United States Atomic Energy Commission*, vol. 1, *The New World, 1939/1946* (University Park, Pa.: Pennsylvania State University Press, 1962), caption of photo facing p. 401.
12. Roy D. Woodruff, paper presented at the 21 April 1992 meeting of the American Physical Society in Washington, D.C., session G3.
13. Ibid.
14. Author's experience preparing uranium targets for physics experiments.
15. Pyrophoric materials spontaneously combust in the air, particularly when they are present in finely divided form such as shavings and dust from machining.
16. See Serber, *The Los Alamos Primer*, 58-59 for a description of a Manhattan Project era diagnostic method. Descriptions such as this illustrate the tension between the need to reveal information in order to be convincing and the need to hold back knowledge which would aid a potential proliferant. More detailed information on diagnostics is, of course, available in the technical literature on explosives and explosions.
17. Hewlett and Anderson, *A History of the United States Atomic Energy Commission*, vol. 1, *The New World, 1939/1946*, 723 (Appendix 2).
18. It is possible that this extreme discrepancy arises because K-25 did not come into full use until long after Y-12 and the Hanford reactors. Hence, the operating budget would have been relatively depressed.
19. The thermal output of the reactor is what determines the rate of plutonium production. In general nuclear reactors are less than 33 percent efficient in producing electricity; hence a nuclear power plant producing 100 megawatts of electricity must produce about 300 megawatts of thermal energy.
20. Leonard S. Spector and Selig Harrison, oral report of the Carnegie mission to North Korea, on-the-record "Face to Face Breakfast," 11 May 1992.
21. H. DeW. Smyth, *Atomic Energy For Military Purposes*, originally published by the

U.S. Government Printing Office, 1945. Now available from Stanford University Press (Stanford, 1989). Pagination varies; use numbered paragraphs for identifying material. Par. 8.20-8.26.

22. See Rhodes, *The Making of the Atomic Bomb*, 603-4, for a description of the appearance and construction of the Hanford, Washington reprocessing plants built during the Manhattan Project. See also Thomas B. Cochran, et al., *Nuclear Weapons Databook*, vol. 2, *U.S. Nuclear Warhead Production* (Cambridge, Mass.: Ballinger, 1987), 139-40, for useful information on the more modern PUREX reprocessing scheme.
23. Hawkins, *Project Y: The Los Alamos Story*, 223.
24. Gram quantities of uranium metal foils may be used as calibration standards; some nuclear physics experiments might be done with uranium or plutonium foils.
25. The design of the Hiroshima bomb was not tested before the weapon was delivered to its target. The scientists involved were completely confident that their product would function properly.
26. Rhodes, *The Making of the Atomic Bomb*, 577. George Kistiakowsky described the primitive Manhattan Project methods for manufacturing explosive charges in a series of (now) unclassified Los Alamos reports.
27. Interviews with inspectors who participated in the inspection of Iraq pursuant to United Nations Security Council Resolution 687 following the 1991 Gulf War.

NOTE ADDED IN PROOF

The cost estimates and the timescales given in the body of the article, as well as judgments about the difficulties a proliferant nation might face, have all been based on the assumption that special nuclear material, ^{235}U and ^{240}Pu, would have to be manufactured by the proliferant itself. Because of the likelihood of a reduced level of control over former Soviet nuclear material resulting from the breakup of the USSR, it is no longer clear that a proliferant with adequate funds will be unable to find a "commercial" source for fissile material. Should a nation determined to build a nuclear weapon quickly be able to purchase the fissile material, the costs for it project would probably plummet, and the time needed to construct its first weapon might decrease from years to months or weeks. This possibility should be borne in mind when considering the advisability of lifting economic sanctions against Iraq, a nation which has demonstrated that it has the technical skills to design nuclear weapons and, almost certainly, to fabricate them given the requisite ingredients; it should also be incorporated into judgments made about other relatively wealthy nations with nuclear ambitions.